Novel Approaches in Biopreservation for Food and Clinical Purposes

Editors

Enriqueta Garcia-Gutierrez
Department of Agricultural Engineering
Institute of Plant Biotechnology
Polytechnic University of Cartagena
Murcia, Spain

Natalia Gomez-Torres
Department of Nutrition and Food Science
Complutense University of Madrid
Spain

Sara Arbulu
Faculty of Chemistry, Biotechnology and Food Science
Norwegian University of Life Sciences
Ås, Norway

CRC Press is an imprint of the
Taylor & Francis Group, an **informa** business

A SCIENCE PUBLISHERS BOOK

Cover credit:

- Strawberries image provided by Dr. Sara Arbulu
- Kefir grains image provided by Dr. Samuel Mortensen
- Dekkera bruxellensis NCYC 823 provided by the National Collection of Yeast Cultures in UK

First edition published 2024
by CRC Press
2385 NW Executive Center Drive, Suite 320, Boca Raton FL 33431

and by CRC Press
4 Park Square, Milton Park, Abingdon, Oxon, OX14 4RN

© 2024 Enriqueta Garcia-Gutierrez, Natalia Gomez-Torres and Sara Arbulu

CRC Press is an imprint of Taylor & Francis Group, LLC

Reasonable efforts have been made to publish reliable data and information, but the author and publisher cannot assume responsibility for the validity of all materials or the consequences of their use. The authors and publishers have attempted to trace the copyright holders of all material reproduced in this publication and apologize to copyright holders if permission to publish in this form has not been obtained. If any copyright material has not been acknowledged please write and let us know so we may rectify in any future reprint.

Except as permitted under U.S. Copyright Law, no part of this book may be reprinted, reproduced, transmitted, or utilized in any form by any electronic, mechanical, or other means, now known or hereafter invented, including photocopying, microfilming, and recording, or in any information storage or retrieval system, without written permission from the publishers.

For permission to photocopy or use material electronically from this work, access www.copyright.com or contact the Copyright Clearance Center, Inc. (CCC), 222 Rosewood Drive, Danvers, MA 01923, 978-750-8400. For works that are not available on CCC please contact mpkbookspermissions@tandf.co.uk

Trademark notice: Product or corporate names may be trademarks or registered trademarks and are used only for identification and explanation without intent to infringe.

Library of Congress Cataloging-in-Publication Data (applied for)

ISBN: 978-1-032-21355-2 (hbk)
ISBN: 978-1-032-21356-9 (pbk)
ISBN: 978-1-003-26799-7 (ebk)
DOI: 10.1201/9781003267997

Typeset in Times New Roman
by Radiant Productions

Foreword

The term "biopreservation" is usually employed to define the sustainable use of microorganisms and/or their metabolites to extend the shelf life, to improve the organoleptic properties and/or to enhance the safety of foods and other biological materials. Biopreserved foods have been consumed by humans for millennia. The Neolithic Revolution meant an increase in the availability of food and, as a consequence, the appearance and development of different techniques for the treatment and conservation of the surplus (cooking, drying, salting, smoking...). Among these techniques, fermentation has occupied a prominent place from then until today, making a virtue of necessity. The first farmers observed that when they stored sugary foods, such as fruits, grains or milk, two things could happen: sometimes unsafe products with a bad smell and taste were generated but, in others cases, derivatives were produced with an acrid but not unpleasant taste, which did not affect health and which were the precursors of cider, wine, beer or cheese. Although they did not know the reason for these processes, they used an empirical system of trial and error for controlling the conditions of time, temperature, atmosphere, etc., that prevented putrefaction and led to what was later called "fermentation".

Fermentations must have been quickly appreciated not only because of the transcendental transformation of highly perishable raw materials into products with a much longer shelf life but, also, because they provided new foods with a wide range of flavors, smells, and textures, and in many cases, more digestible than the starting raw materials. In addition, some of the fermented products must have been used for medicinal purposes for two main reasons: they were very nutritious and, although it was not known at the time, they contained large numbers of microbes with beneficial properties for health.

A practice that proved to be very useful to promote fermentation and prevent putrefaction was the addition of a portion of a fermented product with desirable characteristics to the raw material to be fermented. This practice provoked the selection of microbial communities adapted to the characteristics of the food to be fermented, causing rapid genomic specialization through a variety of mechanisms. Within the new food niches, the metabolic requirements of these microbes became predictable, resulting in the "domestication" of microbial communities with increasingly better fermentative capacities, which developed in the fermentable mass and produced a series of compounds that, on the one hand, prevented colonization by spoilage or pathogenic organisms and, on the other hand, they gave rise to beneficial effects for the people who ingested them.

iv *Novel Approaches in Biopreservation for Food and Clinical Purposes*

Although biopreservation was originally driven by food fermentation, this concept evolved and was successfully extended to other approaches as well as to non-fermented foods, and even to drugs, cells and tissues, as it is shown in this book. New strategies, including –omics approaches, for the selection and characterization of lactic acid bacteria-and other microbes-based cultures are now available, together with a revival in the scientific and industrial interest for some microbial related metabolites or compounds, such as bacteriocins. In addition, plant- and animal-derived compounds and bacteriophages and their endolysins may also play a relevant role in biopreservation. All these strategies may be used by their own or combined among them for summative or synergic effects. New approaches will require new regulatory frames, an issue that is also dealt with in this book.

Although a wide variety of biopreservation strategies are currently available on the market, this area must receive more scientific and technological attention to meet global population- and climate-related challenges. In this frame, novel approaches are required to improve selection and characterization of protective microbes, and to elucidate the mechanisms responsible for their protective effect. This book provides a complete overview of the state of the art and the future directions of biopreservation. The high quality of the editors and authors of this book makes it an indispensable reference in this field.

Juan M. Rodríguez
Dpt. Nutrition and Food Science, Complutense University of Madrid, Spain

Preface

In an interconnected world, we have access to a wide variety of foods from all around the world, regardless of the season or its geographical location. Food matrixes are getting more diverse and complex, while consumers purchasing habits promote the use of more natural preservatives and less processed foods. Biopreservation provides new tools based on microorganism, plants and animal-derived compounds to safeguard the quality and safety of foodstuffs. As a result, there is renewed interest in traditional fermentation, plant extracts or antimicrobial peptides such as bacteriocins. As a multi-disciplinary field, biopreservation draws knowledge from microbiology, chemistry, food science or agriculture that put their efforts together through the entire supply chain, from farm to fork to preserve food in optimal conditions.

The purpose of "Novel approaches in biopreservation for food and clinical purposes" is to complement existing books with state-of-the-art information on food and medical biopreservation methods. This book assumes a minimum level of knowledge in microbiology and food science and aims to give an overview on the most common biopreservation techniques on foodstuffs. These include microorganisms used as protective cultures or as starters for fermentation, bacterial metabolites such as bacteriocins, organic acids and enzymes, herbs, species and essential oils, or the application of bacteriophages and endolysins.

The book is composed of a general introductory chapter that highlights relevant biopreservation techniques and compounds, followed by six chapters that address each of the major food groups, i.e., milk and dairy, vegetables, meat, fish, bread and flours and beverages. The three last chapters describe biopreservation methods in drugs, embryos and gametes, and cell and tissues.

We are grateful to all our coauthors for their dedication to developing informative chapters to broaden our knowledge on the different biopreservatives used in foods and for clinical purposes.

Enriqueta Garcia-Gutierrez
Natalia Gomez-Torres
Sara Arbulu

Contents

Foreword iii

Preface v

1. **Introduction to Biopreservation** 1
   Sara Arbulu and *Enriqueta Garcia-Gutierrez*

2. **Biopreservation of Milk and Dairy Products** 23
   Lucía Fernández, Susana Escobedo, Ana Catarina Duarte, Seila Agún,
   Claudia Rendueles, Pilar García, Ana Rodríguez and *Beatriz Martínez*

3. **Biopreservation of Vegetables** 41
   Bárbara Ramos, Teresa R.S. Brandão, Paula Teixeira and *Cristina L.M. Silva*

4. **Biopreservation in Meat and Meat Products** 66
   Annada Das, Dipanwita Bhattacharya, Pramod Kumar Nanda,
   Santanu Nath and *Arun K. Das*

5. **Biopreservation of Fish** 98
   Vida Šimat, Federica Barbieri, Chiara Montanari, Fausto Gardini,
   Danijela Skroza, Ivana Generalić Mekinić, Fatih Ozogul, Yesim Ozogul and
   Giulia Tabanelli

6. **Biopreservation in Flours and Bread** 130
   Biljana Kovacevik, Sanja Kostadinović Veličkovska, Tuba Esatbeyoglu,
   Aleksandar Cvetkovski, Muhammad Qamar and *João Miguel Rocha*

7. **Biopreservation of Beverages** 205
   Suchi Parvin Biki and *Enriqueta Garcia-Gutierrez*

8. **Biopreservation in Medicines** 226
   Sanjogta Thapa Magar

9. **Biopreservation of Cells and Tissues** 242
   Enriqueta Garcia-Gutierrez

10. **Biopreservation of Gametes and Embryos against**
    Microbiological Risk 255
    José Luis Girela López and *Enriqueta Garcia-Gutierrez*

Index 265

CHAPTER 1

Introduction to Biopreservation

Sara Arbulu[1], and *Enriqueta Garcia-Gutierrez[2]*

Introduction

Biopreservation is a general term that can be defined as the use of various biologically derived substances and techniques to protect biological materials, including food, drugs, cells or tissues from degradation or spoilage while maintaining their efficacy, safety, and quality. The ability to preserve biological materials, whether they be medicines or foodstuffs, is a constant challenge for the agri-food industry and the health sectors. This chapter provides a brief overview of food biopreservation and medical biopreservation methods.

According to the Food and Agriculture Organization, around 14% of the world's food is lost during the period between harvest and reaching the market (FAO, 2019). The United Nations Environment Programme (UNEP) reports that an additional 17% of food is wasted in retail and by households (UNEP, 2021). The consumption of contaminated food is also an important cause of morbidity and mortality worldwide, with an estimation of 600 million ill people consuming contaminated food each year (WHO, 2015). Preservation not only is essential for food safety but plays a crucial role in promoting sustainability by reducing food loss and waste, providing access to food, and preserving natural resources.

Similarly, in the medical field, biopreservation is key in preserving biological materials for research, clinical, and pharmaceutical purposes. It enables long-term storage, reduction of contamination risk, increase availability and extended shelf life.

---

[1] Faculty of Chemistry, Biotechnology and Food Science, Norwegian University of Life Sciences, 1432 Ås, Norway.
[2] Department of Agricultural Engineering, Institute of Plant Biotechnology, Polytechnic University of Cartagena, Murcia, Spain.
* Corresponding author: sara.arbulu@nmbu.no

Biopreservation in foods

All foods, after harvesting start to spoil. Food spoilage results from the chemical, physical or microbiological actions that occur throughout the entire supply chain (Singh et al., 2019). They change the texture, odour, flavour or appearance of food, reducing, in many cases, their nutritional value and making them unacceptable for the consumer. If no preservation method was applied the shelf life of many of the food products we consume would be drastically reduced.

Food preservation is of paramount importance for human development and aims to prevent or reduce spoilage, foodborne illnesses, and nutrient loss while retaining the taste, texture, and nutritional quality of food. A significant proportion of food loss and waste is due to microbiological spoilage, including bacteria, fungi, mould and yeasts (Lorenzo et al., 2018). In this sense, the increasing understanding on microbial dynamics through new technologies such as next generation sequencing or nanotechnology, will revolutionize how we deal with food contamination developing targeted strategies against specific undesirable microorganisms (Angelopoulou et al., 2022; Cook and Nightingale, 2018).

Ensuring food access, quality and safety are, therefore, major challenges for the agri-food sector and preservation is a pillar to achieving these goals.

Preservation methods can be divided into physical, chemical and biological (Amit et al., 2017). Physical methods include thermal processing such as pasteurization, chilling or drying and non-thermal methods, such as irradiation; chemical methods cover mostly the addition of chemical preservatives, while biological methods refer to the use of beneficial microorganisms, their by-products (bacteriocins, enzymes) or naturally derived substances from animals (lysozyme, lactoferrin) plants (essential oils, herbs or spices) or bacteriophages and their lytic enzymes (Moye et al., 2018) as preservation agents (Singh, 2018). Biological preservation methods group together under the umbrella of Biopreservation.

Throughout history humans have dried, chill, freeze, salt, smoke, ferment or pickle foods to increase their shelf life and make food available out of season. Industrialization brought heat treatments, freezing, pasteurizing or canning to a larger scale and chemical additives were slowly introduced into the market. Nowadays, consumers are becoming increasingly interested in fresh, natural, less processed, and additive-free or natural-additive foods. In this context, biopreservation strategies offer new possibilities in the food preservation arena.

Protective cultures versus starter cultures

Protective cultures are live microorganisms that are added to food to inhibit the growth of harmful or spoilage organisms without changing their sensory properties (Varsha and Nampoothiri, 2016). Most of them are lactic acid bacteria (LAB) with a long history of human use and meet the Qualified Presumption of Safety (QPS) and the Generally Regarded As Safe (GRAS) regulations by the European Food Safety Authority (EFSA) and the Food and Drug Administration (FDA), respectively. On the contrary, starter cultures are introduced during the production stage to generate a fermented final product with different nutritional and sensory properties than the initial one (Leroy and De Vuyst, 2004). Fermentation has been used by humans for centuries to preserve all kinds of food, and applied to all kinds of food, including dairy, meat, fish, bread, vegetables and beverages. Fermented foods are often rich in beneficial bacteria and can be a healthy addition to a balanced diet (Leeuwendaal et al., 2022). While the use of starter cultures is well-established for fermented foods, the utilization of protective cultures as food preservatives is more recent (Souza et al., 2022).

The potential of microorganisms as biopreservation agents resides in their capability of producing metabolites with antibacterial and antifungal properties. To qualify as a protective culture, bacteria must meet certain criteria such as being endorsed by regulatory agencies, demonstrating stability, and displaying efficacy against the targeted pathogens. The use of protective cultures has been explored in dairy (Silva et al., 2018), meat (Xu et al., 2021), fish (Cifuentes Bachmann and Leroy, 2015), bakery products (Calasso et al., 2023) and fruit and fresh produce (Linares-Morales et al., 2018). Likewise, there are several commercially available bioprotective cultures, mostly for their application in dairy (Souza et al., 2022).

Microbial-derived products

Organic acids

Organic acids are naturally produced as the end product of carbohydrate fermentation of different organisms including bacteria, yeasts and moulds (Coban, 2020). They are naturally present in fermented foods, but can also be added for flavouring, antioxidant effect or as antimicrobials (Nkosi et al., 2021). As antimicrobials, they effectively inhibit the growth of bacteria and fungi showing bacteriostatic or bactericidal effects. Although their exact mechanism of action is not known, at low pH the undissociated state of the molecule allows them to enter the target cell. Once inside, they dissociate to release charged anions and protons that inhibit metabolic reactions. They have been also proposed to disrupt the membrane and induce stress on intracellular pH homeostasis (Davidson et al., 2012). At the same time, organic acids lower the pH of the medium making harsh conditions for spoiler and pathogen microorganisms to grow (Lund et al., 2020).

The most commonly used organic acids are lactic acid and acetic acid. Homofermentative LAB such as *Leuconostoc*, *Lactococcus*, *Lactobacillus* or *Pediococcus* metabolize sugars via glycolysis to give lactic acid as the final only product (Abedi and Hashemi, 2020). Heterofermentative LAB, such as *Oenococcus*, *Leuconostoc*, and some *Lactobacillus*, metabolize sugars via the pentose phosphoketolase or the phosphogluconate pathways producing not only lactic acid but also, acetic acid, ethanol or carbon dioxide (Abedi and Hashemi, 2020). Acetic acid can also be produced via oxidative fermentation by *Acetobacter* and *Gluconobacter* (Gomes et al., 2018). Several short chain fatty acids, particularly, propionate, butyrate, and formate have been the subject of numerous research studies and commercial applications and can be produced by a wide range of bacterial species (Ricke et al., 2020).

Organic acids can be commercially produced through either chemical synthesis or fermentation processes. The latter are gaining attention and offer several advantages:

4 *Novel Approaches in Biopreservation for Food and Clinical Purposes*

lower production costs, high productivity, the possibility of genetic engineering, co-cultivation or immobilization of the producer microorganisms for an optimized production (Coban, 2020). Additionally, there is growing interest in utilizing agro-industrial waste as a substrate for fermentation. (Mahgoub et al., 2022; Palakawong Na Ayudthaya et al., 2018).

Bacteriocins

Bacteriocins are antimicrobial peptides produced by bacteria that kill or inhibit the growth of other bacteria. They are widely studied as food preservatives and also as potential coadjuvants or alternatives to antibiotics (Simons et al., 2020; Verma et al., 2022). Bacteriocins are widespread in nature, but it is those produced by food-grade LAB that have been more studied for food preservation (da Costa et al., 2019; Silva et al., 2018). Thus, bacteriocin producing bacteria have been found in many LAB isolated from food (Table 1). Traditionally, the discovery of bacteriocins has involved screening for antimicrobial activity in bacterial cultures and purifying of the active compound for further characterization. Luckily, with the rise of sequencing technologies bacteriocins are not only detected by *in vitro* studies, but can also be discovered by using dedicated databases and bioinformatics pipelines such as BAGEL4 (van Heel et al., 2018), Bactibase (Hammami et al., 2010) or BADASS (Costa et al., 2023). After screening whole genome sequences and metagenomes, bacteriocins can be revived in different hosts by genetic engineering or by chemical synthesis (Bédard and Biron, 2018; Collins et al., 2018; Leech et al., 2020; Peña et al., 2022).

Due to their high diversity in terms of size, structure, biosynthesis and mode of action different classification schemes have been proposed over the years. Here, we refer to the classification by Cotter et al. 2013 based on the structure and the modifications they undergo during their biosynthesis (Cotter et al., 2013). Class I bacteriocins, or lantibiotics, includes heat stable small size peptides (< 5 kDa) with extensive post translational modifications including residues such as lanthionine or methyllanthionine. Most lantibiotics exert their mechanism of action by binding to lipid II and pore formation (Dickman et al., 2019). Lantibiotics, represented by nisin A and its variants, are among the most extensively researched bacteriocins.

Class II bacteriocins (< 10 kDa) are unmodified or little modified peptides that can be further subdivided into four subclasses: IIa or pediocin-like bacteriocins, highly active against *Listeria*, IIb or two-peptide bacteriocins that require the two peptide to work together, IIc or circular bacteriocins, which are highly stable due to the covalent union between their N- and C-termini, and unmodified, linear, non-pediocin-like bacteriocins (class IId) that include the leaderless bacteriocins or those secreted through the general sec route. Ultimately, bacteriocins are generally believed to form pores in bacterial membranes, but a number of receptor such as the mannose phosphotransferase system for class IIa bacteriocins (Diep et al., 2007; Kjos et al., 2010), the undecaprenyl-pyrophosphate phosphatase in class IIb (Kjos et al., 2014) or the maltose ABC transporter for circular bacteriocins (Gabrielsen et al., 2012; Pérez-Ramos et al., 2021).

Typically, bacteriocins can be used on food matrixes using different strategies: inoculation of the bacteriocin producing strain on the food for an *in situ* production

Table 1. Bacteriocins identified from foodstuffs.

Bacteriocin class	Bacteriocin	Produced by	Isolated from	References
		Milk and dairy		
Class I	**Bovicin HJ50**	*Streptococcus bovis*	Raw milk	(Xiao et al., 2004, p. 50)
Class I	**Lacticin 3147**	*Lactococcus lactis* subsp. *lactis*	Irish kefir grain	(Ryan et al., 1996)
Class I	**Lacticin 481 (Lactococcin DR)**	*L. lactis* subsp *lactis* CNRZ 481	Dairy isolate	(Donadio et al., 2009)
Class I	**Macedocin**	*Streptococcus macedonicus* ACA-DC 198	Greek cheese	(Georgalaki et al., 2002)
Class I	**Macedovicin**	*S. macedonicus* ACA-DC 198	Greek cheese	(Georgalaki et al., 2013)
Class I	**McdA1**	*S. macedonicus* ACA-DC 198	Greek cheese	(De Vuyst and Tsakalidou, 2008)
Class I	**Nisin A**	*L. lactis*	Milk	(Piper et al., 2011)
Class I	**Nisin Z**	*L. lactis* NIZO 221 86	Dairy isolate	(Mulders et al., 1991)
Class I	**Thermophilin 1277**	*Streptococcus thermophilus* SBT1277	Raw milk	(Kabuki et al., 2007)
Class I	**Enterocin AS-48RJ**	*Enterococcus faecalis* RJ16	Goat cheese	(Abriouel et al., 2005)
Class I	**Lactocyclicin Q**	*Lactococcus* sp. QU12	Cheese	(Sawa et al., 2009)
Class II	**Acidocin 8912**	*Lactobacillus acidophilus*	Dairy product	(Tahara et al., 1992)
Class II	**Acidocin A**	*L. acidophilus* TK9201	Starter fermented milk	(Kanatani et al., 1995)
Class II	**Aureocin A53**	*Staphylococcus aureus* A53	Pasteurised commercial milk	(Netz et al., 2002)
Class II	**Bacteriocin J46**	*L. lactis* subsp. *cremoris* J46	Fermented milk	(Huot et al., 1996)
Class II	**BlpU**	*S. thermophilus* 18311	Yogurt	(Bolotin et al., 2004)
Class II	**Carnocin CP52**	*Carnobacterium piscicola* CP5	Ripened cheese	(Herbin et al., 1997)
Class II	**Enterocin CRL35**	*Enterococcus mundtii* CRL35	Cheese	(Saavedra et al., 2004)
Class II	**Enterocin 96**	*E. faecalis* WHE 96	Munster cheese	(Izquierdo et al., 2009, p. 96)

Table 1 contd. ...

...Table 1 contd.

Bacteriocin class	Bacteriocin	Produced by	Isolated from	References
Milk and dairy				
Class II	**Lactococcin 972**	*L. lactis* subsp. *lactis* IPLA 972	Cheese	(Martínez et al., 1996)
Class II	**Lactococcin A**	*L. lactis* subsp. *cremoris* LMG 2130	Milk	(Holo et al., 1991)
Class II	**Lactococcin B**	*L. lactis* subsp. *cremoris* 9B4	Milk	(Venema et al., 1993)
Class II	**Lactococcin MMFII**	*L. lactis* MMFII	Dairy product	(Ferchichi et al., 2001)
Class II	**LSEI 2163**	*Lactobacillus casei* ATCC 334	Dairy products, emmental cheese	(Kuo et al., 2013, p. 334)
Class II	**LSEI 2386**	*L. casei* ATCC 334	Dairy products, emmental cheese	(Kuo et al., 2013, p. 334)
Class II	**Mesentericin B105**	*Leuconostoc mesenteroides* Y105	Goat milk	(Héchard et al., 1992)
Class II	**Mesentericin Y105 (anti-Listeria)**	*L. mesenteroides* Y105	Goat milk	(Héchard et al., 1992)
Class II	**Propionicin SM1**	*Propionibacterium jensenii* DF1	Swiss raw milk	(Miescher et al., 2000)
Class II	**Propionicin T1**	*Propionibacterium thoenii*	Dairy product	(Faye et al., 2000)
Class II	**Propionicin F**	*Propionibacterium freudenreichii* subsp. *freudenreichii* LMGT2946	Cheese starter	(Brede et al., 2004)
Class II	**Subtilosin (SboX)**	*Bacillus amyloliquefaciens*	Dairy Yogu FarmTM	(Sutyak et al., 2008)
Class II	**BlpD/Thermophilin 9**	*S. thermophilus* LMD-9	Dairy product	(Bolotin et al., 2004; Fontaine et al., 2015; Makarova et al., 2006)
Class III	**Linocin M18**	*Brevibacterium linens* M18	Cheese	(Parkhill et al., 2003)

		Meat and meat foods		
Class I	**Variacin**	*Kocuria varians*	Salami	(O'Mahony et al., 2001)
Class I	**Carnocyclin A**	*Carnobacterium maltaromaticum* UAL307	Fresh pork	(Martin-Visscher et al., 2008)
Class II	**Bavaricin MN**	*L. sake* MN	Meat	(Kaiser and Montville, 1996)
Class II	**Carnobacteriocin A (Piscicolin 61)**	*C. piscicola* LV61	Meat	(Schillinger et al., 1993)
Class II	**Carnobacteriocin B2**	*C. piscicola* LV17	Vacuum-packed meat	(Ahn and Stiles, 1990; McCormick et al., 1996)
Class II	**Carnobacteriocin BM1 (CarnobacteriocinB1)**	*C. piscicola* LV17B	Vacuum-packed meat	(Quadri et al., 1994)
Class II	**Curvacin A**	*Lactobacillus curvatus* LTH1174	Dry sausages	(Tichaczek et al., 1992)
Class II	**Curvaticin FS47**	*L. curvatus* FS47	Retail meat	(Garver and Muriana, 1994)
Class II	**Curvaticin L442**	*L. curvatus* L442	Fermented sausage	(Xiraphi et al., 2006)
Class II	**Enterocin P-like**	*Enterococcus faecium* P 13	Dry fermented sausage	(Cintas et al., 1997)
Class II	**Enterocin A**	*E. faecium* CTC492	Dry fermented sausage	(Aymerich et al., 1996)
Class II	**Enterocin B**	*E. faecium* T136	Dry fermented sausage	(Casaus et al., 1997)
Class II	**Enterocin L50**	*E. faecium* L50	Dry fermented sausage	(Cintas et al., 2000)
Class II	**Enterocin P**	*E. faecium* P13	Dry sausages	(Cintas et al., 1997)
Class II	**Enterocin Q**	*E. faecium* L50	Fermented sausage	(Cintas et al., 2000)
Class II	**Garvieacin Q**	*Lactococcus garvieae* BCC 43578	Nham (fermented pork sausage)	(Tosukhowong et al., 2012, p. 43)
Class II	**Lactocin 705**	*L. casei* CRL 705	Dry fermented sausage	(Palacios et al., 1999)

Table 1 contd. ...

8   *Novel Approaches in Biopreservation for Food and Clinical Purposes*

...Table 1 contd.

		Meat and meat foods		
Bacteriocin class	**Bacteriocin**	**Produced by**	**Isolated from**	**References**
Class II	**Lactococcin A**	*Clostridium perfringens* strain SM101/TypeA	Derivative of NCTC 8798 isolated from meat	(Zhao and Melville, 1998)
Class II	**Leucocin A (LeucocinA-UAL187)**	*Leuconostoc gelidum* UAL 187	Vacuum-packaged meat.	(Martin-Visscher et al., 2008, p. 187)
Class II	**Leucocin B (LeucocinB Ta11a)**	*Leuconostoc carnosum* Ta11a	Vacuum-packaged meat	(Felix et al., 1994)
Class II	**Leucocin_C**	*L. mesenteroides* TA33a	Processed meat	(Felix et al., 1994)
Class II	**Piscicolin 126**	*C. piscicola* JG126	Spoiled ham	(Rw et al., 1996, p. 126)
Class II	**Plantaricin 1.25β**	*L. plantarum* TMW 1.25	Sausage fermentation	(Ehrmann et al., 2000)
Class II	**Plantaricin SA6**	*L. plantarum* SA6	Fermented sausages	(Rekhif et al., 1995)
Class II	**Putative bacteriocin**	*L. sakei* subsp. *sakei* 23K	French sausage	(Chaillou et al., 2005)
Class II	**Sakacin A**	*L. sake* Lb706	Meat product	(Schillinger, 1994)
Class II	**Sakacin G**	*L. sake* 2512	Dry fermented sausages	(Schillinger, 1994)
Class II	**Sakacin P**	*L. curvatus* strain CRL705	Fermented sausages	(Hebert et al., 2012, p. 705)
Class II	**Sakacin P**	*L. sakei* 1151	Fermented sausages	(Urso et al., 2006)
Class II	**Sakacin Q**	*L. curvatus* ACU-1	Dry sausages	(Rivas et al., 2014)
Class II	**Weisselin A**	*Weissella paramesenteroides* DX	Fermented sausages	(Papagianni and Sergelidis, 2013)

Fish

Class				
Class I	Enterocin_W	*E. faecalis* NKR-4-1	Pla-ra, Thai fermented fish	(Sawa et al., 2009)
Class I	Carnolysins A1	*C. maltaromaticum* C2	Brazilian smoked fish	(Tulini et al., 2014)
Class I	Carnolysins A2	*C. maltaromaticum* C2	Brazilian smoked fish	(Tulini et al., 2014)
Class II	Divergicin M35	*C. divergens* M35	Smoked mussels	(Tahiri et al., 2004)
Class II	Enterocin NKR-5-3A	*E. faecium* NRK-5-3	Fermented fish	(Ishibashi et al., 2012)
Class II	Enterocin NKR-5-3D	*E. faecium* NRK-5-3	Fermented fish	(Ishibashi et al., 2012)
Class II	Enterocin NKR-5-3Z	*E. faecium* NRK-5-3	Fermented fish	(Ishibashi et al., 2012)
Class II	Plantaricin F	*L. plantarum* BF001	Catfish fillets	(Fricourt et al., 1994)

Vegetables and fruits

Class				
Class I	Enterocin_W	*E. faecalis* NKR-4-1	Pla-ra, Thai fermented fish	(Sawa et al., 2009)
Class I	Carnolysins A1	*C. maltaromaticum* C2	Brazilian smoked fish	(Tulini et al., 2014)
Class I	Carnolysins A2	*C. maltaromaticum* C2	Brazilian smoked fish	(Tulini et al., 2014)
Class II	Divergicin M35	*C. divergens* M35	Smoked mussels	(Tahiri et al., 2004)
Class II	Enterocin NKR-5-3A	*E. faecium* NRK-5-3	Fermented fish	(Ishibashi et al., 2012)
Class II	Enterocin NKR-5-3D	*E. faecium* NRK-5-3	Fermented fish	(Ishibashi et al., 2012)
Class II	Enterocin NKR-5-3Z	*E. faecium* NRK-5-3	Fermented fish	(Ishibashi et al., 2012)
Class II	Plantaricin F	*L. plantarum* BF001	Catfish fillets	(Fricourt et al., 1994)

Table 1 contd. ...

Bread and flours				
Bacteriocin class	**Bacteriocin**	**Produced by**	**Isolated from**	**References**
Class I	Putative lacticin 3147	*L. lactis* ssp. *lactis* M30	Soudough	(Settanni et al., 2005)
Unclassified	Plantaricin ASM1	*L. plantarum* A-1	Tortilla bread	(Hata et al., 2010)
Class II	*Bavaricin A*	*Lactobacillus bavaricus* MI401	Rye and wheat sourdough	(Larsen et al., 1993)
Beverages				
Class I	**Bozacin B14**	*L. lactis* subsp. *lactis* B14	Boza-Bulgarian traditional cereal beverage	(Ivanova et al., 2000)
Unclassified	**CLK_01**	*Lacticaseibacillus rhamnosus* CLK 101	Fermented juice sample	(Chen et al., 2023)
Unclassified	**Bacteriocin ST13BR**	*L. plantarum* ST13BR	Barley beer	(Todorov and Dicks, 2004)
Class II	**Paracin 1.7**	*Lactobacillus paracasei* HD1-7	Chinese Sauerkraut Juice	(Ge et al., 2016)

(protective cultures), addition of bacteriocins directly to the food in a pure or crude-extract form, or using bacteriocin producing bacteria as starters for fermentation.

Despite the numerous known bacteriocins and their value as antimicrobials, to date, only nisin and pediocin PA-1 produced by *Lactococcus lactis* and *Pediococcus acidilactici*, respectively are commercially available as for food preservation. Nisin effectively inhibits food borne spoilage and pathogen bacteria such as *Listeria* and *Staphylococcus*, and the spore forming *Bacillus* and *Clostridium.* Genetically engineered variants have shown broader spectrum antimicrobial activity (Field et al., 2008, 2012). Nisin is widely applied to dairy products (Silva et al., 2018) and new formulations are been explored to improve is effectivity (Ibarra-Sánchez et al., 2020). As a class IIa bacteriocin, pediocin displays high antimicrobial activity against *Listeria monocytogenes*, but also other food borne pathogens including *Staphylococcus aureus* and LAB such as *Lactococcus, Lactobacillus, Pediococcus, Leuconostoc* or *Enterococcus* (Rodríguez et al., 2002).

Bacterial enzymes

Microbial enzymes are used for improving the technological properties of foodstuffs. Better flavoring and aroma, improved quality of bread, improved efficiency of beer making, tenderization of meat or clarification of fruit juices, are just a few examples (Raveendran et al., 2018). Some of them are also used to increase the shelf life of food products generally by helping in the degradation of toxic products or by breaking down complex molecules to smaller ones that are less susceptible to degradation or microbial spoilage. Additionally, there are enzymes of microbial or animal origin whose primary role is to exert antimicrobial activity and thus, can be harnessed for preservation purposes (Thallinger et al., 2013).

Catalase and glucose oxidase are applied together when the levels of hydrogen peroxide and glucose need to be reduced during food production. Catalase catalyzes the reduction of hydrogen peroxide to water and oxygen while glucose oxidase catalyses the oxidation of glucose to gluconolactone, an antimicrobial compound, and hydrogen peroxide. They have both been reported to be produced at high yields by *Aspergillus niger* (Okpara, 2022; Raveendran et al., 2018).

Laccases serve as a preservative in beer production by eliminating oxygen from the process (Okpara, 2022; Raveendran et al., 2018) and lipases are used in bakery products to increase their shelf life (Dai and Tyl, 2021).

Non-microbial products

There has been a renewed interest in naturally derived antimicrobial compounds from plants and animals (El-Saber Batiha et al., 2021).

Essential oils and spices from plants are usually a mixture of various active compounds including phenolic compounds, terpenes, aliphatic alcohols, aldehydes, ketones, acids, and isoflavonoids (Baindara and Mandal, 2022; Gottardi et al., 2016). Their type and concentration determine their antimicrobial activity and they generally inhibit Gram-positive and Gram-negative bacteria (Tiwari et al., 2009). Antimicrobial peptides from plants (Baindara and Mandal, 2022) or guar gum from leguminous plants (Mudgil et al., 2014) have also been proposed for food preservation.

Among the animal derived compounds, lactoferrin is a milk derived substance with antimicrobial activity against bacteria, virus and fungi that it is used in dairy and meat (Jańczuk et al., 2023); lysozyme, found in many biological fluids has traditionally been extracted from the chicken egg white although heterologous production in microbial hosts is also been explored (He et al., 2020; Zhang et al., 2014). Lysozyme disrupts the peptidoglycan of Gram-positive bacteria and can be used in meat, cheeses or wine (Khorshidian et al., 2022; Pilevar et al., 2022; Zhang and Rhim, 2022). Chitosan is also known in food preservation for its antimicrobial properties. It derives from the polysaccharide chitin, and it is mainly produced from crustacean shells by deacetylating chitin to a positively charged molecule that interacts with the cell wall, cell membranes and cytoplasmic components of both Gram-positive and Gram-negative bacteria (Yilmaz Atay, 2020). Its antifungal activity is due suppressing sporulation and spore germination (Lopez-Moya et al., 2019). It is mainly studied as an edible coating for food protection (Duan et al., 2019; Mihai and Popa, 2015).

Bacteriophages and phage endolysins

Bacteriophages are ubiquitous viruses that infect and replicate in bacteria in a species-specific or even strain-specific manner. Lysogenic viruses integrate their genetic material into the host genome and replicate along with their host, being in a dormant state for a variable amount of time. Conversely, lytic viruses take over the host cell machinery and use it to replicate themselves and generate new viral particles, lysing the host cell in the process (García et al., 2008).

Bacteriophages intended for food preservation should be lytic viruses and stable in the environment where they are going to be applied. Due to their specificity, they can be directed to target only food borne pathogenic bacteria, and can be applied as phage cocktails to target multiple bacteria an minimize the development of resistance (García et al., 2008).

Different phage cocktails have been commercialized to target effectively *Escherichia coli*, *Listeria monocytogenes*, *Salmonella*, *Shigella* and *Campylobacter* in foods (Endersen and Coffey, 2020; Moye et al., 2018).

Phage lysins, also known as endolysins, are enzymes produced by lytic phages that degrade the cell wall peptidoglycan and allow the new viral particles to be released. This ability can be harnesses for food preservation and it is of especial interest to target Gram-positive bacteria since they are susceptible to external lysin action (Schmelcher and Loessner, 2016). Several studies propose the use of endolysins those to control *S. aureus*, *L. monocytogenes* and *Clostridium perfringes* in dairy (Chang, 2020).

Preservation in the medical field

In the medical field, biopreservation involves the use of different strategies to prevent degradation and maintain the stability of pharmaceutical products. This includes the use of antioxidants, stabilizers, or preservatives, usually of chemical origin, to prevent degradation or spoilage, the use of specialized storage conditions such as refrigeration or freezing to maintain drug stability, or the use of advanced packaging materials that provide a protective barrier against environmental factors. The selection of a preservation method depends on the intended use and the nature of the specimen as each technique has its own strengths and weaknesses.

Preservation of medicines

Excipients play a crucial role in the development and formulation of medicines, as well as in their administration as active pharmaceutical ingredients. Initially thought to be inert, nowadays they must pass strict tests to guarantee their adequate use. The most important function of any excipient is to ensure the safety and efficacy of the medicine throughout the formulation, the storage period, and during and after its administration (Abrantes et al., 2016). In this context preservatives such as antioxidants and antimicrobials are of outmost importance. Antimicrobial preservatives prevent the microbial contamination of the drug, particularly for drugs with a high water content, while antioxidants prevent damage by oxidation (Shitole et al., 2022).

Preservation of cell, tissues, gametes and organs

The objective for cells and tissues is to preserve their viability and functionality for research, medicine and biotechnology. It enables the study and analysis of biological structures, development of new drugs and therapies, and facilitates the transplantation of organs and tissues. When it comes to short-term storage, such as for organ transplantation, hypothermic storage at temperatures above freezing point is commonly employed, and there are ongoing efforts to enhance this method by modifying oxygen and temperature variables (Lepoittevin et al., 2022). For long term storage, cryopreservation is the gold standard technique (Bojic et al., 2021), being freezing an vitrification the main approaches (Taylor et al., 2019). Other techniques for cells and tissue preservation include the use of antimicrobials or dry storage (Campos et al., 2012; Chen et al., 2019; Oliver, 2012).

Preservation of gametes and embryos

In fertility preservation of gametes and embryos through cryopreservation, whether it is slow cooling or vitrification, is also the preferred technique (Bojic et al., 2021). Notably, cryopreservation affects cell viability, it induces cellular and epigenetic changes in embryos and changes at the transcriptomic and genomic level (Estudillo et al., 2021).

Perspectives on biopreservation

The food system commonly applies chemical preservatives to guarantee food safety and extend the shelf life of food products. However, health concerns over food additives as well as environmental awareness generate growing interest towards using more natural products. These novel methods include: microorganisms used as protective cultures and their metabolites, plant and animal-based compounds and

14 *Novel Approaches in Biopreservation for Food and Clinical Purposes*

bacteriophages and endolysins. While some of these techniques are already available in the market, further studies are necessary to assess their safety and effectiveness on an industrial scale, thereby enabling more biopreservatives to become available to consumers.

In medical preservation, however, chemical preservatives and physical methods dominate the market of preservation due to their proven effectiveness and widespread availability. More studies are needed to fully understand the mechanisms of action of biopreservatives, to screen for those relevant in medical applications and to meet the regulatory criteria.

References

Abedi, E. and Hashemi, S.M.B. 2020. Lactic acid production—producing microorganisms and substrates sources-state of art. *Heliyon*, 6(10): e04974, doi: 10.1016/j.heliyon.2020.e04974.

Abrantes, C.G., Duarte, D. and Reis, C.P. 2016. An overview of pharmaceutical excipients: safe or not safe? *Journal of Pharmaceutical Sciences*, 105(7): 2019–2026, doi: 10.1016/j.xphs.2016.03.019.

Abriouel, H., Lucas, R., Ben Omar, N., Valdivia, E., Maqueda, M., Martínez-Cañamero, M. and Gálvez, A. 2005. Enterocin AS-48RJ: a variant of enterocin AS-48 chromosomally encoded by *Enterococcus faecium* RJ16 isolated from food. *Systematic and Applied Microbiology*, 28(5): 383–397, doi: 10.1016/j.syapm.2005.01.007.

Ahn, C. and Stiles, M.E. 1990. Plasmid-associated bacteriocin production by a strain of *Carnobacterium piscicola* from meat. *Applied and Environmental Microbiology*, 56(8): 2503–2510, doi: 10.1128/aem.56.8.2503-2510.1990.

Amit, S.K., Uddin, Md.M., Rahman, R., Islam, S.M.R. and Khan, M.S. 2017. A review on mechanisms and commercial aspects of food preservation and processing. *Agriculture & Food Security*, 6(1): 51, doi: 10.1186/s40066-017-0130-8.

Angelopoulou, P., Giaouris, E. and Gardikis, K. 2022. Applications and prospects of nanotechnology in food and cosmetics preservation. *Nanomaterials*, 12(7): 1196, doi: 10.3390/nano12071196.

Aymerich, T., Holo, H., Håvarstein, L.S., Hugas, M., Garriga, M. and Nes, I.F. 1996. Biochemical and genetic characterization of enterocin A from *Enterococcus faecium*, a new antilisterial bacteriocin in the pediocin family of bacteriocins. *Applied and Environmental Microbiology*, 62(5): 1676–1682, doi: 10.1128/aem.62.5.1676-1682.1996.

Baindara, P. and Mandal, S.M. 2022. Plant-derived antimicrobial peptides: novel preservatives for the food industry. *Foods, Multidisciplinary Digital Publishing Institute*, 11(16): 2415, doi: 10.3390/foods11162415.

Bédard, F. and Biron, E. 2018. Recent progress in the chemical synthesis of Class II and S-glycosylated bacteriocins. *Frontiers in Microbiology*, 9: 1048, doi: 10.3389/fmicb.2018.01048.

Bojic, S., Murray, A., Bentley, B.L., Spindler, R., Pawlik, P., Cordeiro, J.L., Bauer, R., *et al.* 2021. Winter is coming: the future of cryopreservation. *BMC Biology*, 19(1): 56, doi: 10.1186/s12915-021-00976-8.

Bolotin, A., Quinquis, B., Renault, P., Sorokin, A., Ehrlich, S.D., Kulakauskas, S., Lapidus, A., *et al.* 2004. Complete sequence and comparative genome analysis of the dairy bacterium *Streptococcus thermophilus*. *Nature Biotechnology*, Nature Publishing Group, 22(12): 1554–1558, doi: 10.1038/nbt1034.

Brede, D.A., Faye, T., Johnsborg, O., Odegård, I., Nes, I.F. and Holo, H. 2004. Molecular and genetic characterization of propionicin F, a bacteriocin from *Propionibacterium freudenreichii*. *Applied and Environmental Microbiology*, 70(12): 7303–7310, doi: 10.1128/AEM.70.12.7303-7310.2004.

Calasso, M., Marzano, M., Caponio, G.R., Celano, G., Fosso, B., Calabrese, F.M., De Palma, D., *et al.* 2023. Shelf-life extension of leavened bakery products by using bio-protective cultures and type-III sourdough. *LWT*, 177: 114587, doi: 10.1016/j.lwt.2023.114587.

Campos, C.O., Bernuci, M.P., Vireque, A.A., Campos, J.R., Silva-de-Sá, M.F., Jamur, M.C. and Rosa-e-Silva, A.C.J.S. 2012. Preventing microbial contamination during long-term *in vitro* culture of human granulosa-lutein cells: an ultrastructural analysis. *ISRN Obstetrics and Gynecology*, 2012: 152781, doi: 10.5402/2012/152781.

Casaus, P., Nilsen, T., Cintas, L.M., Nes, I.F., Hernández, P.E. and Holo, H. 1997. Enterocin B, a new bacteriocin from *Enterococcus faecium* T136 which can act synergistically with enterocin A. *Microbiology, Microbiology Society*, 143(7): 2287–2294, doi: 10.1099/00221287-143-7-2287.

Chaillou, S., Champomier-Vergès, M.-C., Cornet, M., Crutz-Le Coq, A.-M., Dudez, A.-M., Martin, V., Beaufils, S., *et al.* 2005. The complete genome sequence of the meat-borne lactic acid bacterium *Lactobacillus sakei* 23K. *Nature Biotechnology*, 23(12): 1527–1533, doi: 10.1038/nbt1160.

Chang, Y. 2020. Bacteriophage-derived endolysins applied as potent biocontrol agents to enhance food safety. *Microorganisms*, 8(5): 724, doi: 10.3390/microorganisms8050724.

Chen, S., Ren, J. and Chen, R. 2019. 5.11 - Cryopreservation and desiccation preservation of cells. *In*: Moo-Young, M. (Ed.). *Comprehensive Biotechnology (Third Edition)*, Pergamon, Oxford, pp. 157–166, doi: 10.1016/B978-0-444-64046-8.00451-1.

Chen, S.-Y., Yang, R.-S., Ci, B.-Q., Xin, W.-G., Zhang, Q.-L., Lin, L.-B. and Wang, F. 2023. A novel bacteriocin against multiple foodborne pathogens from *Lacticaseibacillus rhamnosus* isolated from juice ferments: ATF perfusion-based preparation of viable cells, characterization, antibacterial and antibiofilm activity. *Current Research in Food Science*, 6: 100484, doi: 10.1016/j. crfs.2023.100484.

Cifuentes Bachmann, D.E. and Leroy, F. 2015. Use of bioprotective cultures in fish products. *Current Opinion in Food Science*, 6: 19–23, doi: 10.1016/j.cofs.2015.11.009.

Cintas, L.M., Casaus, P., Håvarstein, L.S., Hernández, P.E. and Nes, I.F. 1997. Biochemical and genetic characterization of enterocin P, a novel sec-dependent bacteriocin from *Enterococcus faecium* P13 with a broad antimicrobial spectrum. *Applied and Environmental Microbiology*, American Society for Microbiology, 63(11): 4321–4330, doi: 10.1128/aem.63.11.4321-4330.1997.

Cintas, L.M., Casaus, P., Herranz, C., Håvarstein, L.S., Holo, H., Hernández, P.E. and Nes, I.F. 2000. Biochemical and genetic evidence that *Enterococcus faecium* L50 produces enterocins L50A and L50B, the sec-dependent enterocin P, and a novel bacteriocin secreted without an N-terminal extension termed enterocin Q. *Journal of Bacteriology*, 182(23): 6806–6814, doi: 10.1128/ JB.182.23.6806-6814.2000.

Coban, H.B. 2020. Organic acids as antimicrobial food agents: applications and microbial productions. *Bioprocess and Biosystems Engineering*, 43(4): 569–591, doi: 10.1007/s00449-019-02256-w.

Collins, F.W.J., Mesa-Pereira, B., O'Connor, P.M., Rea, M.C., Hill, C. and Ross, R.P. 2018. Reincarnation of Bacteriocins From the *Lactobacillus Pangenomic* Graveyard. *Frontiers in Microbiology*, 9: 1298, doi: 10.3389/fmicb.2018.01298.

Cook, P.W. and Nightingale, K.K. 2018. Use of omics methods for the advancement of food quality and food safety. *Animal Frontiers*, 8(4): 33–41, doi: 10.1093/af/vfy024.

da Costa, R.J., Voloski, F.L.S., Mondadori, R.G., Duval, E.H. and Fiorentini, Â.M. 2019. Preservation of meat products with bacteriocins produced by lactic acid bacteria isolated from meat. *Journal of Food Quality*, Hindawi, 2019: e4726510, doi: 10.1155/2019/4726510.

Costa, S.S., da Silva Moia, G., Silva, A., Baraúna, R.A. and de Oliveira Veras, A.A. 2023. BADASS: Bacteriocin-diversity assessment software. *BMC Bioinformatics*, 24(1): 24, doi: 10.1186/s12859-022-05106-x.

Cotter, P.D., Ross, R.P. and Hill, C. 2013. Bacteriocins—a viable alternative to antibiotics? *Nature Reviews Microbiology*, Nature Publishing Group, 11(2): 95–105, doi: 10.1038/nrmicro2937.

Dai, Y. and Tyl, C. 2021. A review on mechanistic aspects of individual versus combined uses of enzymes as clean label-friendly dough conditioners in breads. *Journal of Food Science*, 86(5): 1583–1598, doi: 10.1111/1750-3841.15713.

Davidson, P.M., Taylor, T.M. and Schmidt, S.E. 2012. Chemical preservatives and natural antimicrobial compounds. *Food Microbiology*, John Wiley & Sons, Ltd, pp. 765–801, doi: 10.1128/9781555818463.ch30.

De Vuyst, L. and Tsakalidou, E. 2008. Streptococcus macedonicus, a multi-functional and promising species for dairy fermentations. *International Dairy Journal*, 18(5): 476–485, doi: 10.1016/j. idairyj.2007.10.006.

Dickman, R., Mitchell, S.A., Figueiredo, A.M., Hansen, D.F. and Tabor, A.B. 2019. Molecular recognition of lipid II by lantibiotics: synthesis and conformational studies of analogues of nisin and Mutacin

16 *Novel Approaches in Biopreservation for Food and Clinical Purposes*

Rings A and B. *The Journal of Organic Chemistry*, American Chemical Society, 84(18): 11493–11512, doi: 10.1021/acs.joc.9b01253.

Diep, D.B., Skaugen, M., Salehian, Z., Holo, H. and Nes, I.F. 2007. Common mechanisms of target cell recognition and immunity for class II bacteriocins. *Proceedings of the National Academy of Sciences*, Proceedings of the National Academy of Sciences, 104(7): 2384–2389, doi: 10.1073/pnas.0608775104.

Donadio, S., Sosio, M., Serina, S. and Mercorillo, D. 2009. Genes and proteins for the biosynthesis of the lantibiotic 107891. 12 February.

Duan, C., Meng, X., Meng, J., Khan, Md.I.H., Dai, L., Khan, A., An, X., *et al.* 2019. Chitosan as A preservative for fruits and vegetables: a review on chemistry and antimicrobial properties. *Journal of Bioresources and Bioproducts*, 4(1): 11–21, doi: 10.21967/jbb.v4i1.189.

Ehrmann, M.A., Remiger, A., Eijsink, V.G.H. and Vogel, R.F. 2000. A gene cluster encoding plantaricin 1.25β and other bacteriocin-like peptides in *Lactobacillus plantarum* TMW1.25. *Biochimica et Biophysica Acta (BBA) - Gene Structure and Expression*, 1490(3): 355–361, doi: 10.1016/S0167-4781(00)00003-8.

El-Saber Batiha, G., Hussein, D.E., Algammal, A.M., George, T.T., Jeandet, P., Al-Snafi, A.E., Tiwari, A., *et al.* 2021. Application of natural antimicrobials in food preservation: recent views. *Food Control*, 126: 108066, doi: 10.1016/j.foodcont.2021.108066.

Endersen, L. and Coffey, A. 2020. The use of bacteriophages for food safety. *Current Opinion in Food Science*, 36: 1–8, doi: 10.1016/j.cofs.2020.10.006.

Estudillo, E., Jiménez, A., Bustamante-Nieves, P.E., Palacios-Reyes, C., Velasco, I. and López-Ornelas, A. 2021. Cryopreservation of gametes and embryos and their molecular changes. *International Journal of Molecular Sciences.* Multidisciplinary Digital Publishing Institute, 22(19): 10864, doi: 10.3390/ijms221910864.

FAO. 2019. The State of Food and Agriculture 2019. Moving forward on food loss and waste reduction.

Faye, T., Langsrud, T., Nes, I.F. and Holo, H. 2000. Biochemical and genetic characterization of propionicin T1, a new bacteriocin from *Propionibacterium thoenii. Applied and Environmental Microbiology*, 66(10): 4230–4236, doi: 10.1128/AEM.66.10.4230-4236.2000.

Felix, J.V., Papathanasopoulos, M.A., Smith, A.A., von Holy, A. and Hastings, J.W. 1994. Characterization of leucocin B-Ta11a: A bacteriocin from *Leuconostoc carnosum* Ta11a isolated from meat. *Current Microbiology*, 29(4): 207–212, doi: 10.1007/BF01570155.

Ferchichi, M., Frère, J., Mabrouk, K. and Manai, M. 2001. Lactococcin MMFII, a novel class IIa bacteriocin produced by *Lactococcus lactis* MMFII, isolated from a Tunisian dairy product. *FEMS Microbiology Letters*, 205(1): 49–55, doi: 10.1111/j.1574-6968.2001.tb10924.x.

Field, D., Connor, P.M.O., Cotter, P.D., Hill, C. and Ross, R.P. 2008. The generation of nisin variants with enhanced activity against specific gram-positive pathogens. *Molecular Microbiology*, 69(1): 218–230, doi: 10.1111/j.1365-2958.2008.06279.x.

Field, D., Begley, M., O'Connor, P.M., Daly, K.M., Hugenholtz, F., Cotter, P.D., Hill, C. *et al.* 2012. Bioengineered Nisin A derivatives with enhanced activity against both gram positive and gram negative pathogens. *PLOS ONE*, Public Library of Science, 7(10): e46884, doi: 10.1371/journal.pone.0046884.

Fontaine, L., Wahl, A., Fléchard, M., Mignolet, J. and Hols, P. 2015. Regulation of competence for natural transformation in streptococci. *Infection, Genetics and Evolution*, 33: 343–360, doi: 10.1016/j.meegid.2014.09.010.

Fricourt, B.V., Barefoot, S.F., Testin, R.F. and Hayasaka, S.S. 1994. Detection and activity of Plantaricin F an antibacterial substance from *Lactobacillus plantarum* BF001 Isolated from processed channel catfish. *Journal of Food Protection*, 57(8): 698–702, doi: 10.4315/0362-028X-57.8.698.

Gabrielsen, C., Brede, D.A., Hernández, P.E., Nes, I.F. and Diep, D.B. 2012. The maltose ABC transporter in *Lactococcus lactis* facilitates high-level sensitivity to the circular bacteriocin garvicin ML", *Antimicrobial Agents and Chemotherapy*, 56(6): 2908–2915, doi: 10.1128/AAC.00314-12.

García, P., Martínez, B., Obeso, J.M. and Rodríguez, A. 2008. Bacteriophages and their application in food safety. *Letters in Applied Microbiology*, 47(6): 479–485, doi: 10.1111/j.1472-765X.2008.02458.x.

Garver, K.I. and Muriana, P.M. 1994. Purification and partial amino acid sequence of curvaticin FS47, a heat-stable bacteriocin produced by *Lactobacillus curvatus* FS47. *Applied and Environmental Microbiology*, 60(6): 2191–2195, doi: 10.1128/aem.60.6.2191-2195.1994.

Ge, J., Sun, Y., Xin, X., Wang, Y. and Ping, W. 2016. Purification and partial characterization of a novel bacteriocin synthesized by *Lactobacillus paracasei* HD1-7 isolated from Chinese Sauerkraut Juice. *Scientific Reports*, 6: 19366, doi: 10.1038/srep19366.

Georgalaki, M., Papadimitriou, K., Anastasiou, R., Pot, B., Van Driessche, G., Devreese, B. and Tsakalidou, E. 2013. Macedovicin, the second food-grade lantibiotic produced by *Streptococcus macedonicus* ACA-DC 198. *Food Microbiology*, 33(1): 124–130, doi: 10.1016/j.fm.2012.09.008.

Georgalaki, M.D., Van den Berghe, E., Kritikos, D., Devreese, B., Van Beeumen, J., Kalantzopoulos, G., De Vuyst, L., *et al.* 2002. Macedocin, a food-grade lantibiotic produced by *Streptococcus macedonicus* ACA-DC 198. *Applied and Environmental Microbiology*, American Society for Microbiology, 68(12): 5891–5903, doi: 10.1128/AEM.68.12.5891-5903.2002.

Gomes, R.J., Borges, M. de F., Rosa, M. de F., Castro-Gómez, R.J.H. and Spinosa, W.A. 2018. Acetic acid bacteria in the food industry: systematics, characteristics and applications. *Food Technology and Biotechnology*, 56(2): 139–151, doi: 10.17113/ftb.56.02.18.5593.

Gottardi, D., Bukvicki, D., Prasad, S. and Tyagi, A.K. 2016. Beneficial effects of spices in food preservation and safety. *Frontiers in Microbiology*, Vol. 7.

Hammami, R., Zouhir, A., Le Lay, C., Ben Hamida, J. and Fliss, I. 2010. BACTIBASE second release: a database and tool platform for bacteriocin characterization. *BMC Microbiology*, 10(1): 22, doi: 10.1186/1471-2180-10-22.

Hata, T., Tanaka, R. and Ohmomo, S. 2010. Isolation and characterization of plantaricin ASM1: anew bacteriocin produced by *Lactobacillus plantarum* A-1. *International Journal of Food Microbiology*, 137(1): 94–99, doi: 10.1016/j.ijfoodmicro.2009.10.021.

He, H., Wu, S., Mei, M., Ning, J., Li, C., Ma, L., Zhang, G., *et al.* 2020. A combinational strategy for effective heterologous production of functional human Lysozyme in *Pichia pastoris*. *Frontiers in Bioengineering and Biotechnology*, 8: 118, doi: 10.3389/fbioe.2020.00118.

Hebert, E.M., Saavedra, L., Taranto, M.P., Mozzi, F., Magni, C., Nader, M.E.F., Font de Valdez, G., *et al.* 2012. Genome sequence of the bacteriocin-producing *Lactobacillus curvatus* Strain CRL705. *Journal of Bacteriology*, American Society for Microbiology, 194(2): 538–539, doi: 10.1128/JB.06416-11.

Héchard, Y., Dérijard, B., Letellier, F. and Cenatiempo, Y. 1992. Characterization and purification of mesentericin Y105, an anti-Listeria bacteriocin from *Leuconostoc mesenteroides*. *Microbiology*, Microbiology Society, 138(12): 2725–2731, doi: 10.1099/00221287-138-12-2725.

van Heel, A.J., de Jong, A., Song, C., Viel, J.H., Kok, J. and Kuipers, O.P. 2018. BAGEL4: a user-friendly web server to thoroughly mine RiPPs and bacteriocins. *Nucleic Acids Research*, 46(W1): W278–W281, doi: 10.1093/nar/gky383.

Herbin, S., Mathieu, F., Brulé, F., Branlant, C., Lefebvre, G. and Lebrihi, A. 1997. Characteristics and genetic determinants of bacteriocin activities produced by *Carnobacterium piscicola* CP5 isolated from cheese. *Current Microbiology*, 35(6): 319–326, doi: 10.1007/s002849900262.

Holo, H., Nilssen, O. and Nes, I.F. 1991. Lactococcin A, a new bacteriocin from *Lactococcus lactis* subsp. cremoris: isolation and characterization of the protein and its gene. *Journal of Bacteriology*, 173(12): 3879–3887, doi: 10.1128/jb.173.12.3879-3887.1991.

Huot, E., Meghrous, J., Barrena-Gonzalez, C. and Petitdemange, H. 1996. Bacteriocin J46, a new bacteriocin produced by *Lactococcus lactis* Subsp. cremoris J46: isolation and characterization of the protein and its gene. *Anaerobe*, 2(3): 137–145, doi: 10.1006/anae.1996.0018.

Ibarra-Sánchez, L.A., El-Haddad, N., Mahmoud, D., Miller, M.J. and Karam, L. 2020. Invited review: advances in nisin use for preservation of dairy products. *Journal of Dairy Science*, 103(3): 2041–2052, doi: 10.3168/jds.2019-17498.

Ishibashi, N., Himeno, K., Fujita, K., Masuda, Y., Perez, R.H., Zendo, T., Wilaipun, P., *et al.* 2012. Purification and characterization of multiple bacteriocins and an inducing peptide produced by *Enterococcus faecium* NKR-5-3 from Thai fermented fish. *Bioscience, Biotechnology, and Biochemistry*, 76(5): 947–953, doi: 10.1271/bbb.110972.

Ivanova, I., Kabadjova, P., Pantev, A., Danova, S. and Dousset, X. 2000. Detection, purification and partial characterization of a novel bacteriocin substance produced by *Lactococous lactis* subsp. Lactis B14 Isolated From Boza|Bulgarian Traditional Cereal Beverage.

Izquierdo, E., Wagner, C., Marchioni, E., Aoude-Werner, D. and Ennahar, S. 2009. Enterocin 96, a novel Class II bacteriocin produced by *Enterococcus faecalis* WHE 96, isolated from munster cheese.

18 *Novel Approaches in Biopreservation for Food and Clinical Purposes*

Applied and Environmental Microbiology, American Society for Microbiology, 75(13): 4273–4276, doi: 10.1128/AEM.02772-08.

Jack, R.W., Wan, J., Gordon, J., Harmark, K., Davidson, B.E., Hillier, A.J., Wettenhall, R.E., Hickey, M.W. and Coventry, M.J. 1996. Characterization of the chemical and antimicrobial properties of piscicolin 126, a bacteriocin produced by *Carnobacterium piscicola* JG126. *Applied and Environmental Microbiology*, 62(8), doi: 10.1128/aem.62.8.2897-2903.1996.

Jańczuk, A., Brodziak, A., Czernecki, T. and Król, J. 2023. Lactoferrin—the health-promoting properties and contemporary application with genetic aspects. *Foods*, Multidisciplinary Digital Publishing Institute, 12(1): 70, doi: 10.3390/foods12010070.

Kabuki, T., Uenishi, H., Watanabe, M., Seto, Y. and Nakajima, H. 2007. Characterization of a bacteriocin, Thermophilin 1277, produced by *Streptococcus thermophilus* SBT1277. *Journal of Applied Microbiology*, 102(4): 971–980, doi: 10.1111/j.1365-2672.2006.03159.x.

Kaiser, A.L. and Montville, T.J. 1996. Purification of the bacteriocin bavaricin MN and characterization of its mode of action against *Listeria monocytogenes* Scott A cells and lipid vesicles. *Applied and Environmental Microbiology*, 62(12): 4529–4535.

Kanatani, K., Oshimura, M. and Sano, K. 1995. Isolation and characterization of acidocin A and cloning of the bacteriocin gene from *Lactobacillus acidophilus*. *Applied and Environmental Microbiology*, 61(3): 1061–1067.

Khorshidian, N., Khanniri, E., Koushki, M.R., Sohrabvandi, S. and Yousefi, M. 2022). An overview of antimicrobial activity of lysozyme and its functionality in cheese. *Frontiers in Nutrition*, Vol. 9.

Kjos, M., Salehian, Z., Nes, I.F. and Diep, D.B. 2010. An extracellular loop of the mannose phosphotransferase system component IIC Is responsible for specific targeting by class IIa bacteriocins. *Journal of Bacteriology*, 192(22): 5906–5913, doi: 10.1128/JB.00777-10.

Kjos, M., Oppegård, C., Diep, D.B., Nes, I.F., Veening, J.-W., Nissen-Meyer, J. and Kristensen, T. 2014. Sensitivity to the two-peptide bacteriocin lactococcin G is dependent on UppP, an enzyme involved in cell-wall synthesis. *Molecular Microbiology*, 92(6): 1177–1187, doi: 10.1111/mmi.12632.

Kuo, Y.-C., Liu, C.-F., Lin, J.-F., Li, A.-C., Lo, T.-C. and Lin, T.-H. 2013. Characterization of putative class II bacteriocins identified from a non-bacteriocin-producing strain *Lactobacillus casei* ATCC 334. *Applied Microbiology and Biotechnology*, 97(1): 237–246, doi: 10.1007/s00253-012-4149-2.

Larsen, A.G., Vogensen, F.K. and Josephsen, J. 1993. Antimicrobial activity of lactic acid bacteria isolated from sour doughs: purification and characterization of bavaricin A, a bacteriocin produced by *Lactobacillus bavaricus* MI401. *The Journal of Applied Bacteriology*, 75(2): 113–122, doi: 10.1111/j.1365-2672.1993.tb02755.x.

Leech, J., Cabrera-Rubio, R., Walsh, A.M., Macori, G., Walsh, C.J., Barton, W., Finnegan, L., *et al.* 2020. Fermented-food metagenomics reveals substrate-associated differences in taxonomy and health-associated and antibiotic resistance determinants. *MSystems*, 5(6): e00522-20, doi: 10.1128/mSystems.00522-20.

Leeuwendaal, N.K., Stanton, C., O'Toole, P.W. and Beresford, T.P. 2022. Fermented foods, health and the gut microbiome. *Nutrients*, 14(7): 1527, doi: 10.3390/nu14071527.

Lepoittevin, M., Giraud, S., Kerforne, T., Barrou, B., Badet, L., Bucur, P., Salamé, E., *et al.* 2022. Preservation of organs to be transplanted: an essential step in the transplant process. *International Journal of Molecular Sciences*, 23(9): 4989, doi: 10.3390/ijms23094989.

Leroy, F. and De Vuyst, L. 2004. Lactic acid bacteria as functional starter cultures for the food fermentation industry. *Trends in Food Science & Technology*, 15(2): 67–78, doi: 10.1016/j.tifs.2003.09.004.

Linares-Morales, J.R., Gutiérrez-Méndez, N., Rivera-Chavira, B.E., Pérez-Vega, S.B. and Nevárez-Moorillón, G.V. 2018. Biocontrol processes in fruits and fresh produce, the use of lactic acid bacteria as a sustainable option. *Frontiers in Sustainable Food Systems*, Vol. 2.

Lopez-Moya, F., Suarez-Fernandez, M. and Lopez-Llorca, L.V. 2019. Molecular mechanisms of chitosan interactions with fungi and plants. *International Journal of Molecular Sciences*, 20(2): 332, doi: 10.3390/ijms20020332.

Lorenzo, J.M., Munekata, P.E., Dominguez, R., Pateiro, M., Saraiva, J.A. and Franco, D. 2018. Main groups of microorganisms of relevance for food safety and stability. *Innovative Technologies for Food Preservation*, pp. 53–107, doi: 10.1016/B978-0-12-811031-7.00003-0.

Lund, P.A., De Biase, D., Liran, O., Scheler, O., Mira, N.P., Cetecioglu, Z., Fernández, E.N., *et al.* 2020. Understanding how microorganisms respond to Acid pH Is central to their control and successful exploitation. *Frontiers in Microbiology*, 11: 556140, doi: 10.3389/fmicb.2020.556140.

Mahgoub, S.A., Kedra, E.G.A., Abdelfattah, H.I., Abdelbasit, H.M., Alamoudi, S.A., Al-Quwaie, D.A., Selim, S., *et al.* 2022. Bioconversion of some agro-residues into organic acids by cellulolytic rock-phosphate-solubilizing *Aspergillus japonicus*. *Fermentation*, Multidisciplinary Digital Publishing Institute, 8(9): 437, doi: 10.3390/fermentation8090437.

Makarova, K., Slesarev, A., Wolf, Y., Sorokin, A., Mirkin, B., Koonin, E., Pavlov, A., *et al.* 2006. Comparative genomics of the lactic acid bacteria. *Proceedings of the National Academy of Sciences*, Proceedings of the National Academy of Sciences, 103(42): 15611–15616, doi: 10.1073/pnas.0607117103.

Martínez, B., Suárez, J.E. and Rodríguez, A. 1996. Lactococcin 972: a homodimeric lactococcal bacteriocin whose primary target is not the plasma membrane. *Microbiology*, Microbiology Society, 142(9): 2393–2398, doi: 10.1099/00221287-142-9-2393.

Martin-Visscher, L.A., van Belkum, M.J., Garneau-Tsodikova, S., Whittal, R.M., Zheng, J., McMullen, L.M. and Vederas, J.C. 2008. Isolation and characterization of carnocyclin a, a novel circular bacteriocin produced by *Carnobacterium maltaromaticum* UAL307. *Applied and Environmental Microbiology*. American Society for Microbiology, 74(15): 4756–4763, doi: 10.1128/AEM.00817-08.

McCormick, J.K., Worobo, R.W. and Stiles, M.E. 1996. Expression of the antimicrobial peptide carnobacteriocin B2 by a signal peptide-dependent general secretory pathway. *Applied and Environmental Microbiology*, 62(11): 4095–4099, doi: 10.1128/aem.62.11.4095-4099.1996.

Miescher, S., Stierli, M.P., Teuber, M. and Meile, L. 2000. Propionicin SM1, a bacteriocin from *Propionibacterium jensenii* DF1: isolation and characterization of the protein and its gene. *Systematic and Applied Microbiology*, 23(2): 174–184, doi: 10.1016/S0723-2020(00)80002-8.

Mihai, A.L. and Popa, M.E. 2015. Chitosan coatings, a natural and sustainable food preservation method. *Journal of Biotechnology*, 208: S81, doi: 10.1016/j.jbiotec.2015.06.250.

Moye, Z.D., Woolston, J. and Sulakvelidze, A. 2018. Bacteriophage applications for food production and processing. *Viruses*, 10(4): 205, doi: 10.3390/v10040205.

Mudgil, D., Barak, S. and Khatkar, B.S. 2014. Guar gum: processing, properties and food applications—A Review. *Journal of Food Science and Technology*, 51(3): 409–418, doi: 10.1007/s13197-011-0522-x.

Mulders, J.W., Boerrigter, I.J., Rollema, H.S., Siezen, R.J. and de Vos, W.M. 1991. Identification and characterization of the lantibiotic nisin Z, a natural nisin variant. *European Journal of Biochemistry*, 201(3): 581–584, doi: 10.1111/j.1432-1033.1991.tb16317.x.

Netz, D.J.A., Bastos, M. do C. de F. and Sahl, H.-G. 2002. Mode of action of the antimicrobial peptide Aureocin A53 from *Staphylococcus aureus*. *Applied and Environmental Microbiology*, American Society for Microbiology, 68(11): 5274–5280, doi: 10.1128/AEM.68.11.5274-5280.2002.

Nkosi, D.V., Bekker, J.L. and Hoffman, L.C. 2021. The use of organic acids (Lactic and Acetic) as a microbial decontaminant during the slaughter of meat animal species: a review. *Foods*, 10(10): 2293, doi: 10.3390/foods10102293.

Okpara, M.O. 2022. Microbial enzymes and their applications in food industry: a mini-review. *Advances in Enzyme Research*, Scientific Research Publishing, 10(1): 23–47, doi: 10.4236/aer.2022.101002.

Oliver, A.E. 2012. Dry state preservation of nucleated cells: progress and challenges. *Biopreservation and Biobanking*, 10(4): 376–385, doi: 10.1089/bio.2012.0020.

O'Mahony, T., Rekhif, N., Cavadini, C. and Fitzgerald, G.F. 2001. The application of a fermented food ingredient containing 'variacin', a novel antimicrobial produced by Kocuria varians, to control the growth of *Bacillus cereus* in chilled dairy products. *Journal of Applied Microbiology*, 90(1): 106–114, doi: 10.1046/j.1365-2672.2001.01222.x.

Palacios, J., Vignolo, G., Farías, M.E., de Ruiz Holgado, A.P., Oliver, G. and Sesma, F. 1999. Purification and amino acid sequence of lactocin 705, a bacteriocin produced by *Lactobacillus casei* CRL 705. *Microbiological Research*, 154(2): 199–204, doi: 10.1016/S0944-5013(99)80015-9.

Palakawong Na Ayudthaya, S., van de Weijer, A.H.P., van Gelder, A.H., Stams, A.J.M., de Vos, W.M. and Plugge, C.M. 2018. Organic acid production from potato starch waste fermentation by rumen microbial communities from Dutch and Thai dairy cows. *Biotechnology for Biofuels*, 11: 13, doi: 10.1186/s13068-018-1012-4.

20 Novel Approaches in Biopreservation for Food and Clinical Purposes

Papagianni, M. and Sergelidis, D. 2013. Effects of the presence of the curing agent sodium nitrite, used in the production of fermented sausages, on bacteriocin production by Weissella paramesenteroides DX grown in meat simulation medium. *Enzyme and Microbial Technology*, 53(1): 1–5, doi: 10.1016/j.enzmictec.2013.04.003.

Parkhill, J., Sebaihia, M., Preston, A., Murphy, L.D., Thomson, N., Harris, D.E., Holden, M.T.G., *et al.* 2003. Comparative analysis of the genome sequences of *Bordetella pertussis, Bordetella parapertussis and Bordetella bronchiseptica. Nature Genetics*, Nature Publishing Group, 35(1): 32–40, doi: 10.1038/ng1227.

Peña, N., Bland, M.J., Sevillano, E., Muñoz-Atienza, E., Lafuente, I., Bakkoury, M.E., Cintas, L.M., *et al.* 2022. *In vitro* and *in vivo* production and split-intein mediated ligation (SIML) of circular bacteriocins. *Frontiers in Microbiology*, 13: 1052686, doi: 10.3389/fmicb.2022.1052686.

Pérez-Ramos, A., Madi-Moussa, D., Coucheney, F. and Drider, D. 2021. Current knowledge of the mode of action and immunity mechanisms of LAB-Bacteriocins. *Microorganisms*, 9(10): 2107, doi: 10.3390/microorganisms9102107.

Pilevar, Z., Abhari, K., Tahmasebi, H., Beikzadeh, S., Afshari, R., Eskandari, S., Bozorg, M.J.A., *et al.* 2022. Antimicrobial properties of lysozyme in meat and meat products: possibilities and challenges. *Acta Scientiarum. Animal Sciences*, Editora da Universidade Estadual de Maringá - EDUEM, 44: e55262, doi: 10.4025/actascianimsci.v44i1.55262.

Piper, C., Hill, C., Cotter, P.D. and Ross, R.P. 2011. Bioengineering of a Nisin A-producing *Lactococcus lactis* to create isogenic strains producing the natural variants Nisin F, Q and Z. *Microbial Biotechnology*, 4(3): 375–382, doi: 10.1111/j.1751-7915.2010.00207.x.

Quadri, L.E., Sailer, M., Roy, K.L., Vederas, J.C. and Stiles, M.E. 1994. Chemical and genetic characterization of bacteriocins produced by *Carnobacterium piscicola* LV17B. *Journal of Biological Chemistry*, 269(16): 12204–12211, doi: 10.1016/S0021-9258(17)32702-3.

Raveendran, S., Parameswaran, B., Ummalyma, S.B., Abraham, A., Mathew, A.K., Madhavan, A., Rebello, S., *et al.* 2018. Applications of microbial enzymes in food industry. *Food Technology and Biotechnology*, 56(1): 16–30, doi: 10.17113/ftb.56.01.18.5491.

Rekhif, N., Atrih, A. and Lefebvrexy, G. 1995. Activity of plantaricin SA6, a bacteriocin produced by *Lactobacillus plantarum* SA6 isolated from fermented sausage. *Journal of Applied Bacteriology*, 78(4): 349–358, doi: 10.1111/j.1365-2672.1995.tb03417.x.

Ricke, S.C., Dittoe, D.K. and Richardson, K.E. 2020. Formic acid as an antimicrobial for poultry production: a review. *Frontiers in Veterinary Science*, Vol. 7.

Rivas, F.P., Castro, M.P., Vallejo, M., Marguet, E. and Campos, C.A. 2014. Sakacin Q produced by *Lactobacillus curvatus* ACU-1: functionality characterization and antilisterial activity on cooked meat surface. *Meat Science*, 97(4): 475–479, doi: 10.1016/j.meatsci.2014.03.003.

Rodríguez, J.M., Martínez, M.I. and Kok, J. 2002. Pediocin PA-1, a wide-spectrum bacteriocin from lactic acid bacteria. *Critical Reviews in Food Science and Nutrition*, 42(2): 91–121, doi: 10.1080/10408690290825475.

Ryan, M.P., Rea, M.C., Hill, C. and Ross, R.P. 1996. An application in cheddar cheese manufacture for a strain of *Lactococcus lactis* producing a novel broad-spectrum bacteriocin, lacticin 3147.*Applied and Environmental Microbiology*, 62(2): 612–619, doi: 10.1128/aem.62.2.612-619.1996.

Saavedra, L., Minahk, C., de Ruiz Holgado, A.P. and Sesma, F. 2004. Enhancement of the enterocin CRL35 activity by a synthetic peptide derived from the NH2-terminal sequence. *Antimicrobial Agents and Chemotherapy*, 48(7): 2778–2781, doi: 10.1128/AAC.48.7.2778-2781.2004.

Sawa, N., Zendo, T., Kiyofuji, J., Fujita, K., Himeno, K., Nakayama, J. and Sonomoto, K. 2009. Identification and characterization of lactocyclicin Q, a novel cyclic bacteriocin produced by *Lactococcus* sp. strain QU 12. *Applied and Environmental Microbiology*, American Society for Microbiology, 75(6): 1552–1558, doi: 10.1128/AEM.02299-08.

Schillinger, U., Stiles, M.E. and Holzapfel, W.H. 1993). Bacteriocin production by *Carnobacterium piscicola* LV 61. *International Journal of Food Microbiology*, 20(3): 131–147, doi: 10.1016/0168-1605(93)90106-q.

Schillinger, U. 1994. Sakacin a produced by *Lactobacillus sake* Lb 706. pp. 419–434. *In:* De Vuyst, L. and Vandamme, E.J. (Eds.). *Bacteriocins of Lactic Acid Bacteria: Microbiology, Genetics and Applications*, Springer US, Boston, MA, doi: 10.1007/978-1-4615-2668-1_16.

Schmelcher, M. and Loessner, M.J. 2016. Bacteriophage endolysins: applications for food safety. *Current Opinion in Biotechnology*, 37: 76–87, doi: 10.1016/j.copbio.2015.10.005.

Settanni, L., Massitti, O., Van Sinderen, D. and Corsetti, A. 2005. *In situ* activity of a bacteriocin-producing *Lactococcus lactis* strain. Influence on the interactions between lactic acid bacteria during sourdough fermentation. *Journal of Applied Microbiology*, 99(3): 670–681, doi: 10.1111/j.1365-2672.2005.02647.x.

Shitole, S., Shinde, S., Waghmare, S. and Kamble, H. 2022. A Review On: Preservatives Used in Pharmaceuticals and Impacts on Health, 5(7).

Silva, C.C.G., Silva, S.P.M. and Ribeiro, S.C. 2018. Application of bacteriocins and protective cultures in dairy food preservation. *Frontiers in Microbiology*, Vol. 9.

Simons, A., Alhanout, K. and Duval, R.E. 2020. Bacteriocins, antimicrobial peptides from bacterial origin: overview of their biology and their impact against multidrug-resistant bacteria.*Microorganisms*, 8(5): 639, doi: 10.3390/microorganisms8050639.

Singh, P.K., Singh, R.P., Singh, P. and Singh, R.L. 2019). Chapter 2 - food hazards: physical, chemical, and biological. pp. 15–65. *In:* Singh, R.L. and Mondal, S. (Eds.). *Food Safety and Human Health*, Academic Press, doi: 10.1016/B978-0-12-816333-7.00002-3.

Singh, V.P. 2018. Recent approaches in food bio-preservation—a review. *Open Veterinary Journal*, 8(1): 104–111, doi: 10.4314/ovj.v8i1.16.

Souza, L.V., Martins, E., Moreira, I.M.F.B. and de Carvalho, A.F. 2022. Strategies for the development of bioprotective cultures in food preservation. *International Journal of Microbiology*, Hindawi, 2022: e6264170, doi: 10.1155/2022/6264170.

Sutyak, K. e., Wirawan, R. e., Aroutcheva, A. a. and Chikindas, M. l. 2008. Isolation of the *Bacillus subtilis* antimicrobial peptide subtilosin from the dairy product-derived *Bacillus amyloliquefaciens*. *Journal of Applied Microbiology*, 104(4): 1067–1074, doi: 10.1111/j.1365-2672.2007.03626.x.

Tahara, T., Kanatani, K., Yoshida, K., Miura, H., Sakamoto, M. and Oshimura, M. 1992. Purification and some properties of acidocin 8912, a novel bacteriocin produced by *Lactobacillus acidophilus* TK8912. *Bioscience, Biotechnology, and Biochemistry*, 56(8): 1212–1215, doi: 10.1271/bbb.56.1212.

Tahiri, I., Desbiens, M., Benech, R., Kheadr, E., Lacroix, C., Thibault, S., Ouellet, D., *et al.* 2004. Purification, characterization and amino acid sequencing of divergicin M35: a novel class IIa bacteriocin produced by *Carnobacterium divergens* M35. *International Journal of Food Microbiology*, 97(2): 123–136, doi: 10.1016/j.ijfoodmicro.2004.04.013.

Taylor, M.J., Weegman, B.P., Baicu, S.C. and Giwa, S.E. 2019. New approaches to cryopreservation of cells, tissues, and organs. *Transfusion Medicine and Hemotherapy*, 46(3): 197–215, doi: 10.1159/000499453.

Thallinger, B., Prasetyo, E.N., Nyanhongo, G.S. and Guebitz, G.M. 2013. Antimicrobial enzymes: an emerging strategy to fight microbes and microbial biofilms. *Biotechnology Journal*, 8(1): 97–109, doi: 10.1002/biot.201200313.

Tichaczek, P.S., Nissen-Meyer, J., Nes, I.F., Vogel, R.F. and Hammes, W.P. 1992. Characterization of the bacteriocins curvacin A from *Lactobacillus curvatus* LTH1174 and Sakacin P from L. sake LTH673. *Systematic and Applied Microbiology*, 15(3): 460–468, doi: 10.1016/S0723-2020(11)80223-7.

Tiwari, B.K., Valdramidis, V.P., O' Donnell, C.P., Muthukumarappan, K., Bourke, P. and Cullen, P.J. 2009. Application of natural antimicrobials for food preservation. *Journal of Agricultural and Food Chemistry*, American Chemical Society, 57(14): 5987–6000, doi: 10.1021/jf900668n.

Todorov, S.D. and Dicks, L.M.T. 2004. Screening of lactic-acid bacteria from South African barley beer for the production of bacteriocin-like compounds. *Folia Microbiologica*, 49(4): 406–410, doi: 10.1007/BF02931601.

Tosukhowong, A., Zendo, T., Visessanguan, W., Roytrakul, S., Pumpuang, L., Jaresitthikunchai, J. and Sonomoto, K. 2012. Garvieacin Q, a novel class II bacteriocin from *Lactococcus garvieae* BCC 43578. *Applied and Environmental Microbiology*, American Society for Microbiology, 78(5): 1619–1623, doi: 10.1128/AEM.06891-11.

Tulini, F.L., Lohans, C.T., Bordon, K.C.F., Zheng, J., Arantes, E.C., Vederas, J.C. and De Martinis, E.C.P. 2014. Purification and characterization of antimicrobial peptides from fish isolate *Carnobacterium maltaromaticum* C2: Carnobacteriocin X and carnolysins A1 and A2. *International Journal of Food Microbiology*, 173: 81–88, doi: 10.1016/j.ijfoodmicro.2013.12.019.

22 *Novel Approaches in Biopreservation for Food and Clinical Purposes*

United Nations Environment Programme (2021). Food Waste Index Report 2021. Nairobi.

Urso, R., Rantsiou, K., Cantoni, C., Comi, G. and Cocolin, L. 2006. Sequencing and expression analysis of the sakacin P bacteriocin produced by a *Lactobacillus sakei* strain isolated from naturally fermented sausages. *Applied Microbiology and Biotechnology*, 71(4): 480–485, doi: 10.1007/s00253-005-0172-x.

Varsha, K.K. and Nampoothiri, K.M. 2016. Appraisal of lactic acid bacteria as protective cultures. *Food Control*, 69: 61–64, doi: 10.1016/j.foodcont.2016.04.032.

Venema, K., Abee, T., Haandrikman, A.J., Leenhouts, K.J., Kok, J., Konings, W.N. and Venema, G. 1993. Mode of action of lactococcin B, a thiol-activated bacteriocin from *Lactococcus lactis. Applied and Environmental Microbiology*, 59(4): 1041–1048.

Verma, D.K., Thakur, M., Singh, S., Tripathy, S., Gupta, A.K., Baranwal, D., Patel, A.R., *et al.* 2022. Bacteriocins as antimicrobial and preservative agents in food: Biosynthesis, separation and application. *Food Bioscience*, 46: 101594, doi: 10.1016/j.fbio.2022.101594.

WHO. 2015. Food safety, available at: https://www.who.int/news-room/fact-sheets/detail/food-safety (accessed 13 March 2023).

Xiao, H., Chen, X., Chen, M., Tang, S., Zhao, X. and Huan, L. 2004). Bovicin HJ50, a novel lantibiotic produced by *Streptococcus bovis* HJ50. *Microbiology*, Microbiology Society, 150(1): 103–108, doi: 10.1099/mic.0.26437-0.

Xiraphi, N., Georgalaki, M., Driessche, G.V., Devreese, B., Beeumen, J.V., Tsakalidou, E., Metaxopoulos, J., *et al.* 2006. Purification and characterization of curvaticin L442, a bacteriocin produced by *Lactobacillus curvatus* L442. *Antonie Van Leeuwenhoek*, 89(1): 19–26, doi: 10.1007/s10482-005-9004-3.

Xu, M.M., Kaur, M., Pillidge, C.J. and Torley, P.J. 2021. Evaluation of the potential of protective cultures to extend the microbial shelf-life of chilled lamb meat. *Meat Science*, 181: 108613, doi: 10.1016/j.meatsci.2021.108613.

Yilmaz Atay, H. 2020. Antibacterial activity of chitosan-based systems. *Functional Chitosan*, pp. 457–489, doi: 10.1007/978-981-15-0263-7_15.

Zhang, H., Fu, G. and Zhang, D. 2014. Cloning, characterization, and production of a novel lysozyme by different expression hosts. *Journal of Microbiology and Biotechnology*, 24(10): 1405–1412, doi: 10.4014/jmb.1404.04039.

Zhang, W. and Rhim, J.-W. 2022. Functional edible films/coatings integrated with lactoperoxidase and lysozyme and their application in food preservation. *Food Control*, 133: 108670, doi: 10.1016/j.foodcont.2021.108670.

Zhao, Y. and Melville, S.B. 1998. Identification and characterization of sporulation-dependent promoters upstream of the enterotoxin gene (cpe) of *Clostridium perfringens. Journal of Bacteriology*, 180(1): 136–142, doi: 10.1128/JB.180.1.136-142.1998.

CHAPTER **2**

Biopreservation of Milk and Dairy Products

Lucía Fernández, Susana Escobedo, Ana Catarina Duarte, Seila Agún, Claudia Rendueles, Pilar García, Ana Rodríguez and *Beatriz Martínez**

Introduction

The dairy sector has a remarkable importance in the economy of European countries, as is illustrated by the available production and consumption data. For instance, the European Union (EU) is the largest producer of raw milk worldwide, with a production of 160.1 million tonnes of raw milk in 2020 (https://www.statista.com/). Also, this region exhibits one of the highest intakes of milk in the world and had a per capita consumption of milk-derived products of approximately 53 kilograms between 2017 and 2021. As a matter of fact, despite the current economic slowdown, the EU is expected to remain the largest dairy supplier in the world market, with a predicted share of approximately 30% of the global dairy trade in 2031. Nonetheless, when looking at the coming decade, experts foresee that the dairy market will likely be influenced by the need to meet sustainability, climate change and health concerns, which will, in turn, involve the development of new production strategies and consumption patterns (CE, 2021). Unfortunately, milk and its derived products are nutrient-rich foods that support the growth of pathogenic and spoilage microorganisms that may lead to health risks to consumers and economic losses, respectively. Indeed, food of animal origin, including dairy products, was implicated in 65.7% of the 3,086 foodborne outbreaks reported in the EU in 2020. Moreover, these outbreaks were caused by some of the most dangerous food pathogenic bacteria such as *Campylobacter*, *Salmonella enteritidis*, *Listeria monocytogenes*,

---

Instituto de Productos Lácteos de Asturias, CSIC. Paseo Rio Linares s/n. 33300 Villaviciosa, Asturias, Spain.
* Corresponding author: bmf1@ipla.csic.es

Shiga toxin-producing *Escherichia coli* (STEC) and toxin-producing *Staphylococcus aureus* (EFSA and ECDC, 2021). These bacteria get into food from several sources including raw materials, the processing industry and incorrect transport or storage. To reduce bacterial contamination, heat treatment is usually applied to milk intended for consumption as liquid milk and for the production of most dairy products. However, the number of consumers who prefer raw milk and dairy products made with raw milk is growing, which may negatively affect food safety and, consequently, public health (Costard et al., 2017). Moreover, the need to prevent the transmission of zoonotic bacteria along the dairy food chain, from primary production to the final product, requires the implementation of global approaches, bearing in mind the need for minimising antibiotic use in primary production and avoiding the unintended delivery of biocides/antibiotics to the environment. Additionally, it is necessary to rethink the current production and preservation systems, paying special attention to reducing both food waste and outbreaks. In this context, biopreservation is perceived by most consumers as a healthy and environmentally-friendly strategy that helps to maintain the nutritional value and organoleptic properties of dairy products without compromising their safety.

Biopreservation is defined as the use of natural microbiota and/or naturally occurring antimicrobials to preserve foods and beverages and extend their shelf-life (Stiles, 1996). For instance, fermentation, one of the oldest methods of milk biopreservation, creates an unfavourable environment for the development of pathogenic and spoilage microorganisms through the growth of lactic acid bacteria (LAB). LAB are currently the main components of dairy starters and are very adept at outcompeting competitors by deploying a large arsenal of antimicrobial compounds (bacteriocins, organic acids, etc.) as will be later described in this chapter. Moreover, there is an increasing interest in the use of bacteriophages and their peptidoglycan-degrading enzymes as pathogen biocontrol agents along the dairy food chain (Fernández et al., 2017). However, the introduction of new biopreservation agents is subjected to the same legal constraints as any other food additive, which often results in notorious delays in their reaching the market. Despite such hurdles, several biopreservation formulations are already commercially available, paving the way towards a healthier and more sustainable dairy chain.

Microbial composition of milk and milk products

The high nutrient content and neutral pH of milk make it a favourable environment for the growth of many microorganisms, allowing the development of a complex and highly variable community. This microbial population includes bacteria that participate in the production of milk-derived products, as well as others that may lead to spoilage or even disease. More recently, we have also learned that some dairy microbes may even have a beneficial effect on human health. Therefore, given their multiple roles, it seems clear that getting to know the microscopic inhabitants of milk throughout the dairy chain is very important from both an industrial and a clinical perspective. Indeed, dairy products are the foods for which we have the most available data on microbial composition (de Filippis et al., 2017).

Studies regarding the milk microbiota may be carried out using culture-dependent or, especially more recently, culture-independent approaches. While the former were important for providing our initial data on this field, it is only since the implementation of the latter techniques, particularly those involving high throughput sequencing, that we have realized the huge complexity of the communities that populate this environment. Nonetheless, culture-independent methods do not indicate whether the identified bacteria correspond to viable cells that will have an actual impact on community dynamics. To overcome this issue, some strategies use propidium monoazide (PMA) to avoid PCR amplification of DNA from dead cells (Emerson et al., 2017). On the other hand, culture-based methods also enable the isolation and preservation of potentially interesting strains. Therefore, a combination of different techniques remains the best option for studying milk microbial communities. It must be noted, however, the importance of including methods with a high taxonomic resolution (species level or below), given the need to differentiate between closely related bacteria that may be dairy starters, non-starters, spoilers or even pathogens (Parente et al., 2020).

From the moment it is synthesised in the mammary gland and further down the dairy chain, milk is exposed to varying environmental conditions that will unavoidably modulate the microbiota composition. Here, we will focus mainly on bacteria although fungi and viruses also inhabit the milk environment. Inside the udder cells, milk is supposed to be sterile, but it starts coming into contact with microbes as it goes down the teat canal and, especially, when it reaches the teat skin. The specific composition of this microbiota varies greatly depending on several factors such as the season, geographical location, feeding and milking practices. Nonetheless, if correct hygienic practices are followed, the bacterial load would be expected to be fairly low at this stage (less than 1×10^4 CFU/ml). The microbiota of freshly obtained raw milk from healthy animals is rich in LAB, such as members of the genera *Lactococcus*, *Lactobacillus*, *Streptococcus* and *Leuconostoc*, many of which exhibit technological properties and are used as dairy starters. Other genera frequently isolated in raw milk include *Enterococcus*, *Corynebacterium* and *Arthrobacter*. However, it must be noted that clinical and subclinical mastitis generally result in a less diverse microbiota, frequently enriched in pathogenic bacteria, such as *S. aureus*, *Streptococcus uberis, Escherichia coli* or *Streptococcus agalactiae* (Andrews et al., 2019). Notably, a study observed that the dysbiosis caused by a mammary glands infection may linger long after the symptoms have gone into remission (Falentin et al., 2016). It is important to emphasise that the potential presence of pathogens makes the consumption of raw milk dangerous to consumers.

According to several studies, the microbiota of bulk tank milk is quite different from that of teat milk, probably because of the contamination from milking equipment together with the effect of cold temperatures during storage (Falardeau et al., 2019). Again, the composition of these communities varies due to multiple factors, including time of year, geographical location, or farming practices amongst others, but the main factors that determine community structure at this stage are probably storage temperature and duration (Doyle et al., 2017). Indeed, storage temperatures as low as 4ºC may favour the growth of psychrotrophic bacteria like *Pseudomonas*, *Acinetobacter* and *Psychrobacter*, which will partly replace the original population of raw milk, including technologically relevant bacteria. Additionally, many psychrotrophs can produce heat resistant enzymes, like proteases and lipases that persist in milk during subsequent processing and lead to spoilage (off flavours, physico-chemical instability).

Once it reaches the dairy processing plant, milk is generally subjected to a heat treatment of varying intensity and duration, such as pasteurization. This process leads to a dramatic decrease in the bacterial population, drastically reducing the number of psychrotrophic and mesophilic bacteria. However, certain microbes like spore-formers (*Bacillus, Anoxybacillus, Turicibacter, Clostridium*) and thermotolerant microbes (*Thermus*) can withstand high temperatures. For instance, endospores can germinate after the treatment, grow and potentially synthesise spoilage enzymes. The main spoilers are aerobic psychrotrophic endospore formers of the *Bacillus cereus* group, although other *Bacillus* can also be involved. Heat resistant enzymes produced during cold storage by other bacteria like *Pseudomonas* can also have an impact on dairy spoilage at this stage. Beyond that, it is also important to consider the potential for post-pasteurization contamination by different microbes present in the processing environment, including those from biofilms formed on different surfaces.

Last but not least, it is worth mentioning the significant role of some moulds as part of the microbiota of some cheese varieties. While yeasts and moulds may be in milk at different stages throughout the dairy chain, sometimes contributing to spoilage, it is during cheese ripening that they may become key players. Indeed, proteolysis and lipolysis by moulds are responsible for the aromas and flavours of many important types of cheese. Perhaps the best known are *Penicillium roqueforti* and *Penicillium camemberti*, which participate in the ripening of blue and camembert-style cheeses, respectively. In turn, yeasts such as *Kluyveromyces marxianus* participate actively in several fermented milks (e.g., kefir). *Debaryomyces hansenii* and *Geotrichum candidum* are commonly found in surface-ripened cheeses.

To sum up, the microbiota of milk and its derived products is highly complex and variable. Importantly, the composition and abundance of these microorganisms is critical regarding the quality and safety of the resulting dairy products. As a result, the more we understand the population dynamics in these environments, the more successful we will be in manufacturing products that have all the necessary organoleptic properties and will be safe for consumers, while being profitable for producers.

Biopreservation strategies for enhanced food safety and extended shelf life

Dairy starters and protective cultures

In the past, milk fermentation was empirically employed as a biopreservation strategy, taking advantage of the spontaneous fermentation caused by the natural microbiota of milk. Nowadays, fermentation is used not only to extend shelf-life, but also to obtain desirable nutritional and organoleptic properties, using selected and often standardised starter cultures. Nonetheless, dairy manufacturers now have access to both defined and undefined formulations of LAB starters, which allow the production of a wide range of fermented dairy products. This variety is further increased by the use of diverse raw materials, particularly different types of milk, and fermentation conditions, which are influenced by external factors such as regional culinary traditions, climate or socioeconomic conditions (Shiferaw and Augustin, 2020; Bourdichon et al., 2021).

Lactococcus lactis and *Lactococcus cremoris* are the main components of the mesophilic dairy starters used for cheese manufacture, while *Streptococcus thermophilus* and *Lactobacillus delbrueckii* are commonly used for yogurt production. The main function of dairy starters during cheese and yogurt manufacturing is the production of lactic acid through the fermentation of lactose, the main carbohydrate in milk, leading to a decrease in pH that is required for the coagulation of caseins. Other desirable technological traits of starter cultures are the production of proteolytic activity, volatile compounds, and texturing molecules, such as exopolysaccharides, as well as resistance to bacteriophages. The highly active metabolism of LAB together with their fast acidifying activity and the production of bacteriocins and other metabolites with antimicrobial activity, as described in the following sections, are behind the antagonistic attributes of this group of bacteria. Additionally, the proteolytic activity of LAB can also release antimicrobial peptides encrypted within milk proteins (Bougherra et al., 2017).

Protective cultures are microorganisms employed to control the growth of undesirable bacteria in food without altering its organoleptic properties. Often, the same microorganisms used as starter cultures can also be used for this purpose and the distinction between starter and protective cultures is ambiguous. Indeed, the mode of action of protective cultures is similar to that of starters and is based on their antagonistic effects with other microorganisms by competition for nutrients or space, or by the production of antimicrobial metabolites. Besides preventing the proliferation of undesirable microorganisms, protective cultures must not pose any potential health risk for consumers or lead to the development of negative organoleptic effects, and must be adapted to the product, as well as resistant to the manufacturing and refrigeration conditions (Silva et al., 2018). The effectiveness of protective cultures in dairy products is well documented and their efficacy is not limited to pathogenic bacteria, but also extends to spoilage bacteria, yeasts and moulds.

Low molecular weight microbial metabolites with antimicrobial activity

Organic acids

The fermentation of carbohydrates is usually coupled with substrate-level phosphorylation and, in homofermentative LAB, the main end product is lactic acid. However, under certain conditions such as carbon limitation or excess, the bacterial metabolism shifts from homolactic to mixed-acid fermentation, in which formic acid, acetic acid and CO_2 are produced in addition to lactic acid. Heterofermentative LAB convert glucose to lactic acid, CO_2 and ethanol (Mayo et al., 2010).

In all cases, organic acid production results in a decrease of the extracellular pH, which is deleterious for most bacteria. However, it must be noted that only the undissociated form of organic acids can diffuse across the cytoplasmic membrane of bacterial cells. Acid dissociation subsequently occurs inside the cell since the

28 *Novel Approaches in Biopreservation for Food and Clinical Purposes*

cytoplasmic pH is about neutral. The dissociated form then exerts its antimicrobial action by lowering the intracellular pH due to the release of H^+ ions. This results in the disruption of the transmembrane proton motive force, which is essential for the viability of the bacterial cell. The pH decrease may also negatively affect acid-sensitive enzymes and cause damage to DNA, proteins or other structural and functional components. Ultimately, all these effects can lead to death of the bacterial cell.

Diacetyl

Diacetyl (2,3-butanedione) is produced by citrate-fermenting LAB via pyruvate. This compound displays antimicrobial activity against bacteria, yeasts and moulds, although high concentrations (200–300 µg/mL) are generally required to observe this effect. Regarding its mechanism of action, it is thought that diacetyl inactivates different enzymes involved in the utilization of arginine in Gram-negative bacteria. However, the use of diacetyl may be limited by its aromatic properties. Indeed, concentrations as low as 2–4 µg/mL result in a characteristic buttery flavour and aroma that may negatively affect the sensory properties of the food product, clearly reducing the potential of this molecule as a biopreservative (Helander et al., 1997).

Hydrogen peroxide

Certain LAB (mainly lactobacilli) are able to release hydrogen peroxide (H_2O_2) in the presence of oxygen through the action of flavoprotein oxidases, NADH oxidases and superoxide dismutase. Production by some LAB species ranges between 0.05 and 1 mM and can reach up to 1.8 mM under intensive aeration (Strus et al., 2006). The antimicrobial effect of H_2O_2 is due to its strong oxidizing effect on the bacterial cell. The oxidation of the sulfhydryl groups of essential enzymes results in their denaturalization and the peroxidation of membrane lipids increases membrane permeability.

In milk, the enzyme lactoperoxidase catalyzes the oxidation of the endogenous thiocyanate (SCN^-) to hypothiocyanate ($OSCN^-$) in the presence of H_2O_2, intensifying its inhibitory activity. Indeed, the use of the lactoperoxidase system has been proposed as a strategy to extend the shelf-life of raw milk in areas where refrigeration is not available (FAO/WHO, 2005).

Reuterin

Reuterin is a multi-component system consisting of 3-HPA (3-hydroxypropionaldehyde), HPA-hydrate, HPA-dimer and acrolein. It is released by *Lactobacillus reuteri* (currently *Limosilactobacillus reuteri*) during fermentation of glycerol under anaerobic conditions. The antimicrobial activity of this substance is mainly attributed to acrolein, which accumulates due to the spontaneous dehydration of 3-HPA in aqueous buffers (Engels et al., 2016). The mode of action of reuterin is based on the depletion of free SH-groups in reduced glutathione (GSH) and proteins leading to an imbalance of the cellular redox status and, consequently, to cell death.

Reuterin shows a broad inhibitory spectrum including Gram-positive and Gram-negative bacteria, yeasts, moulds and protozoa, and is active at a wide range of pHs (2–8). Several studies have reported the activity of reuterin, alone or in combination with other antimicrobials, in milk and dairy products. For instance, the combination of reuterin with nisin and the lactoperoxidase system (LP) showed a synergistic effect in "cuajada" (a Spanish dairy product) on *S. aureus* and *L. monocytogenes* (Arqués et al., 2008). More recently, the reuterin-producing strain *L. reuteri* INIA P572 has been used as a protective culture in semi-hard cheese manufacture for controlling *L. monocytogenes* Ohio serotype 4b and *E. coli* O157:H7 (Langa et al., 2018) and *Clostridium tyrobutyricum* (Ávila et al., 2017). However, in order to promote the production of this substance and, consequently, observe its antimicrobial activity, glycerol must be added to milk.

Antifungal compounds

Some LAB are able to produce substances that often act synergistically with organic acids to prevent the growth of moulds and yeasts. The chemical space of these compounds is large and comprises fatty acids (caproic acid, 3-hydroxy-5-cis-dodecenoic acid, 2-hydroxy-4-methylpentanoic acid, etc.), phenyl-lactic acid (PLA) (3-phenyllactic acid, 4-hydroxy-phenyllactic acids, etc.), and proteinaceous compounds, including cyclic dipeptides such as cyclo(L-Phe-L-Pro), cyclo(L-Phe-*trans*-4-OH-L-Pro) and cyclo-(Gly-L-Leu), among others (Siedler et al., 2019; Chen et al., 2021). They are mostly produced via secondary metabolism or bioconversion. For example, PLA is often produced by *Lactiplantibacillus plantarum* (formerly, *Lactobacillus plantarum*) in the presence of phenylalanine, which is transaminated to phenylpiruvic acid and further reduced to PLA.

The antifungal activity of many of these compounds relies mainly on cell permeabilisation and enzymatic inhibition, and can be achieved *in situ* in cottage cheese, yogurt, sour cream and semi-hard cheeses (Chen et al., 2021). This would lead to the potential inhibition of yeasts and mould species, including mycotoxin producers such as *Penicillium expansum* and *Aspergillus flavus*.

Bacteriocins

Bacteriocins are a large and diverse family of ribosomally-synthesised antimicrobial peptides produced by bacteria that play a major role in displacing bacterial competitors (Cotter et al., 2005; Martínez et al., 2016; Heilbronner et al., 2021). In the case of LAB, bacteriocins participate extensively in their antagonistic potential, enhancing safety in fermented dairy products (Bourdichon et al., 2021). For this reason, this subsection will mainly focus on LAB bacteriocins.

Bacteriocins are very diverse in terms of their chemical structure and functional properties, making it difficult to develop a universal bacteriocin classification (Acedo et al., 2018). Nevertheless, small (less than 10 kDa) amphipathic peptides with high isoelectric points are overrepresented within those bacteriocins with better prospects for food biopreservation. This is because they can withstand rather harsh conditions such as high temperature and extreme pHs, being compatible with several food processes. Moreover, there are bacteriocins with a wide spectrum of inhibition, i.e., attacking relevant foodborne and food spoilage Gram-positive bacteria, which makes them well suited for food biopreservation purposes (e.g., the lantibiotic nisin and the circular AS-48). By contrast, others specifically target a narrower set of bacteria, being sometimes restricted to members of the same genus or even species as the bacteriocin producer (e.g., Lcn972 and several plantaricins).

There are basically two main bacteriocin classes, namely class I, encompassing post-translationally modified peptides, and class II, which includes regular peptides with no or just minor modifications. Perhaps the most common examples of the post-translational modifications found in class I bacteriocins are the presence of dehydrated amino acids and lanthionine or β-methyl lanthionine, with the last two being characteristic of the well-known lantibiotics group, which includes nisin and lacticin 3147. There are other modifications, such as head-to-tail cyclization (e.g., the circular bacteriocin AS-48), glycosylation (e.g., glycocin F) and several other enzymatically-driven modifications that give rise to an enormous structural variability (Montalbán-López et al., 2021). Class II bacteriocins are also subdivided into several subclasses, among which bacteriocins belonging to class IIa (or pediocin-like bacteriocins) stand out for their potent anti-*Listeria* activity. The biosynthetic machinery of LAB bacteriocins, including dedicated immunity functions, is either chromosomally-encoded or in plasmids and other mobilizable genetic elements. Secretion is, in some instances, achieved through specialized transporters and their production may be regulated by quorum sensing (Martínez et al., 2016).

The mode of action of LAB bacteriocins is largely due to their physicochemical properties, especially their hydrophobicity and cationic nature. Most, if not all, disrupt the cell envelope functions of the target bacteria by targeting two main action sites, namely the cytoplasmatic membrane and the biosynthesis of the cell wall (Pérez-Ramos et al., 2021). As a result of their action, many of these bacteriocins form pores in the cytoplasmic membrane of susceptible bacteria, causing leakage of intracellular components, altering the ionic balance, and corrupting energy-dependent cell processes, irremediably leading to cell death. In contrast, other bacteriocins inhibit cell wall biosynthesis by binding to the cell wall precursor lipid II (Martínez et al., 2008), while others, such as nisin, combine these two modes of action and exhibit a potent inhibitory activity (Brötz et al., 1998). Generally speaking, LAB bacteriocins are quite potent antimicrobials that act at the nanomolar range. However, their *in situ* activity in complex food matrices, such as dairy products, may be compromised.

The use of LAB bacteriocins for the biopreservation of dairy products has attracted much attention following the success in multiple food settings of nisin, the first discovered bacteriocin (Ibarra-Sánchez et al., 2020). Moreover, bacteriocin-producing LAB strains may be included in the formulation of dairy starters so that they produce bacteriocins during milk fermentation, or even during maturation in the case of cheeses (Rilla et al., 2003, 2004). This strategy, like the use of protective cultures, helps to avoid incurring in additional production costs and labelling issues compared to the addition of purified bacteriocin preparations, while ensuring the control of undesirable bacteria. Alternatively, some fermentates, such as powdered formulations of a food-compatible substrate fermented by bacteriocin producers, can also be utilised as food ingredients in some countries. This strategy simultaneously extends the shelf-life of the product and provides additional health benefits linked to the presence of some bacterial metabolites (Mathur et al., 2020).

The efficacy of LAB bacteriocins in milk and dairy products for controlling pathogens of concern, such as *L. monocytogenes*, *S. aureus*, *Bacillus cereus*, and

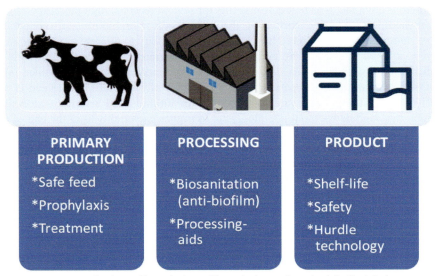

Fig. 1. Applications of bacteriocins and bacteriophages along the daily food chain.

spoilage bacteria has been well documented (López-Cuellar et al., 2016; Martínez et al., 2016; Silva et al., 2018; Ibarra-Sánchez et al., 2020). However, there is also evidence reporting their potential application at other stages along the dairy food chain. For instance, bacteriocins can be used in the primary sector to prevent mastitis in dairy cows, or added during processing to control non-starter bacteria or to accelerate cheese ripening (Fig. 1). Moreover, there are many current research efforts that aim to further improve the success of bacteriocin-based strategies in the dairy sector, including encapsulation, incorporation into packaging films or edible coatings and combination of bacteriocins with additional hurdles.

Bacteriophages and phage-encoded endolysins

Bacteriophages are viruses that specifically infect bacteria. They are obligate parasites that need a bacterial host to multiply and complete their life cycle which concludes, in most cases, with the death of the bacterial cell to release new viral particles. Consequently, phage infection implies an antimicrobial activity that found an early application in the treatment of infectious diseases caused by bacteria (phage therapy). Although the implementation of phage therapy was brought to a halt following the discovery of antibiotics, the interest in using bacteriophages as antimicrobials has recovered importance due to the rapid emergence of antibiotic resistant bacteria. Moreover, this interest has been extended to other fields such as agriculture, veterinary medicine and food safety (Fernández et al., 2018). Traditionally, in dairy settings, bacteriophages have had a negative impact, as they can infect starter LAB cultures leading to fermentation failures and economic losses (Fernández et al., 2017). However, they are now considered a promising alternative for fighting foodborne bacteria, as well as potential eco-friendly antimicrobials to substitute conventional disinfectants for application in dairy plants. Bacteriophages offer numerous advantages as biocontrol agents for several reasons: high target specificity (i.e., harmless to dairy starters or the consumer's microbiota); self-replication; continuous adaption to altered host systems and capacity to tolerate food processing environmental stresses. They generally do not generate resistance transmittable to other microorganisms, even though some may promote antibiotic resistance gene transfer by transducing events or after the release of bacterial DNA (Keen et al., 2017). In addition, phages encode proteins with antimicrobial activity: endolysins, which can degrade bacterial peptidoglycan to trigger bacterial lysis at the end of the lytic infection cycle, are also able to cause cell lysis when added externally (lysis from without). Therefore, they might be used in the same context as whole phages. Also known as enzybiotics (Fischetti, 2010), endolysins exhibit a rapid host killing capacity, are active against persister cells, and no bacterial resistance to endolysins has been reported to date. However, unlike phages, endolysins do not increase in number during treatment and can be inactivated over time. Combination with more stable antimicrobials, such as antibiotics, disinfectants, bacteriocins (Rendueles et al., 2022) or even bacteriophages (Duarte et al., 2021) can help overcoming their short-lived activity. Other strategies are the design of chimeric proteins resulting from catalytic and cell wall binding domain shuffling (Gutiérrez et al., 2018) and fusion with a cationic peptide (artilysins) (Briers et al., 2014) or with phage receptor-binding proteins (Zampara et al., 2020) to expand their lytic activity against Gram-negative bacteria.

The use of phages and their derived proteins to promote food quality and safety can be done at different stages along the dairy food chain as shown in (Fig. 1). In primary production, phage-based products are already available to fight pathogens affecting crops of animal feed products, preventing diseases that would compromise their availability (Buttimer et al., 2017). Several studies have assessed the efficacy of using bacteriophages for both, the prevention and treatment of dairy cattle infectious diseases such as mastitis caused by *S. aureus* (Song et al., 2021). Phages can also be used for biosanitation as disinfectants in dairy facilities to improve cleaning and disinfection procedures. In that context, phages can help to reduce the risk of bacterial contamination, with the added advantages that they can be easily inactivated before being released to the environment, and that they allow a reduction in the use of chemical disinfectants. Furthermore, numerous studies have aimed to use bacteriophages to disrupt biofilms formed from inadequate cleaning of the equipment in dairy plants. Biofilms are structures where bacterial cells are protected by a reticulated polymeric matrix that enhances the bacterium's resistance. For example, the staphylococcal phages phiIPLA-RODI and phiIPLA-C1C exhibit the capacity to control biofilm formation (Gutiérrez et al., 2015). Still, phages could never totally replace the use of disinfectants due to their high specificity, due to the diversity of microorganisms found in different settings, although they can certainly help to improve their effectiveness. Nevertheless, reduction of (cross-)contamination during industrial processing to prevent milk-borne pathogenic or spoilage bacteria has also been approached either by applying phages on food surfaces or by adding them directly to raw milk. Lytic phages, Φ88 and Φ35, were successfully used to completely remove *S. aureus* during curd manufacturing and maturation of fresh and hard-type cheeses (Bueno et al., 2012). Similarly, the application of *Listeria* phages in combination with a bacteriocin (coagulin C23) was proved to extend the shelf life of milk contaminated with *L. monocytogenes* (Rodriguez Rubio et al., 2015). Phages have been reported to lyse hosts at low temperatures, limiting the growth of pathogenic and spoilage bacteria on refrigerated foods and further controlling their proliferation.

Overall, bacteriophages and their derived proteins play a significant role in the success of dairy production. However, the future of phages in food safety is further dependent on the regulatory agencies that still display reticence regarding their use along the food chain, making it necessary to conduct further research to ensure that procedures are effective, safe and easily available. There is also a need to educate farmers, producers, and the general public about the advantages of phages and endolysins as well as the necessity of searching for new methods to adapt lab-scale results to large-scale bacteriophage production and purification or the design of phage mixtures effective against a large number of target strains (Fernández et al., 2018).

Regulatory framework

The application of dairy preservatives is subjected to the same legal requirements as other food preservatives, and most of them have maximum use concentrations allowed in certain foods. From a regulatory perspective, even though many natural methodologies have been recently suggested for the purpose of dairy preservation, only a few have completed the long path that extends from the research lab to the dairy industry to finally reach the market. This situation is due to the strict criteria established by the competent authorities to approve each product in order to ensure their safety and efficacy. Moreover, regulations and standards regarding the control of food additives vary from country to country, and many of them have adopted strict requirements for classifying the different types of compounds that are permitted for addition to foods. Albeit preservatives are treated in a different manner by the most revered food organizations (EFSA-European Food Safety Authority and FDA-The Food and Drug Administration), the aim of both agencies is to ensure food safety with transparency.

In the EU, biopreservatives are legislated as food additives and they are regulated by Regulation (EC) No.1333/2008, which requires a positive safety evaluation by EFSA's Panel on Food additives and Flavourings (FAF). This panel reviews all the available scientific data for a candidate compound, including its potential toxicity, chemical and biological properties and human daily intake in order to draw a conclusion. Once deemed safe by EFSA, each additive must be listed indicating the conditions of use, technological need and maximum use levels established by the Codex Alimentarius Commission. In 2003, EFSA developed the qualified presumption of safety (QPS) status to harmonise generic safety pre-appraisal of microorganisms used in food, such as protective and starter cultures (Leuschner et al., 2010). Since 2012, EFSA's expert FAF Panel performs re-evaluation of the authorised food additives based on new scientific evidence or technical information (Regulation EU 257/2010).

In the USA, food additives are regulated by the FDA according to the Food Additives Amendment to the Federal Food, Drug, and Cosmetic Act of 1958, and is responsible for assessing their safety and efficacy. In particular, the approval of natural antimicrobials considered as biopreservatives in the USA relies on the Generally Recognized as safe (GRAS) assessment. Under the FDA and the Federal Food, Drug and Cosmetic Act, GRAS is a regulatory classification term used to identify compounds with safety evidence that is generally available. GRAS compounds range from usual food ingredients to engineered substances and they cannot be used in food without a positive safety evaluation by the FDA. However, once a given substance attains this recognition, it will not require premarket approval by the FDA as it will not be considered a food additive.

Nisin is a landmark example about the different regulatory pathways that must be followed by biopreservatives, since it is the only bacteriocin licensed as a food preservative (E 234) almost worldwide (Soltani et al., 2021). Although the first commercial preparation of nisin was made in 1953 by Aplin and Barret in the United Kingdom, it was not until 1969 that nisin was first approved as a food preservative by the Joint Food and Agriculture Organization (FAO) and the World Health Organization (WHO) of the United Nations. EU approval took place in 1983 and a few years later, in 1988, nisin was labelled as GRAS by the FDA (83/463/EEC, 1983; 53 FR 11247, 1988). The safety of nisin (E 234) as a food additive was evaluated in 2006 by the EFSA Panel on Food Additives, Flavourings, Processing Aids and Materials in Contact with Food and reviewed in 2017 to extend its use in unripened cheese (EFSA, 2017). The legal limit concentration of nisin as a food additive allowed by FAO/WHO Codex Committee in milk and milk products is 12.5 mg/kg product, while the US FDA permits the use of concentrations of up to 250 mg/kg (Ibarra-Sánchez et al., 2020).

Another regulated biopreservative is natamycin, an antifungal compound with a long history of safe use in food products. In most countries, natamycin has been approved for its use in cheeses, especially for surface treatment of the rind of semi-hard and semi-soft cheeses. The FAO/WHO Expert Committee established the acceptable daily intake (ADI) at 0.3 mg/kg of body weight per day in 1969, amount that has been reviewed in 1976 and 2002, being later reviewed and confirmed to be safe in 2009 by EFSA (EFSA, 2009). In the EU, natamycin (E 235) application is restricted to the surface of cheeses and the maximum permitted level is 1 mg/dm^2 surface according to the EU Directive 1333/2008. In the USA, the FDA approved its used as a food additive in cheese in 1982 with a maximum level of 20 mg/kg (21 CFR 172.155), and granted GRAS status in 2014 for use in yogurt at a maximum level of 5 mg/kg (FDA, 2014).

Concerning bacteriophages, the EFSA BIOHAZ Panel (2009) concluded that they are harmless for consumers. However, they are not included on the QPS list as they must be evaluated on a case-by-case basis to confirm the absence of any virulence factors or/and antimicrobial resistance genes. Likewise, the FDA evaluates each phage-based product in order to grant the GRAS status. Conversely, endolysins have not been approved as biopreservatives by any food authority yet, although they may be regulated as food additives when enough scientific evidence about their safety becomes available.

Commercially available biopreservation solutions

As already mentioned above, the oldest products used as biopreservatives are nisin and natamycin. As a result, several commercial products containing these bacteriocins are currently available on the market (Table 1). Because of their widespread use, biotech companies offer nisin and natamycin preparations in different formats, sources and substrates. Industrial production of nisin is mainly carried out by fermentation using *L. lactis*, whereas natamycin, a fungicide of the polyen macrolide group, is produced by submerged aerobic fermentation by natural strains of *Streptomyces natalensis*.

With the recent increase in antibiotic resistance and foodborne diseases, the demand for new strategies to overcome this problem is growing. Nowadays, the use of phage-based antimicrobials as disinfectants is still not generally well regarded by society. However, there is always an exception to every rule, and some countries have already approved a variety of phage products for use in biocontrol, biopreservation or disinfection (Table 1). For example, there are some commercial products already

Table 1. Examples of products currently available on the market for biocontrol, biopreservation and disinfection (non-exhaustive list).

	Product	Target	Field of application	Website[1]
Bacteriocins	Nisin	Wide range of Gram-positive bacteria and their spores	Fermented dairy products and others	http://www.chinanisin.com https://www.dsm.com/ https://www.dupontnutritionandbiosciences.com https://www.handary.com http://www.nisin.in http://www.pimaripro.com/ http://www.siveele.com
Others	Natamycin (Pimaricin)	Yeasts and moulds	Dairy and meat products	
Bacteriophages	ListShield™	*L. monocytogenes*	Fermented dairy products, meat and fish products	http://www.intralytix.com
	Phage Guard Listex	*L. monocytogenes*	Fermented dairy products, meat and fish products	https://phageguard.com
	PhageLab	*E. coli* and *Salmonella*	Breeding, poultry, livestock and aquaculture industries	https://phage-lab.com
Protective cultures	FreshQ® Holdbac® Delvo®Guard	Yeast and moulds	Fermented dairy products	https://www.chr-hansen.com https://www.dupontnutritionandbiosciences.com https://www.dsm.com
	BIOSAFE™ Dairy Safe™	Clostridia	Fermented dairy products	https://www.chr-hansen.com https://www.dsm.com

[1] Most companies produce both nisin and natamycin

36 *Novel Approaches in Biopreservation for Food and Clinical Purposes*

available on the market such as ListShield™ or PhageGuard Listex. These two products have been proposed as disinfectants that prevent *Listeria* contamination in processing plants and on ripened cheese surfaces, while leaving the necessary bacterial starter cultures unharmed and, as a result, without affecting the organoleptic properties of the final products.

There is also a company, PhageLab, which manufactures phage-based products for the control of bacterial growth in breeding facilities belonging to the poultry, cattle and aquaculture industries, with the aim of reducing the use of antibiotics in the primary sector. In this context, the main products have been developed to eliminate the presence of *E. coli* in pigs or *Salmonella* in birds, and to prevent infectious diarrhea caused by *E. coli* and *Salmonella* spp. in cattle. However, to the best of our knowledge, no phage-based product has been developed for the treatment of mastitis in dairy cows yet.

There are also several commercial formulations to be used as protective cultures. Besides their intrinsic antimicrobial nature, the metabolic activity of some bacteria can be used as an indicator of microbial risk. Indeed, under certain abuse conditions of temperature or handling, indicator bacteria will grow and produce gas or acid to warn consumers. Some such products are meant to reduce the risk of spoilage caused by yeast and mould contamination, ensuring optical freshness of the product, while others have been specifically designed against clostridia present in milk to prevent food poisoning (Table 1). Depending on the product, protective cultures can be commercialized in one of two formats: freeze-dried or frozen. Generally, freeze-dried cultures are used to avoid increasing water activity, contributing in this way to the better taste of the product as well as to its shelf life.

Conclusions

Milk and dairy products are an essential part of our diet with significant shares in the food business as well. Therefore, it is of paramount importance to guarantee that dairy products fulfil today's quality standards and consumer preferences, an objective that we believe can be achieved by using biopreservation strategies. From the once-empirical use of lactic fermentation to the knowledge-guided strategies that incorporate lessons learnt from nature, such as the use of LAB metabolites and phages, the natural enemies of bacteria, the field of biopreservation keeps evolving. Now, it is time to fill in the gap between the lab bench and the market, something that necessarily implies a long way towards approval.

Funding

Research at DairySafe is currently supported by grant PID2019-105311RB-I00 (MCIN/AEI/10.13039/501100011033), grant PID2020-119697RB-I00 (MCIN/AEI/10.13039/501100011033) and grant AYUD/2021/52120 (Program of Science, Technology and Innovation 2021–2023 and FEDER EU, Principado de Asturias, Spain). C.R is a fellow of the program "Ayudas Severo Ochoa" of the Principality of Asturias (BP20 006). S.A. is funded by grant PRE2020-093719 funded by MCIN/AEI/10.13039/501100011033 and by "ESF Investing in your future". A.C.D. is a

Biopreservation of Milk and Dairy Products   37

fellow from the European Union's Horizon 2020 research and innovation program under the Marie Skłodowska-Curie Grant Agreement No. 813439.

Acknowledgements

The authors wish to thank all the past and present members of the DairySafe team (IPLA-CSIC) who have devoted their efforts to implement food biopreservation strategies suitable for milk and dairy products.

References

Acedo, J.Z., Chiorean, S., Vederas, J.C. and van Belkum, M.J. 2018. The expanding structural variety among bacteriocins from Gram-positive bacteria. *FEMS Microbiol. Rev.*, 42: 805–828.

Andrews, T., Neher, D., Weicht, T. and Barlow, J. 2019. Mammary microbiome of lactating organic dairy cows varies by time, tissue site, and infection status. *PloS One*, 14: e0225001.

Arqués, J.L., Rodríguez, E., Nuñez, M. and Medina, M. 2008. Antimicrobial activity of nisin, reuterin, and the lactoperoxidase system on *Listeria monocytogenes* and *Staphylococcus aureus* in cuajada, a semisolid dairy product manufactured in Spain. *J. Dairy Sci.*, 91(1): 70–75.

Ávila, M., Gómez-Torres, N., Delgado, D., Gaya, P. and Garde, S. 2017. Industrial-scale application of *Lactobacillus reuteri* coupled with glycerol as a biopreservation system for inhibiting *Clostridium tyrobutyricum* in semi-hard ewe milk cheese. *Food Microbiology*, 66: 104–109.

Bougherra, F., Dilmi-Bouras, A., Balti, R., Przybylski, R., Adoui, F., Elhameur, H., et al. 2017. Antibacterial activity of new peptide from bovine casein hydrolyzed by a serine metalloprotease of *Lactococcus lactis* subsp. *lactis* BR16. *J. Funct. Foods*, 32: 112–122.

Bourdichon, F., Arias, E., Babuchowski, A., Bückle, A., Dal Bello, F., Dubois, A., Fontana, A., Fritz, D., Kemperman, R., Laulund, S., McAuliffe, O., Hanna Miks, M., Papademas, P., Patrone, V., Sharma, D.K., Sliwinski, E., Stanton, C., Von Ah, U., Yao, S. and Morelli, L. 2021. The forgotten role of food cultures. *FEMS Microbiology Letters*, 368(14): 1–15.

Briers, Y., Walmagh, M., Van Puyenbroeck, V., Cornelissen, A., Cenens, W., Aertsen, A., Oliveira, H., Azeredo, J., Verween, G., Pirnay, J.P., Miller, S., Volckaert, G., Lavigne, R. 2014. Engineered endolysin-based "Artilysins" to combat multidrug-resistant gram-negative pathogens. MBio 5. https://doi.org/10.1128/MBIO.01379- 14 e01379-e01314.

Brötz, H., Josten, M., Wiedemann, I., Schneider, U., Götz, F., Bierbaum, G. and Sahl, H.G. 1998. Role of lipid-bound peptidoglycan precursors in the formation of pores by nisin, epidermin and other lantibiotics. *Mol. Microbiol.*, 30: 317–327.

Bueno, E., García, P., Martínez, B. and Rodríguez, A. 2012. Phage inactivation of *Staphylococcus aureus* in fresh and hard-type cheeses. *Int. J. Food Microbiol.*, 158(1): 23–7. doi: 10.1016/j.ijfoodmicro.2012.06.012.

Buttimer, C., McAuliffe, O., Ross, R.P., Hill, C., O'Mahony, J. and Coffey, A. 2017. Bacteriophages and bacterial plant diseases. *Front Microbiol.*, 8: 34. doi:10.3389/fmicb.2017.00034.

Chen, H., Yan, X., Du, G., Guo, Q., Shi, Y., Chang, J., Wang, X., Yuan, Y. and Yue, T. 2021. Recent developments in antifungal lactic acid bacteria: application, screening methods, separation, purification of antifungal compounds and antifungal mechanisms. *Critical Reviews in Food Science and Nutrition*, 1–15. https://doi.org/10.1080/10408398.2021.1977610.

Commission Regulation (EU) No 257/2010 of 25 March 2010 setting up a program for the re-evaluation of approved food additives in accordance with Regulation (EC) No 1333/2008 of the European Parliament and of the Council on food additives. OJ L 80, 26.3.2010, p. 19–27.

Costard, S., Espejo, L., Groenendaal, H. and Zagmutt, F.J. 2017. Outbreak-related disease burden associated with consumption of unpasteurized cow's milk and cheese, United States, 2009–2014. *Emerging Infectious Diseases*, 23(6): 957–964.

Cotter, P.D., Hill, C. and Ross, R.P. 2005. Bacteriocins: developing innate immunity for food. *Nat. Rev. Microbiol.*, 3: 777–788.

38   *Novel Approaches in Biopreservation for Food and Clinical Purposes*

De Filippis, F., Parente, E. and Ercolini, D. 2017. Metagenomics insights into food fermentations. *Microbial Biotechnology*, 10: 91–102.

Doyle, C.J., Gleeson, D., O'Toole, P.W. and Cotter, P.D. 2017. High-throughput metataxonomic characterization of the raw milk microbiota identifies changes reflecting lactation stage and storage conditions. *International Journal of Food Microbiology*, 255: 1–6.

Duarte, A.C., Fernández, L., De Maesschalck, V., Gutiérrez, D., Campelo, A.B., Briers, Y., Lavigne, R., Rodríguez, A. and García, P. 2021. Synergistic action of phage phiIPLA-RODI and lytic protein CHAPSH3b: a combination strategy to target *Staphylococcus aureus* biofilms. *npj Biofilms Microbiomes*, 7: 39. https://doi.org/10.1038/S41522- 021-00208-5.

EC. 2021. EU agricultural outlook for markets, income and environment, 2021–2031. European Commission, DG Agriculture and Rural Development, Brussels.

EFSA (European Food Safety Authority). 2009. Scientific Opinion on the use of Natamycin (E 235) as a food additive. EFSA Panel on Food Additives and Nutrient Sources added to Food (ANS). *EFSA Journal*, 7(12): 1412.

EFSA (European Food Safety Authority). 2017. Safety of nisin (E 234) as a food additive in the light of new toxicological data and the proposed extension of use. *EFSA Journal*, 15(12): 5063. DO: 10.2903/j.efsa.2017.5063.

EFSA and ECDC. 2021. European food safety authority and European Centre for disease prevention and control. The European Union One Health 2020 Zoonoses Report. *EFSA Journal*, 19(12): 6971.

EFSA BIOHAZ Panel (EFSA Panel on Biological Hazards). 2009. Scientific opinion on the maintenance of the list of QPS microorganisms intentionally added to food or feed (2009 update). *EFSA Journal*, 7(12): 1431, 92 pp. doi:10.2903/j.efsa.2009.1431.

Emerson, J.B., Adams, R.I., Roman, C.M.B., Brooks, B., Coil, D.A., Dahlhausen, K. et al. 2017. Schrödinger's microbes: tools for distinguishing the living from the dead in microbial ecosystems. *Microbiome*, 5. Article 86.

Engels, C., Schwab, C., Zhang, J., *et al.* 2016. Acrolein contributes strongly to antimicrobial and heterocyclic amine transformation activities of reuterin. *Sci. Rep.*, 6: 36246. https://doi.org/10.1038/srep36246.

European Economy Community. 1983. EEC Commission Directive 83/463/EEC.

Falardeau, J., Keeney, K., Trmcic, A., Kitts, D. and Wang, S. 2019. Farm-to-fork profiling of bacterial communities associated with an artisan cheese production facility. *Food Microbiology*, 83: 48–58.

Falentin, H., Rault, L., Nicolas, A., Bouchard, D.S., Lassalas, J., Lamberton, P., et al. 2016. Bovine teat microbiome analysis revealed reduced alpha diversity and significant changes in taxonomic profiles in quarters with a history of mastitis. *Frontiers in Microbiology*, 7. Article 954.

FAO/WHO (Food and Agricultural Organization of the United Nations/World Health Organization). 2005. Benefits and potential risks of the lactoperoxidase system of raw milk preservation. FAO/WHO Techical Meeting (Rome, Nov 28 –Dec 2, 2005).

FDA. 1988. Federal Register. Nisin preparation: affirmation of GRAS status as a direct human food ingredient. 21 CFR Part 184, Fed. Reg. 1988; 53: 11247–11251.

FDA. 2014. Agency Response Letter GRAS Notice No. GRN 000517. GRAS Notice Inventory. November 21. 1413 Accessed June 27, 2017.

Fernández, F., Escobedo, S., Gutiérrez, D., Portilla, S., Martínez, B., García, P. and Rodríguez, A. 2017. Bacteriophages in the dairy environment: from enemies to allies. *Antibiotics* (Basel), 6(4): 27.

Fernández, L., Gutiérrez, D., Rodríguez, A., García, P. 2018. Application of bacteriophages in the agro-food sector: a long way toward approval. *Front. Cell. Infect. Microbiol.*, 8: 296. https://doi.org/10.3389/FCIMB.2018.00296.

Fischetti, V.A. 2010. Bacteriophage endolysins: a novel anti-infective to control Gram-positive pathogens. *Int. J. Med. Microbiol.*, 300: 357–362. doi: 10.1016/j.ijmm.2010.04.002.

Gutiérrez, D., Vandenheuvel, D., Martínez, B., Rodríguez, A., Lavigne, R. and García, P. 2015. Two Phages, phiIPLA-RODI and phiIPLA-C1C, Lyse Mono- and Dual-Species Staphylococcal Biofilms. *Appl. Environ. Microbiol.*, 81(10): 3336–48. doi: 10.1128/AEM.03560-14.

Gutiérrez, D., Fernéndez, L., Rodríguez, A. and García, P. 2018. Are phage lytic proteins the secret weapon to kill *Staphylococcus aureus*? *MBio.*, 9: e01923-17. 10.1128/mBio.01923-17.

Heilbronner, S., Krismer, B., Brötz-Oesterhelt, H. and Peschel, A. 2021. The microbiome-shaping roles of bacteriocins. *Nat. Rev. Microbiol.*, 19: 726–739.

Helander, I.M., von Wright, A. and Mattila-Sandhom, T.M. 1997. Potential of lactic Acid Bacteria and novel antimicrobials against Gram-negative bacteria. *Trends In Food Microbiology and Biotechnology*, 8: 146–150.

Ibarra-Sánchez, L.A., El-Haddad, N., Mahmoud, D., Miller, M.J. and Karam, L. 2020. Invited review: advances in nisin use for preservation of dairy products. *J. Dairy Sci.*, 103: 2041–2052.

Keen, E.C., Bliskovsky, V.V., Malagon, F., Baker, J.D., Prince, J.S., Klaus, J.S. and Adhya, S.L. 2017. Novel "Superspreader" bacteriophages promote horizontal gene transfer by transformation. *mBio.*, 8(1): e02115-16. doi: 10.1128/mBio.02115-16.

Langa, S., Martín-Cabrejas, I., Montiel, R., Peirotén, A., Arqué, J.L. and Medina, M. 2018. Protective effect of reuterin-producing *Lactobacillus reuteri* against *Listeria monocytogenes* and *Escherichia coli* O157:H7 in semi-hard cheese. *Food Control*, 84: 284–289.

Leuschner, R.G.K., Robinson, T.P., Hugas, M., et al. 2010. Qualified presumption of safety (QPS): a generic risk assessment approach for biological agents notified to the European Food Safety Authority (EFSA). *Trends Food Sci. Tech.*, 21:425–35.

López-Cuellar, M.D.R., Rodríguez Hernández, A.-I. and Chavarría Hernández, N. 2016. LAB bacteriocin applications in the last decade. *Biotechnol. Biotechnol. Equip.*, 30: 1–12.

Martínez, B., Böttiger, T., Schneider, T., Rodríguez, A., Sahl, H.G. and Wiedemann, I. 2008. Specific interaction of the unmodified bacteriocin Lactococcin 972 with the cell wall precursor lipid II. *Appl. Env. Microbiol.*, 74: 4666–4670.

Martínez, B., Rodríguez, A. and Suárez, E. 2016. Antimicrobial peptides produced by bacteria: the bacteriocins bt - new weapons to control bacterial growth. pp. 15–38. *In*: Villa, T.G. and Vinas, M. (Eds.). *New Weapons to Control Bacterial Growth*. Cham: Springer International Publishing.

Mathur, H., Beresford, T.P. and Cotter, P.D. 2020. Health benefits of lactic acid bacteria (Lab) fermentates. *Nutrients*, 12: 1–16.

Mayo, B., Aleksandrzak-Piekarczyk, T., Fernández, M., Kowalczyk, M., Alvarez-Martín, P. and Bardowski, J. 2010. Updates in the metabolism of lactic acid bacteria. pp. 3–33. *In*: Fernanda Mozzi, Raúl R. Raya and Graciela M. Vignolo (Eds.). Biotechnology of Lactic Acid Bacterial. Novel Applications. Wiley-Blackwell.

Montalbán-López, M., Scott, T.A., Ramesh, S., Rahman, I.R., van Heel, A.J., Viel, J.H., et al. 2021. New developments in RiPP discovery, enzymology and engineering. *Nat. Prod. Rep.*, 38: 130–239.

Parente, E., Ricciardi, A. and Zotta, T. 2020. The microbiota of dairy milk: a review. *International Dairy Journal*, 107: 104714.

Pérez-Ramos, A., Madi-Moussa, D., Coucheney, F., and Drider, D. 2021. Current knowledge of the mode of action and immunity mechanisms of LAB-Bacteriocins. *Microorganisms*, 9: 2107. doi: 10.3390/microorganisms9102107.

Rendueles, C., Duarte, A.C., Escobedo, S., Fernández, L., Rodríguez, A., García, P. and Martínez, B. 2022. Combined use of bacteriocins and bacteriophages as food biopreservatives. A review. *Int. J. Food Microbiol.*, 368: 109611. doi: 10.1016/j.ijfoodmicro.2022.109611.

Rilla, N., Martínez, B., Delgado, T., and Rodríguez, A. 2003. Inhibition of *Clostridium tyrobutyricum* in Vidiago cheese by *Lactococcus lactis* ssp. lactis IPLA 729, a nisin Z producer. *Int. J. Food Microbiol.*, 85: 23–33.

Rilla, N., Martínez, B. and Rodríguez, A. 2004. Inhibition of a methicillin-resistant *Staphylococcus aureus* strain in Afuega'l Pitu cheese by the nisin Z-producing strain *Lactococcus lactis* subsp. lactis IPLA 729. *J. Food Prot.*, 67: 928–933.

Rodríguez-Rubio, L., García, P., Rodríguez, A., Billington, C., Hudson, J.A. and Martínez, B. 2015. Listeriaphages and coagulin C23 act synergistically to kill *Listeria monocytogenes* in milk under refrigeration conditions. *Int. J. Food Microbiol.*, 205: 68–72. doi: 10.1016/j.ijfoodmicro.2015.04.007.

Shiferaw, N. and Augustin, M.A. 2020. Fermentation for tailoring the technological and health related functionality of food products. *Critical Reviews in Food Science and Nutrition*, 60(17): 2887–2913.

Siedler, S., Balti, R. and Neves, A.R. 2019. Bioprotective mechanisms of lactic acid bacteria against fungal spoilage of food. *Curr. Opin. Biotechnol.*, 56: 138–146.

Silva, C.C.G., Silva, S.P.M. and Ribeiro, S.C. 2018. Application of bacteriocins and protective cultures in dairy food preservation. *Frontiers in Microbiology*, 9: 594.

40 *Novel Approaches in Biopreservation for Food and Clinical Purposes*

Soltani, S., Hammami, R., Cotter, P.D., Rebuffat, S., Said, L.B., Gaudreau, H., Bédard, F., Biron, E., Drider, D. and Fliss, I. 2021. Bacteriocins as a new generation of antimicrobials: toxicity aspects and regulations. *FEMS Microbiol. Rev.* doi: 10.1093/femsre/fuaa039.

Song, J., Ruan, H., Chen, L., Jin, Y., Zheng, J., Wu, R. and Sun, D. 2021. Potential of bacteriophages as disinfectants to control of *Staphylococcus aureus* biofilms. *BMC Microbiol.* 2021 Feb 20; 21(1): 57. doi: 10.1186/s12866-021-02117-1.

Stiles, M.E. 1996. Biopreservation by lactic acid bacteria. *Antonie Leeuwenhoek Journal*, 70: 331–345.

Strus, M., Brzychczy-Włoch, M., Gosiewski, T., Kochan, P. and Heczko, P.B. 2006. The *in vitro* effect of hydrogen peroxide on vaginal microbial communities. *FEMS Immunology and Medical Microbiology*, 48(1): 56–63.

The European Parliament and the Council of the European Union. 2008. Regulation (EC) no 1333/2008 of the European Parliament and of the Council of 16 December 2008 on food additives. Off J Eur Union L 354, 31.12.2008. https://www.fsai.ie/uploadedFiles/Consol_Reg1333_2008.pdf.

Zampara, A., Sørensen, M.C.H., Grimon, D., Antenucci, F., Vitt, A.R., Bortolaia, V., Briers, Y. and Brøndsted, L. 2020. Exploiting phage receptor binding proteins to enable endolysins to kill Gram-negative bacteria. *Sci. Rep.*, 10: 12087. https://doi.org/ 10.1038/S41598-020-68983-3.

CHAPTER 3

Biopreservation of Vegetables

Bárbara Ramos, Teresa R.S. Brandão, Paula Teixeira and
*Cristina L.M. Silva**

Introduction

The natural microbiota of raw vegetables are usually non-pathogenic for humans and may be present at the time of consumption (Ahvenainen, 1996; Food and Drug Administration, 2008). However, during growth, harvest, transportation and further processing and handling, vegetables can be contaminated with pathogens from human, animal or environmental sources (Ahvenainen, 1996; Brandl, 2006; Froder et al., 2007). As a result, these products can be a vehicle of transmission of bacterial, parasitic and viral pathogens, capable of causing human illness.

Nowadays, a wide range of technologies are available to eliminate pathogens from the food chain (Ramos et al., 2013). However, pathogens may be more resistant to the decontamination procedures than the indigenous microbiota, and by removing this group of microorganisms, a natural barrier for pathogenic growth is also removed. As a result, disinfection may provide conditions which favour survival and growth of the pathogens. Furthermore, reduction of the natural microbiota could be risky if further hurdles are not applied to the produce, especially if they are contaminated with a pathogen after processing (Allende et al., 2008). Due to the pathogen mechanisms to survive and the increasing consumption of minimally processed and ready to eat vegetables, outbreaks associated with these products are still frequent (Gandhi and Chikindas, 2007; Naghmouchi et al., 2007; Ramos et al., 2014; Ramos et al., 2013). *Salmonella* spp., *Escherichia coli*, *Shigella* spp. and *Listeria monocytogenes* are amongst the major pathogens associated to outbreaks caused by the consumption of

---

CBQF – Centro de Biotecnologia e Química Fina – Laboratório Associado, Escola Superior de Biotecnologia, Universidade Católica Portuguesa/Porto, Rua Arquiteto Lobão Vital, Apartado 2511, 4202-401 Porto, Portugal.

* Corresponding author: clsilva@porto.ucp.pt

42 *Novel Approaches in Biopreservation for Food and Clinical Purposes*

contaminated vegetables. The produce most associated with outbreaks is salad, since it has all kind of mixed vegetables (Ramos et al., 2013). In addition, WHO (World Health Organization) categorized lettuce and salads (all varieties), leafy vegetables (spinach, cabbage, raw watercress) and fresh herbs of highest priority in terms of fresh produce safety from a global perspective (FAO/WHO, 2008; Goodburn and Wallace, 2013).

An increasing number of consumers prefer minimally processed foods, prepared without chemical preservatives. Many of these ready-to-eat and novel food types represent new food systems with respect to health risks and spoilage association. Relying on improved understanding and knowledge of the complexity of microbial interactions, one alternative approach is biopreservation or biological control (biocontrol) to prevent growth of pathogens and spoilage microorganisms in produce (Holzapfel et al., 1995; Kostrzynska and Bachand, 2006; Ramos et al., 2013). By biopreservation, storage life is extended and food safety improved through the use of native microbiota and/or their metabolites. Different mechanisms, like production of inhibitory compounds, competition for nutrients, space or even colonization sites, are responsible for pathogen inhibition by biocontrol agents. The use of protective cultures of generally recognized as safe (GRAS) microorganisms, such as lactic acid bacteria (LAB), have been developed over the last few decades to increase the safety and shelf-life of fresh and minimally processed vegetables (Engelhardt et al., 2015; Ramos et al., 2013). In addition, several other microorganisms from the natural microbiota, including strains of *Pseudomonas fluorescens, Pseudomonas syringae, Pseudomonas viridiflava, Gluconobacter asaii* and *Enterobacter asburie* have been proposed as biocontrol agents in these foods (Siroli et al., 2015a). Bacteriocins, ribosomally synthesized antimicrobial peptides, are able to kill or inhibit the growth of other bacteria and are considered to be safe natural biopreservatives. The direct application of bacteriocins in fresh cut products has been tested in recent years with promising results (Allende et al., 2007; Molinos et al., 2005; Randazzo et al., 2009; Siroli et al., 2015a). Bacteriophage (phage) prophylaxis is also a possible natural method to be used as a biopreservative. Phages are bacterial viruses that invade specific bacterial cells, disrupt bacterial metabolism, and cause the bacterium to lyse without compromising the viability of other flora in the habitat. They are the most abundant microorganisms in our environment and are present in high numbers in water and foods. Promising results using phage biocontrol have been reported for several pathogens, including *Salmonella* spp., *L. monocytogenes* and *E. coli* O157:H7 (Oliveira et al., 2015).

This is an overview of all past year's research to find effective biopreservation approaches. This review particularly focuses on the use of protective cultures, bacteriocins and bacteriophages to control pathogens in fresh and minimally processed vegetables.

Protective cultures

A distinction can be made between protective and starter cultures. In fact, the same culture can be used with different purposes under different conditions. For a protective culture the antimicrobial activity is the key effect whilst the metabolic activity and technological potential (e.g., exopolysaccharides production and contribution to flavour development) is a secondary effect. For a starter culture the functional objective is the inverse (Holzapfel et al., 1995). In short, antagonistic cultures that are only added to inhibit pathogens and/or prolong the shelf life, while changing the sensory properties of the food product as little as possible, are termed protective cultures. Protective cultures are mainly selected due to their potential to produce bioactive metabolites with antimicrobial activity, which must not alter or negatively interfere with the food matrix (Matamoros et al., 2009).

In order to grow in the fresh and minimally processed vegetables, protective cultures should tolerate the naturally occurring antimicrobial compounds present in many vegetables and be able to efficiently use the nutrients available (Table 1). Temperature is also a key factor, as these kind of food products relay on the cold storage to maintain its quality and safety. So, protective cultures should be able to grow and to cause their inhibitory activity on the target microorganisms at low temperatures (Gálvez et al., 2012).

Lactic acid bacteria are particularly interesting candidates for biological control. They are naturally present in food and are often strong competitors. Their use and the use of their metabolites are generally recognized as safe (GRAS) and benefit from the healthy image of many dairy products (Deegan et al., 2006; Holzapfel et al., 1995; Zacharof and Lovitt, 2012).

Table 1. Desirable properties of protective cultures to be applied as biopreservation agent to fresh foods (adapted from Holzapfel et al., 1995).

Target Microorganism
High inhibitory activity
Antimicrobial activity at low temperatures
Competitiveness against autochthonous organisms
Health
No health risk
No production of toxins
No production of biogenic amines or other metabolites detrimental to health
Non pathogenic
Product
Adapt to product/substrate
No negative sensory effects
No negative nutritional effects
Predictability of metabolic activity under given conditions
Specific enzymatic activity
Function as 'indicator' under abuse conditions
Compatible with other control systems, and therefore can be applied together (hurdle technologies)

LAB cultures and antimicrobial mechanisms

Application of a protective culture for antimicrobial protection of food should be considered only as an additional measure to good manufacturing, processing, storage and distribution practices. Lactic acid bacteria cultures have been used to preserve meat, fish, dairy products and fermented vegetables and fruit juices (Galvez et al., 2006; Rodgers, 2001; Siroli et al., 2015a). In addition, protective cultures of LAB have been developed over the last decades to increase the safety and shelf-life of fresh and minimally processed vegetables (Siroli et al., 2015a; Siroli et al., 2015b; Trias et al., 2008; Vescovo et al., 1996). The preservative effect of LAB is due to the production of one or more active metabolites, such as organic acids (lactic, acetic, formic, propionic acids) that intensify their action by reducing the pH of the media, and other substances, like fatty acids, acetoin, hydrogen peroxide, diacetyl, antifungal compounds (propionate, phenyl-lactate, hydroxyphenyl-lactate, cyclic dipeptides and 3-hydroxy fatty acids), bacteriocins (nisin, reuterin, reutericyclin, pediocin, lacticin, enterocin and others) and bacteriocin-like inhibitory substances—BLIS (Galvez et al., 2006; Holzapfel et al., 1995; Matamoros et al., 2009; Reis et al., 2012).

Organic acids

Organic acids such as lactic, acetic and propionic acids are end products of LAB fermentation. They have antimicrobial activity towards Gram (+) and Gram (−) bacteria due to the interference in the potential of the cell membrane, inhibition of the active transport, pH reduction and to the inhibition of metabolic functions (O'Bryan et al., 2015; Reis et al., 2012).

Hydrogen peroxide (H_2O_2)

Production of H_2O_2 by LAB can prevent the growth of foodborne pathogens and can also be beneficial in food preservation. The antimicrobial effect derives from the increasing of the membrane permeability, by denaturation of enzymes and peroxidation of the membrane lipids. Lactic acid bacteria that produce H_2O_2 have been shown to inhibit the growth of pathogenic microorganisms at refrigeration temperatures (Reis et al., 2012).

Diacetyl

Some LAB produce diacetyl from the excessing pyruvate. Diacetyl is a nonpolar, volatile diaketone that can inhibit Gram (−) bacteria by obstructing the arginine utilization (O'Bryan et al., 2015).

Bacteriocins

Bacteriocins are defined as a group of heterogeneous, bioactive peptides or proteins with antimicrobial activity against other bacteria (Beshkova and Frengova, 2012). A substantial number of Gram (+) and Gram (−) bacteria produces bacteriocins during their growth. Usually, they have a low molecular weight (rarely over 10 kDa) and are active against strains of species related to the producing bacteria (Zacharof and Lovitt, 2012). The fact that bacteriocins, active against numerous foodborne and human pathogens, are produced by GRAS microorganisms and are readily degraded by proteolytic host systems, makes them attractive candidates for biotechnological applications (Settanni and Corsetti, 2008).

Further aspects of bacteriocins will be discussed below.

Applications of protective cultures to fresh and minimally processed vegetables

The first attempts to control pathogens in fresh produce using biological agents were based on bacteriocinogenic LAB strains. However, other bacteria have been demonstrating the potential to be effective biopreservative cultures. A summary of bacteriogenic culture applications in fresh and minimally processed vegetables is presented in Table 2.

Lactic acid bacteria (LAB)

The potential of bacteriocin producing LAB to inhibit microbial populations of ready-to-eat salads was demonstrated by Vescovo et al. (1995). In particular, coliforms and enterococci were strongly reduced or eliminated from the product after inoculation with LAB cultures (Vescovo et al., 1995). From five LAB strains tested in salads, *Lacticaseibacillus casei* IMPCLC34 was the most effective in reducing the total mesophilic bacteria and the coliform group as well as *Aeromonas hydrophila*, *Salmonella typhimurium* and *Staphylococcus aureus*, while the *Listeria monocytogenes* counts remained constant (Vescovo et al., 1996).

The bacteriocin producing *Lactococcus lactis* subsp. *lactis*, isolated from bean sprouts, was able to reduce *L. monocytogenes* counts on ready to eat Caesar salad by 1 to 1.4 logs during storage for 10 days at 7 and 10°C (Cai et al., 1997).

Torriani et al. (1997) showed that the addition of *Lb. casei* IMPC LC34 and of 3% of its culture permeate to mixed salads reduced the total mesophilic bacteria counts, and suppressed coliforms, enterococci, and *A. hydrophila* after 6 days of storage at 8°C. In addition, they revealed the ability of *L. plantarum* IMPC LP4 to extend the shelf life of shredded carrots, by controlling the growth of their microbiota, particularly *Leuconostoc* spp. (Torriani et al., 1999).

The inhibitory activity of *L. lactis* against *L. monocytogenes* was assessed on alfalfa sprouts. When the co-inoculation onto the seeds was made at the beginning of the sprouting process, the maximum inhibition of *L. monocytogenes* was approximately 1 log unit (Palmai and Buchanan, 2002). The authors reported a decrease in the effectiveness of the biopreservation agent in real systems compared with model systems, by the action of the sprout microbiota.

In a model reproducing the characteristics of fresh vegetables, Scolari et al. (2004) showed that although the growth of *L. plantarum* and *S. aureus* was affected by temperature, the pathogenic strain independently of its inoculum size, was always inhibited by *L. plantarum*. They suggested that a proper combination of specific LAB and storage temperature should improve the safety of the vegetable products. Likewise, Scolari and Vescovo (2004) performed various tests in scarola salad leaves that indicated the remarkable inhibitory effect of *L. casei* towards *S. aureus*, *A. hydrophila*, *E. coli* and *L. monocytogenes*.

One of the largest screenings for antagonistic bacteria was performed by Trias et al. (2008b) among seven hundred samples of fresh fruit and vegetables. Lactic acid bacteria were isolated and tested for the capacity to inhibit *E. coli*, *L. monocytogenes*, *Pseudomonas aeruginosa*, *S. typhimurium*, and *S. aureus* in Iceberg lettuce cuts.

46   *Novel Approaches in Biopreservation for Food and Clinical Purposes*

Table 2. Applications of bacteriogenic cultures to fresh and minimally processed vegetables.

Biopreservation agent	Target microorganism	Food	References
Pseudomonas fluorescens	*Listeria monocytogenes*	Endive leaves	Carlin et al., 1996
Lacticaseibacillus casei IMPCLC34	*Aeromonas hydrophila* *Salmonella typhimurium* *Staphylococcus aureus* *Listeria monocytogenes*	Vegetable salads	Vescovo et al., 1996
Lactococcus lactis subsp. *lactis*	*Listeria monocytogenes*	Ready to eat Caesar salad	Cai et al., 1997
Lacticaseibacillus casei IMPC LC34	Coliforms Enterococci *Aeromonas* hydrophila	Mixed salads	Torriani et al., 1997
Pseudomonas fluorescens *Pseudomonas viridiflava*	*Listeria monocytogenes*	Potato tuber slices	Liao and Sapers, 1999
Lactiplantibacillus plantarum IMPC LP4	*Leuconostoc* spp.	Shredded carrots	Torriani et al., 1999
Pseudomonas fluorescens and yeast	*Salmonella Chester* *Listeria monocytogenes*	Green pepper disks	Liao and Fett, 2001
Lactococcus lactis	*Listeria monocytogenes*	Alfalfa sprouts	Palmai and Buchanan, 2002
Gram (–) microbiota	*Staphylococcus aureus* *Escherichia coli* *Listeria monocytogenes* *Salmonella montevideo*	Model system	Schuenzel and Harrison, 2002
Lactiplantibacillus plantarum	*Staphylococcus aureus*	Fresh vegetables model system	Scolari et al., 2004
Lacticaseibacillus casei	*Staphylococcus aureus* *Aeromonas hydrophila* *Escherichia coli* *Listeria monocytogenes*	Scarola salad leaves	Scolari and Vescovo, 2004
Pseudomonas fluorescens	*Salmonella* spp.	Sprouts	Matos and Garland, 2005
Pseudomonas fluorescens 2–79	*Salmonella enterica*	Sprouts	Fett, 2008
Pseudomonas jessenii	*Salmonella* Senftenberg	Ready-to-eat sprouts	Weiss et al., 2007
Pseudomonas fluorescens 2–79	*Salmonella* spp.	Alfalfa seeds	Liao, 2008
Leuconostoc spp. *Lactiplantibacillus plantarum*, *Weissella* spp. *Lactococcus lactis*	*Salmonella Typhimurium* *Escherichia coli* *Listeria monocytogenes*	Iceberg lettuce cuts	Trias et al., 2008b

Table 2 contd. ...

...Table 2 contd.

Biopreservation agent	Target microorganism	Food	References
Leuconostoc mesenteroides Leuconostoc citreum	*Listeria monocytogenes*	Iceberg lettuce	Trias et al., 2008a
Pseudomonas fluorescens 2–79	*Listeria monocytogenes Yersinia enterocolitica*	Bell pepper disks	Liao, 2009
Pediococcus acidilactici CCA3	*Listeria monocytogenes*	Minimally processed kale	Costa et al., 2009
Enterobacter asburiae JX1 alone, or in combination with a Bacteriophage cocktail	*Salmonella* spp.	Mung beans Alfalfa seeds	Ye et al., 2010
Bovamine® (commercially available LAB culture product)	*Escherichia coli* O157:H7	Baby spinach	Gragg and Brashears, 2010
LactiGuard™ (commercially available LAB culture product)	*Escherichia coli* O157:H7 *Clostridium sporogenes*	Spinach	Brown et al., 2011
Bacteriogenic LAB	*Listeria innocua*	Fresh cut onions	Yang et al., 2012
Lacticaseibacillus paracasei LMGP22043	*Listeria monocytogenes Salmonella enterica* subsp. *enterica Escherichia coli*	Ready to eat artichoke products	Valerio et al., 2013
LactiGuard™ (commercially available LAB culture product)	*Escherichia coli* O157:H7 *Salmonella enterica*	Spinach	Calix-Lara et al., 2014
Lactococcus lactis CBM21	*Listeria monocytogenes Escherichia coli* Total mesophilic species	Minimally processed lamb's lettuce	Siroli et al., 2014
Lactiplantibacillus plantarum V7B3 *Lacticaseibacillus casei* V4B4	*Escherichia coli Listeria monocytogenes*	Lamb's lettuce	Siroli et al., 2015
Pseudomonas spp. strain M309	*Salmonella enterica Escherichia coli Listeria monocytogenes*	Lettuce disks (*in vivo* assay)	Oliveira et al., 2015
Pseudomonas graminis CPA-7 *Pseudomonas* spp. strain M309	*Listeria monocytogenes Salmonella* spp.	Fresh cut lettuce	Oliveira et al., 2015
Pediococcus pentosaceus DT016	*Listeria monocytogenes*	Iceberg lettuce Rocket salad Spinach Parsley	Ramos et al., 2020

48 *Novel Approaches in Biopreservation for Food and Clinical Purposes*

The selected strains, predominantly belonged to *Leuconostoc* spp. and *L. plantarum*, and a few corresponded to *Weissella* spp. and *L. lactis*. The antagonist's strains reduced the *S. typhimurium* and *E. coli counts by* 1 to 2 log CFU/wound or g and inhibited the growth of *L. monocytogenes* in lettuce cuts. On the other hand, the strains did not cause negative effects on the general aspect of lettuce tissues. In a subsequent work, Trias et al. (2008a) tested ten *Leuconostoc mesenteroides* and one *Leuconostoc citreum* strains isolated from fresh produce for their antagonistic capacity against *L. monocytogenes*. The inhibition effect was due to organic acids, hydrogen peroxide and bacteriocins production. In this study, the effect of relative dose of pathogen and LAB on *L. monocytogenes* inactivation was assessed, and the importance of the protective culture inoculum on biopreservation approaches was emphasized.

A study focusing on the inhibition of *L. monocytogenes,* from minimally processed kale, revealed that *Pediococcus acidilactici* CCA3, a strain isolated from kale, was able to inhibit the pathogen (2.3 log units at 15°C) and did not alter the product sensorial characteristics during the shelf life period (Costa et al., 2009).

Yang et al. (2012) revealed the potential of bacteriogenic LAB isolated from dairy products to control *Listeria innocua* on fresh cut onions.

The capacity of commercial LAB food antimicrobials to inhibit pathogens in produce has also been demonstrated. Bovamine® (Nutrition Physiology Corporation, Guymon, OK) effectively inhibited *E. coli* O157:H7 on baby spinach surfaces kept at refrigeration temperature (Gragg and Brashears, 2010). Similarly, LactiGuard™ (Guardian Food Technologies, LLC, Overland Park, KS), applied to spinach, reduced by 1.4 and 1.1 log units the counts of *E. coli* O157:H7 and *Clostridium sporogenes*, respectively (Brown et al., 2011). In another experiment, the application of LactiGuard™ significantly reduced *E. coli* O157:H7 and *Salmonella enterica* populations on spinach by 1.6 and 1.9 log CFU/g, respectively (Calix-Lara et al., 2014).

Valerio et al. (2013) reported the capacity of *L. paracasei* LMGP22043 to inhibit the growth of *L. monocytogenes*, *S. enterica* subsp. *enterica* and *E. coli* in ready to eat artichoke products.

Siroli et al. (2014) revealed the effectiveness of the nisin producing strain *L. lactis* CBM21, inoculated in the washing solution of minimally processed lamb's lettuce, to inhibit *L. monocytogenes*, *E. coli* and the total mesophilic species and significantly increase the product shelf-life. Moreover, Siroli et al. (2015b) showed that applying *L. plantarum* V7B3 and *L. casei* V4B4 to lettuce during the washing phase can increase its safety and shelf-life. In fact, *L. plantarum* V7B3 increased *E. coli* death kinetics and reduced the viability of *L. monocytogenes.* In addition, combining the selected strains with natural antimicrobials produced a further increase in the shelf life of these products, without affecting the organoleptic qualities.

Ramos et al. (2020) demonstrated the high capacity of *P. pentosaceus* DT016, previously isolated from lettuce (Ramos et al., 2016), to control *L. monocytogenes* proliferation on fresh lettuce, rocket salad, parsley and spinach kept under cold storage.

Other bacteria

Gram (–) bacteria dominates the microbiota of most vegetables. They have a high potential to be use as biopreservation agents to reduce pathogen growth and survival in produce. These organisms have the advantage of being part of the natural microbial community that is already established on the target produce, which may facilitate their colonization and survival when applied in appropriate numbers (Gálvez et al., 2012; Ramos et al., 2013; Siroli et al., 2015a).

Carlin et al. (1996) found that *P. fluorescens* inhibited the growth of *L. monocytogenes* by 1 log unit, on endive leaves stored at refrigeration (10°C). Liao and Sapers (1999) in a study on potato tuber slices, verified that *P. fluorescens* and *Pseudomonas viridiflava* have antagonist activity towards *L. monocytogenes* growth.

The native microbiota of green bell peppers, Romaine lettuce, baby carrots, alfalfa and clover were screened for their ability to inhibit the growth of *Salmonella chester*, *L. monocytogenes*, *E. coli* and *Erwinia carotovora* subsp. *carotovora*. Six isolates with the capacity for inhibition of at least one pathogen were selected and identified as *Bacillus* spp. (three strains), *P. aeruginosa* (one strain), *P. fluorescens* and a yeast. The application of *P. fluorescens* and the yeast to green pepper disks resulted in a reduction of *S. chester* and *L. monocytogenes* populations of 1 and 2 log units, respectively (Liao and Fett, 2001).

Schuenzel and Harrison (2002) screened the microbiota of ready to eat salads for antibacterial activity against *S. aureus*, *E. coli*, *L. monocytogenes* and *Salmonella montevideo*. Of the 1.180 isolates screened for inhibitory activity, 37 (3.22%) were found to have various degrees of inhibitory activity against at least one pathogen, and several isolates showed inhibitory activity against all four pathogens.

Fluorescent pseudomonads, especially *P. fluorescens*, play an important role in the survival and growth of *Salmonella* spp. in sprouts (Matos and Garland, 2005). In previous trials, *P. fluorescens* isolated from plants have shown inhibitory activity towards a wide range of spoilage and pathogenic bacteria (Liao, 2006). Fett (2006) found that *P. fluorescens* 2–79 was effective in inhibiting the growth of *S. enterica* in sprouts. In addition, other studies reported that treatment with *P. fluorescens* 2–79 reduced the growth of salmonellae by 2–3 log units on alfalfa seeds (Liao, 2008), and reduced other human pathogens on bell pepper disks (Liao, 2009).

Weiss et al. (2007) verified that *Pseudomonas jessenii* suppressed the growth of *Salmonella senftenberg* LTH5703 on ready to eat sprouts, and therefore it was considered to be used as a protective culture for produce.

The natural microbiota of fresh cut iceberg lettuce and baby spinach was evaluated to isolate and identify antagonist bacteria towards *E. coli* O157:H7. Evidence of naturally occurring microorganisms and of possible antagonistic activity toward *E. coli* O157:H7 on fresh lettuce (295 isolates) and spinach (200 isolates) was documented, displaying that produce microbiota can have inhibitory activities towards foodborne pathogens (Johnston et al., 2009).

Enterobacter asburiae is commonly associated with plants and has been tested as a biocontrol strain for inhibiting the growth of enteric pathogens, such as *Salmonella* and *E. coli* O157:H7 (Cooley et al., 2003). *Enterobacter asburiae* JX1, isolated from mung bean sprouts, exhibited antibacterial activity against a

50 *Novel Approaches in Biopreservation for Food and Clinical Purposes*

broad range of *Salmonella* serovars. The antagonistic bacteria in combination with lytic bacteriophages (F01, P01, P102, P700, P800 and FL 41) was tested to control the growth of *Salmonella* spp. on sprouting mung beans and alfalfa seeds. The combination of the biopreservative agents was effective in controlling the pathogen growth on sprouting alfalfa seeds and on mung beans, reducing the pathogen levels by 5.7 to 6.4 log CFU/mL (Ye et al., 2010).

Other authors assessed the effect of *P. fluorescens* on the *E. coli* O157:H7 fate on baby spinach. The biopreservation agent reduced the pathogen loads by 0.5–2.1 log CFU/g of spinach and proved to yield moderate reductions of *E. coli* O157:H7 populations on spinach, when the ratios of *P. fluorescens* to pathogen were similar (Olanya et al., 2013).

Oliveira et al. (2015) isolated and tested *Pseudomonas* sp. strain M309 against *S. enterica*, *E. coli* and *L. monocytogenes* on lettuce disks. M309 strain was highly effective at controlling *S. enterica* and *E. coli* O157:H7 growth on lettuce disks (*in vivo* assay). Furthermore, they tested various biopreservative agents, including M309 strain and *Pseudomonas graminis* CPA-7 (Alegre et al., 2013), against *Salmonella* and *L. monocytogenes* on fresh cut lettuce. The addition of M309 strain and CPA-7 strain did not result in a significant reduction of *Salmonella* population. However, CPA-7 strain reduced *L. monocytogenes* numbers in 1.5 log units after 6 days at 10°C. Results for other biopreservative agents (nisin and bacteriophages) are referred in the corresponding section (Bacteriocins and Bacteriophages).

Bacteriocins

Bacteriocins are bacterial ribosomally synthesised peptides or proteins with antimicrobial activity, and are generally recognized as "natural" compounds (Galvez et al., 2006; García et al., 2010; Settanni and Corsetti, 2008). Bacteriocins are bactericidal with some exceptions. Inhibitory activity of the bacteriocin producing strains are mostly confined to Gram (+) bacteria. Most of the bacteriocins described to date act by inserting into the bacterial cytoplasmic membrane (Balciunas et al., 2013; Gálvez et al., 2012). Other bacteriocins do not interact with the bacterial cytoplasmic membrane, they bind to lipid II and arrest cell wall synthesis (e.g., mersacidin) or act by inhibiting septum formation, halting cell division (e.g., lactococcin 972). Colicins are the most diverse in mode of action, for example, they can act as pore-formers, DNAses, RNAses, and peptidoglycan synthesis inhibitors (Balciunas et al., 2013; Bennik et al., 1999).

Bacteriocins are often confused in the literature with antibiotics. This would limit their use in food applications from a legal standpoint (Cleveland et al., 2001). They can be distinguished from antibiotics on the basis of synthesis, mode of action, antimicrobial spectrum, toxicity and resistance mechanisms. The major difference is that bacteriocins restrict their activity to strains of species related to the producing species, and particularly to strains of the same species. In addition, bacteriocins are produced in the primary phase of growth, though antibiotics are usually secondary metabolites (Deegan et al., 2006; Settanni and Corsetti, 2008; Zacharof and Lovitt, 2012).

Biopreservation of Vegetables 51

Table 3. Bacteriocin classification (according to Heng and Tagg, 2006).

Class I-Lantibiotic peptides
Ia-Linear, e.g., nisin, lacticin 481, plantaricin C
Ib-Globular, e.g., mersacidin
Ic-Multi-component, e.g., lacticin 3147, plantaricin W
Class II- Unmodified peptides (<10 kDa)
IIa-Pediocin-like, e.g., pediocin PA-1
IIb-Miscellaneous, e.g., enterocin L50
IIc-Multi-component, e.g., lactococcin G, plantaricin S
Class III- Large peptides (> 30 kDa)
IIIa-Bacterolytic, e.g., enterolysin A, lysostaphin
IIIb-Non-lyptic, e.g., helveticin J., colicins
Class IV- Cyclic peptides
e.g., AS-48, gassericin A, acidocin B

Bacteriocins comprise a very heterogeneous group regarding their primary structure, composition and physicochemical properties (Deegan et al., 2006). A classification into four classes has been proposed by Heng and Tagg (2006) (Table 3). These classes also include several subclasses, according to bacteriocin structure.

Class I includes the lantibiotics family. Class II includes small, heat-stable peptide bacteriocins and is by far the largest class among Gram (+) bacteriocins. Of particular relevance for food biopreservation is the potent antilisterial activity displayed by the pediocin-like bacteriocins produced by *Pediococcus* spp. Class III includes bacteriolytic and non-lytic large proteins. Class IV includes cyclic peptides. The LAB bacteriocins that have been applied in food biopreservation belong to Class Ia, II and IV (Gálvez et al., 2012; García et al., 2010; Heng and Tagg, 2006).

In recent years, a large number of bacteriocin producing LAB, including *Lactobacillus*, *Lactococcus*, *Leuconostoc* and *Pediococcus* spp., have been isolated from a variety of foods (Abrams et al., 2011; Albano et al., 2007; Cleveland et al., 2001; Gao et al., 2015; Todorov et al., 2011). However, the use of bacteriocins as food additives demands an exhaustive evaluation for toxicological effects before legal acceptance. For that reason, nisin and pediocin PA-1 are the only bacteriocins commercially exploited to date (Sobrino-López and Martín-Belloso, 2008).

Nisin, produced by *Lactococcus lactis*, is the only bacteriocin approved for food applications, being considered to be safe by the Food and Agriculture Organization/ World Health Organization (FAO/WHO) in 1969, and is a biopreservative ingredient in the European food additive list (E234) (Balciunas et al., 2013). Pediocin PA-1, produced by *Pediococcus acidilactici*, is commercially available and marketed as ALTA™ 2431 (Kerry Bioscience, Carrigaline, Co. Cork, Ireland) (García et al., 2010; Settanni and Corsetti, 2008).

Application of bacteriocins to fresh and minimally processed vegetables

Bacteriocins can be incorporated into foods as a concentrated, though not purified, preparation made with food-grade technique. The direct addition of purified bacteriocins obviously provides a more controllable preservative tool in such products. Some desirable properties of the bacteriocins are listed in Table 4.

52 Novel Approaches in Biopreservation for Food and Clinical Purposes

Table 4. Desirable properties of bacteriocins to be applied as biopreservative agents to fresh foods.

Microorganism
Bactericidal mode of action
Acting on the bacterial cytoplasmic membrane
No cross resistance with antibiotics
Feasible genetic manipulation

Health
Non toxic
GRAS status
Inactivated by digestive proteases
Little influence on the gut microbiota
Effective at low concentrations

Product
pH tolerant
High temperature tolerant
Stable at storage conditions
Extend the shelf life
No negative sensory effects
No negative nutritional effects
Low cost
Extra protection during temperature abuse conditions
Reduced risk of pathogenic cross-contamination
Reduce the use of chemical preservatives
Less severe than heat treatments
Marketing of "novel" foods (less acidic, with a lower salt content, and with a higher water content)
Compatible with other control systems, and therefore can be applied together (hurdle technologies)
No medical application

The application of bacteriocins to fresh and fresh cut vegetables has been tested in recent years (Table 5).

Bennik et al. (1999) isolated a bacteriocin producing *Enterococcus mundtii* from minimally processed vegetables, and reported the potential of mundticin as a biopreservative agent applied to mung bean sprouts in a washing step or a coating procedure.

Bari et al. (2005) assessed the effect of nisin and pediocin individually, or in combination with sodium lactate, citric acid, phytic acid, potassium sorbate and EDTA, as potential sanitizers against *L. monocytogenes* on fresh cut cabbage, broccoli and mung bean sprouts. They found that when tested alone all compounds resulted in a reduction between 2–4 log units of *L. monocytogenes* from the vegetables. However, the combination of nisin plus pediocin plus phytic acid was the most effective in reducing the pathogen population.

In a vegetable food model system, the antibacterial efficiency of bacteriocins, from *Lactobacillus* isolates from appam batter (LABB) and vegetable pickle (LABP), and nisin were evaluated, individually and in combination, against *L. monocytogenes* and *S. aureus*. The bacteriocin LABB was the most effective in inhibiting the two pathogens, compared to either nisin or LABP, while the combination of LABB with nisin resulted in further reductions of both pathogens (Jamuna et al., 2005).

Table 5. Bacteriocin applications to fresh and minimally processed vegetables.

Biopreservation agent	Application	Target microorganism	Food	References
Mundticin	Washing solution Coating	*Listeria monocytogenes*	Mung bean sprouts	(Bennik et al., 1999)
Colicin Hu194	Additive Coating	*Escherichia coli* O157:H7	Alfalfa seeds	(Nandiwada et al., 2004)
Nisin Pediocin alone, and in combination with other antimicrobial compounds	Washing solution	*Listeria monocytogenes*	Fresh cut cabbage Broccoli Mung bean sprouts	(Bari et al., 2005)
Bacteriocin LABB Bacteriocin LABP Nisin	Washing solution	*Listeria monocytogenes* *Staphylococcus aureus*	Vegetable model system	(Jamuna et al., 2005)
Enterocin AS-48, and in combination with other antimicrobial compounds	Washing solution	*Listeria monocytogenes*	Alfalfa sprouts Soybean sprouts Green asparagus	(Molinos et al., 2005)
Nisin Coagulin Nisin and coagulin cocktail	Washing solution	*Listeria monocytogenes*	Fresh cut lettuce	(Allende et al., 2007)
Enterocin AS-48 (25 µg/ml) with polyphosphoric acid (0.1 to 2.0%)	Washing solution	*Salmonella enterica* *Escherichia coli* O157:H7 *Shigella* spp. *Enterobacter aerogenes* *Yersinia enterocolitica* *Aeromonas hydrophila* *Pseudomonas fluorescens*	Soybean sprouts	(Molinos et al., 2008a)
Enterocin AS-48, and in combination with other antimicrobial compounds	Washing solution	*Bacillus cereus* *Bacillus weihenstephanensis*	Alfalfa sprouts Soybean sprouts Green asparagus	(Molinos et al., 2008b)

Table 5 contd. ...

...Table 5 contd.

Biopreservation agent	Application	Target microorganism	Food	References
Bacteriocin RUC9	Washing solution	*Listeria monocytogenes*	Minimally processed iceberg lettuce	(Randazzo et al., 2009)
Enterocin 416K1, and in combination with chitosan	Additive	*Listeria monocytogenes*	Zucchini Corn Radishes Mixed salad Carrots	(Anacarso et al., 2011)
Nisin	Washing solution	*Listeria monocytogenes*	Fresh cut lettuce	(Oliveira et al., 2015)
Pediocin DT016	Washing solution	*Listeria monocytogenes*	Iceberg lettuce Rocket salad Spinach Parsley	(Ramos et al., 2020)

Molinos et al. (2005) assessed the effect of immersion solutions containing enterocin AS-48 for decontamination of vegetables by *L. monocytogenes*. In particular, treatments with the bacteriocin alone or in combination with chemical preservatives were tested on alfalfa sprouts, soybean sprouts and green asparagus. For the sprouts treated with enterocin AS-48, the *L. monocytogenes* viable counts were reduced below detection limits at days 1 to 7 at 6°C and 15°C, and in green asparagus at 15°C. Treatment with solutions containing enterocin AS-48 and chemicals, such as lactic acid, sodium lactate, sodium nitrite, sodium nitrate, trisodium phosphate, trisodium trimetaphosphate, sodium thiosulphate, *n*-propyl *p*-hydroxybenzoate, *p*-hydoxybenzoic acid methyl ester, hexadecylpyridinium chloride, peracetic acid, or sodium hypochlorite, reduced *L. monocytogenes* viable counts below the detection limits. Significantly increased antimicrobial activity was found for AS-48 in combination with potassium permanganate, acetic acid, citric acid, sodium propionate, and potassium sorbate.

Allende et al. (2007) tested the effect of washing with bacteriocin solutions (nisin+, coagulin+ and a nisin–/ coagulin+ mixture) on the survival and proliferation of *L. monocytogenes* on fresh cut lettuce stored at 4°C. The washing step immediately decreased the viability of *L. monocytogenes* by 1.2–1.6 log units, however during storage the control over the pathogen growth was minimum.

In subsequent studies, enterocin AS-48 was assessed for the decontamination of soybean sprouts. Combinations of enterocin AS-48 (25 µg/ml) and polyphosphoric acid, in a concentration range of 0.1 to 2.0%, significantly reduced or inhibited growth of the populations of *S. enterica*, *E. coli* O157:H7, *Shigella* spp., *Enterobacter aerogenes*, *Yersinia enterocolitica*, *A. hydrophila* and *P. fluorescens* in sprout samples stored at 6°C and 15°C (Molinos et al., 2008a). The bacteriocin washing was also tested to inhibit *Bacillus* spp. on alfalfa, soybeans sprouts and green asparagus. The treatment with enterocin reduced viable cell counts of *Bacillus cereus* and *Bacillus weihenstephanensis* by 1.0–1.5 and by 1.5–2.4 log units, respectively. The bacteriocin was effective in reducing the remaining viable population below detection levels during storage at 6°C. Application of washing treatments containing enterocin AS-48 in combination with several other antimicrobials and sanitizers was also tested. The combinations of AS-48 and sodium hypochlorite, peracetic acid or hexadecylpyridinium chloride provided the best results. After application of the combined treatments, *B. cereus* and *B. weihenstephanensis* were not detected or remained at very low concentrations in the sprouts treated, along the storage period at 15°C (Molinos et al., 2008b).

Bacteriocin RUC9, produced by a wild strain of *L. lactis,* was tested as a washing solution of minimally processed Iceberg lettuce to control *L. monocytogenes* during storage at 4°C. The treatment resulted in a reduction of 2.7 log units of *L. monocytogenes* counts after 7 days, however it was not effective in removing completely the pathogen from the produce (Randazzo et al., 2009).

Anacarso et al. (2011) investigated the effect of adding Enterocin 416K1, alone or in combination with chitosan, to zucchini, corn, radishes, mixed salad and carrots on the *L. monocytogenes* population. When both antibacterial substances were used, the minimal reduction achieved in *Listeria* population was at least 2 log units. In addition, the *L. monocytogenes* reduction achieved with the enterocin 416k1 alone

56 *Novel Approaches in Biopreservation for Food and Clinical Purposes*

was almost comparable with the antibacterial activity observed for the combination of bacteriocin and chitosan.

In the study by Oliveira et al. (2015), where nisin washing was evaluated, a 1.8 log unit reduction on the *L. monocytogenes* numbers present on fresh cut lettuce during storage at 10°C was observed.

Recently, Ramos et al. (2020) revealed that a washing step using a pediocin DT016 based solution prevented the proliferation of *L. monocytogenes* on fresh lettuce, rocket salad, parsley and spinach. Moreover, when comparing with a available commercial hypochlorite sodium solution, the pediocin solution resulted in lower pathogen loads, by at least 2.7 log unit on all the fresh vegetables.

The bacteriocins from Gram (–) bacteria can be useful in the control of enteric pathogens. Numerous Gram (–) species produce bacteriocins, but those produced by *E. coli* strains (or colicins) are the best studied (Gálvez et al., 2011). Semi-crude colicin Hu194, produced by *E. coli* strain Hu194, was applied in alfalfa seeds contaminated with *E. coli* O157:H7 strains. The bacteriocin treatment successfully reduced the pathogen viable counts ($\approx$ 5 log CFU/g reduction) from the alfalfa seeds (Nandiwada et al., 2004).

Bacteriophages

Bacteriophages or phages are the microorganisms most abundant in Earth (10^{31} particles) and are present in numerous foods. Bacteriophages are viruses that specifically infect bacterial cells, being harmless to humans, animals and plants (García et al., 2010). They are considered safe and represent a potential natural intervention to reduce human pathogens from vegetables (Anany et al., 2015; Żaczek et al., 2015). Bacteriophages have an average size of 20 to 200 nm and possess two main components: genetic material in the form of DNA or RNA as a core, and a surrounding protein or lipoprotein shell (capsid). The core nucleic is connected with a tail that interacts with various bacterial surface receptors via the tip of the tail fibers. This interaction shows an affinity that is specific to certain group of bacteria or even to a particular strain (Anany et al., 2015; Hagens and Loessner, 2010).

The phages are classified into 13 families based on their shape, size and type of nucleic acid, and presence/absence of envelope or lipids in their structure. Most of them are tailed bacteriophage, which accounts for 96% of all phages present on earth, belonging to the order *Caudovirales*. According to the morphological features of the tail, they are classified into three families: the *Myoviridae* (long contractile tail), the *Siphoviridae* (long non contractile tail) and the *Podoviridae* (short non contractile tail) (Deresinski, 2009; García et al., 2010).

As bacteria's parasites, bacteriophages start the infection with adsorption to the suitable host cell, followed by injection of their genetic material into the bacterial cytoplasm (Sharma and Sharma, 2012). The bacteriophages can be divided in two groups, lytic (virulent) or lysogenic (temperate) bacteriophages, depending on their life cycle.

For lytic phages, the bacteriophage genes are transcribed by the host cell machinery, the virion particles are assembled and the lysis of the peptidoglycan layer releases virion particles from the cell, which allow the particles to infect other bacterial hosts (Anany et al., 2015; Sharma and Sharma, 2012). On the other hand, for temperate bacteriophages the phage genome remains in a repressed state in the host genome and it is replicated as part of the bacterial chromosome until lytic cycle is induced. Therefore, temperate phages are not suitable for biocontrol (Anany et al., 2015).

Bacteriophages lytic action is specific to groups or species of bacteria. This allows their use to target pathogenic bacteria, without substantially changing the microbial ecology or microbiota of the produce commodity. This specificity at strain level can be a limitation for application of bacteriophages. Nevertheless, several studies have shown the efficacy of mixtures containing different bacteriophages and broad host range bacteriophages to attack a high number of bacterial strains, including the most virulent strains found in foods (Gálvez et al., 2014; Hagens and Loessner, 2010).

Applications of bacteriophages to fresh and minimally processed vegetables

There are several bacteriophage preparations commercially available for food safety applications, such as ListShield™ and EcoShield™ (Intralytix, Inc., USA), Agriphage™ (Omnilytics, Inc., USA), and Listex™ P100 and Salmonelex™ (Micreos Food Safety, The Netherlands) (Boyacioglu et al., 2013; Mahony et al., 2011). The approval of using phage preparations in food products by U.S. Food and Drug Administration (USFDA) provided the impetus for further investigation into their applications. A summary of bacteriophages applications in fresh and minimally processed vegetables is presented in Table 6.

Pao et al. (2004) isolated two bacteriophages capable of lysing *Salmonella*, phage A and phage B, from sewage water and assessed their potential to control *Salmonella* spp. on sprouting seeds. They found that bacteriophages application resulted in a 1.5 log suppression of *Salmonella* growth in the soaking water of broccoli seeds. When only phage A was applied a 1.4 log suppression of the pathogen growth was achieved on mustard seeds.

Abuladze et al. (2008) showed that bacteriophages may be useful for reducing contamination of various vegetables by *E. coli* O157:H7. In particular, ECP-100 (now EcoShield™ (Intralytix, Inc., USA)), a bacteriophage cocktail containing three *Myoviridae* phages lytic for *E. coli* O157:H7, was applied to decontaminate tomato, spinach and broccoli. Treatments with the ECP-100 preparations resulted in *E. coli* reductions ranging from 94% (tomato) to 100% (spinach).

In a broad set of experiments, a virulent broad host range phages revealed to be very effective for specific biocontrol of *L. monocytogenes* in ready to eat foods. Bacteriophages A511 and P100 reduced the pathogen load by 5 log units in sliced cabbage and lettuce leaves (Guenther et al., 2009).

Kocharunchitt et al. (2009) suggested the existence of a temporary, acquired, non-specific phage resistance phenomenon by bacteria. Briefly, two *Salmonella* bacteriophages (SSP5 and SSP6) were evaluated for their potential to control *Salmonella oranienburg on* alfalfa seeds. Addition of phage SSP6 to alfalfa seeds, previously contaminated with the pathogen, caused approximately 1 log unit reduction of viable *Salmonella*. However, thereafter the phage had no inhibitory effect on the pathogen growth.

58 *Novel Approaches in Biopreservation for Food and Clinical Purposes*

Table 6. Bacteriophages applications to fresh and minimally processed vegetables.

Biopreservation agent	Application	Target microorganism	Food	References
Phage A and phage B	Additive	*Salmonella* spp.	Broccoli seeds, Mustard seed	Pao et al., 2004
ECP-100 (cocktail of three lytic phages ECML-4, ECML-117, ECML-134)	Spraying solution	*Escherichia coli* O157:H7	Broccoli, Spinach	Abuladze et al., 2008
Bacteriophages A511 and P100	Additive	*Listeria monocytogenes*	Fresh cut cabbage, Fresh cut lettuce	Guenther et al., 2009
Bacteriophage SSP 5, Bacteriophage SSP6	Washing solution	*Salmonella Oranienburg*	Alfalfa seeds	Kocharunchitt et al., 2009
ECP-100 (cocktail of 3 lytic phages)	Spraying solution	*Escherichia coli O157:H7*	Fresh cut lettuce	Sharma et al., 2009
BEC8 (mixture of eight lytic bacteriophages), and in combination with trans-cinnamaldehyde	Additive	*Escherichia coli* O157:H7	Baby spinach, Baby romaine lettuce leaves	Viazis et al., 2011
EcoShield™ (cocktail of three lytic phages)	Spraying solution	*Escherichia coli* O157:H7	Lettuce	Carter et al., 2012
EcoShield™ (cocktail of three lytic phages)	Spraying solution	*Escherichia coli* O157:H7	Fresh spinach, Romaine lettuce	Boyacioglu et al., 2013
EcoShield™ (cocktail of three lytic phages)	Additive	*Escherichia coli* O157:H7	Iceberg lettuce leaves	Ferguson et al., 2013
Bacteriophage cocktail, (UAB_Phi 20, UAB_Phi78, and UAB_Phi87)	Washing solution	*Salmonella enterica serovar Typhimurium*, *Salmonella enterica serovar Enteritidis*	Romaine Lettuce	Spricigo et al., 2013
Bacteriophage cocktail, and in combination with levulinic acid	Washing solution	*Escherichia coli O157:H7*, *Shigella* spp., *Salmonella* spp.	Broccoli	Magnone et al., 2013
Listex P100, Salmonelex	Washing solution	*Listeria monocytogenes*, *Salmonella* spp.	Fresh cut lettuce	Oliveira et al., 2015

ECP-100 was an effective treatment to reduce *E. coli* 0157:H7 on fresh cut lettuce stored at refrigeration temperature (4°C) (Sharma et al., 2009).

As referred previously, using a combination of bacteriophages with *E. asburiae* JX1, the levels of *Salmonella* spp. associated with mung bean sprouts were only detected by enrichment (Ye et al., 2010).

In experiments by Viazis et al. (2011), organic baby spinach and baby romaine lettuce leaves artificially contaminated with *E. coli* O157:H7 were treated with BEC8, a mixture of eight lytic bacteriophages, at different multiplicity of infection levels (MOI 1, 10 and 100) and under various conditions. The treatment reduced the number of *E. coli* O157:H7 cells in the produce and it was observed that higher MOI, temperature and incubation period resulted in greater bacterial inactivation. The authors also demonstrated that phage treatment combined with the essential oil trans-cinnamaldehyde was more effective than the phages alone, and the most environment friendly way to reduce bacterial contamination from food.

Several studies assessed the effect of EcoShield™ (Intralytix, Inc., USA) on vegetables artificially contaminated with *E. coli* O157:H7. Carter et al. (2012) studied the capacity of this bacteriophage cocktail in lower concentrations to inhibit the pathogen from lettuce. It was concluded that the phage treatment was effective in reducing the bacterial count by 1–2 logs on lettuce and in maintaining the pathogen levels during the storage period at 4°C. However, the reduction in the phage concentration slightly reduced its efficacy. Boyacioglu et al. (2013) and Ferguson et al. (2013) considered the application of this preparation in fresh cut leafy greens. Boyacioglu et al. (2013) contaminated fresh spinach and romaine lettuce with *E. coli* O157:H7 and sprayed the fresh cut leaves with an EcoShield™ solution. In another experiment Ferguson et al. (2013) studied the potential effect of EcoShield™ in preventing cross contamination of Iceberg lettuce, by treatment of the produce with bacteriophage solution and later contamination with *E. coli* O157:H7. In both cases, phage application significantly reduced the pathogen population in produce. Boyacioglu et al. (2013) observed the effect of the phage treatment as early as 30 min after EcoShield™ spraying, and this was maintained over the storage period at 4°C. In studies by Ferguson et al. (2013) preventive phage application did not immediately cause reduction in *E. coli* cells, but was most successful after several days of storage at 4°C. The presented experiments indicate that EcoShield™ has the potential to inhibit growth of *E. coli* O157:H7 in ready to eat vegetables and can be used as a biocontrol tool in the food industry.

Spricigo et al. (2013) reported the effectiveness of a bacteriophage cocktail (UAB_Phi 20, UAB_Phi78, and UAB_Phi87) in reducing *Salmonella enterica* serovar Typhimurium and *S. enterica* serovar Enteritidis loads from lettuce. A significant bacterial reduction (2.2 log CFU/g) was obtained in the lettuce samples dipped in a solution containing the bacteriophage cocktail for 60 min at room temperature.

Washings with lytic bacteriophage cocktails, levulinic acid and a combination of both treatments were investigated for their effectiveness against the foodborne pathogens: *E. coli* O157:H7, *Shigella* spp. and *Salmonella* on broccoli. The combination of both treatments was effective in reducing the pathogens from produce

60 *Novel Approaches in Biopreservation for Food and Clinical Purposes*

and was not influenced negatively by the presence of high loads of organic matter (Magnone et al., 2013).

Perera et al. (2015) observed that ListShield™ treatment of lettuce reduced *L. monocytogenes* cells by 1.1 log units.

In the experiments by Oliveira et al. (2015) it was additionally demonstrated that Listex™ P100 and Salmonelex™ treatments were not efficient in reducing, respectively, *L. monocytogenes* and *Salmonella* spp. populations from fresh cut lettuce. This highlighted that effective biocontrol strategies may need to be combined with other technologies.

Conclusions

Application of live cultures to vegetables calls for innovative multidisciplinary inputs from various fields of food science. Protective culture applications offer advantages, such as shelf-life extension and food safety improvements. This review highlights the importance of the careful isolation and selection of biopreservation strain(s). Protective cultures efficacy is affected by the inoculation level, the microbiota, the physic-chemical and structure of the products, and the storage conditions. In the future, combined protective culture preparations could be possible.

Bacteriocins, applied directly or as a washing treatment, have shown to be an effective approach to improve microbial safety and reduce the chemical treatment in minimally processed vegetable processing. However, bacteriocins are affected by numerous factors, e.g., they may be efficient only in a narrow pH range, which excludes their utilization in many food products. Thus, a single bacteriocin-based technique is specific to a single food matrix, and its application on different matrices needs to be tested. The combination of bacteriocins with other antimicrobial compounds, and with other preservative techniques, can overcome this limitation and further improve the safety of the products.

In the last few years, several authors have assessed the potential of using bacteriophages in biocontrol of human foodborne pathogens. One important advantage of their use, is that they may enable targeted elimination of a specific pathogenic bacteria in foods without affecting the microbiota of the foods.

Bacteriophages applications in the vegetable processing needs to be further studied. For example, bacteriophages can act synergistically with antagonistic bacteria against selected human pathogenic bacteria, this is an interesting approach that has very seldom been exploited. The lytic enzymes produced by bacteriophages could also be exploited as antimicrobial agents, minimizing the impact of phage specificity. The combination of different biopreservation agents represents a promising, chemical-free approach for controlling the growth of foodborne pathogens in vegetables.

Commercial applications of biopreservation methods require a wider availability of protective cultures, bacteriocins and bacteriophages with limited or known sensory effects. In addition, more information about their health and toxicological effects and safety is required before legal acceptance. Therefore, additional studies using biopreservation approaches, or combined methods, to extend and enhance the safety of this kind of products are crucial.

References

Abrams, D., Barbosa, J., Albano, H., Silva, J., Gibbs, P.A. and Teixeira, P. 2011. Characterization of bacPPK34 a bacteriocin produced by *Pediococcus pentosaceus* strain K34 isolated from "Alheira". *Food Control*, 22: 940–946.

Abuladze, T., Li, M., Menetrez, M.Y., Dean, T., Senecal, A. and Sulakvelidze, A. 2008. Bacteriophages reduce experimental contamination of hard surfaces, tomato, spinach, broccoli, and ground beef by *Escherichia coli* O157:H7. *Applied and Environmental Microbiology*, 74: 6230–6238.

Ahvenainen, R. 1996. New approaches in improving the shelf life of minimally processed fruit and vegetables. *Trends in Food Science & Technology*, 7: 179–187.

Albano, H., Todorov, S.D., van Reenen, C.A., Hogg, T., Dicks, L.M.T. and Teixeira, P. 2007. Characterization of two bacteriocins produced by *Pediococcus acidilactici* isolated from "Alheira", a fermented sausage traditionally produced in Portugal. *International Journal of Food Microbiology*, 116: 239–247.

Alegre, I., Viñas, I., Usall, J., Teixidó, N., Figge, M.J. and Abadias, M. 2013. Control of foodborne pathogens on fresh-cut fruit by a novel strain of *Pseudomonas graminis*. *Food Microbiology*, 34: 390–399.

Allende, A., Martínez, B., Selma, V., Gil, M.I., Suárez, J.E. and Rodríguez, A. 2007. Growth and bacteriocin production by lactic acid bacteria in vegetable broth and their effectiveness at reducing *Listeria monocytogenes in vitro* and in fresh-cut lettuce. *Food Microbiology*, 24: 759–766.

Allende, A., Selma, M.V., López-Gálvez, F., Villaescusa, R. and Gil, M.I. 2008. Role of commercial sanitizers and washing systems on epiphytic microorganisms and sensory quality of fresh-cut escarole and lettuce. *Postharvest Biology and Technology*, 49: 155–163.

Anacarso, I., de Niederhäusern, S., Iseppi, R., Sabia, C., Bondi, M. and Messi, P. 2011. Anti-listerial activity of chitosan and Enterocin 416K1 in artificially contaminated RTE products. *Food Control*, 22: 2076–2080.

Anany, H., Brovko, L.Y., El-Arabi, T., Griffiths and M.W. 2015. 4—Bacteriophages as antimicrobials in food products: history, biology and application. pp. 69–87. *In*: Taylor, T.M. (Ed.). *Handbook of Natural Antimicrobials for Food Safety and Quality*. Woodhead Publishing, Oxford.

Balciunas, E.M., Martinez, F.A.C., Todorov, S.D., Franco, B., Converti, A. and Oliveira, R.P.D. 2013. Novel biotechnological applications of bacteriocins: a review. *Food Control*, 32: 134–142.

Bari, M.L., Ukuku, D.O., Kawasaki, T., Inatsu, Y., Isshiki, K. and Kawamoto, S. 2005. Combined efficacy of nisin and pediocin with sodium lactate, citric acid, phytic acid, and potassium sorbate and EDTA in reducing the *Listeria monocytogenes* population of inoculated fresh-cut produce. *Journal of Food Protection*, 68: 1381–1387.

Bennik, M.H.J., van Overbeek, W., Smid, E.J. and Gorris, L.G.M. 1999. Biopreservation in modified atmosphere stored mungbean sprouts: the use of vegetable-associated bacteriocinogenic lactic acid bacteria to control the growth of *Listeria monocytogenes*. *Letters in Applied Microbiology*, 28: 226–232.

Beshkova, D. and Frengova, G. 2012. Bacteriocins from lactic acid bacteria: microorganisms of potential biotechnological importance for the dairy industry. *Engineering in Life Sciences*, 12: 419–432.

Boyacioglu, O., Sharma, M., Sulakvelidze, A. and Goktepe, I. 2013. Biocontrol of *Escherichia coli* O157. *Bacteriophage*, 3: e24620.

Brandl, M.T. 2006. Fitness of human enteric pathogens on plants and implications for food safety. *Annual Review of Phytopathology*, 44: 367–392.

Brown, A.L., Brooks, J.C., Karunasena, E., Echeverry, A., Laury, A. and Brashears, M.M. 2011. Inhibition of *Escherichia coli* O157:H7 and *Clostridium sporogenes* in spinach packaged in modified atmospheres after treatment combined with chlorine and lactic acid bacteria. *Journal of Food Science*, 76: M427–432.

Cai, Y., Ng, L.K. and Farber, J.M. 1997. Isolation and characterization of nisin-producing *Lactococcus lactis* subsp. *lactis* from bean-sprouts. *Journal of Applied Microbiology*, 83: 499–507.

Calix-Lara, T.F., Rajendran, M., Talcott, S.T., Smith, S.B., Miller, R.K., Castillo, A., Sturino, J.M. and Taylor, T.M. 2014. Inhibition of *Escherichia coli* O157:H7 and *Salmonella enterica* on spinach and identification of antimicrobial substances produced by a commercial Lactic Acid Bacteria food safety intervention. *Food Microbiology*, 38: 192–200.

62 Novel Approaches in Biopreservation for Food and Clinical Purposes

Carlin, F., Nguyen-The, C. and Morris, C.E. 1996. Influence of background microflora on *Listeria monocytogenes* on minimally processed fresh broad-leaved endive (*Cichorium endivia* var latifolia). *Journal of Food Protection*, 59: 698–703.

Carter, C.D., Parks, A., Abuladze, T., Li, M., Woolston, J., Magnone, J., Senecal, A., Kropinski, A.M. and Sulakvelidze, A. 2012. Bacteriophage cocktail significantly reduces *Escherichia coli* O157. *Bacteriophage*, 2: 178–185.

Cleveland, J., Montville, T.J., Nes, I.F. and Chikindas, M.L. 2001. Bacteriocins: safe, natural antimicrobials for food preservation. *International Journal of Food Microbiology*, 71: 1–20.

Cooley, M.B., Miller, W.G. and Mandrell, R.E. 2003. Colonization of *Arabidopsis thaliana* with *Salmonella enterica* and enterohemorrhagic *Escherichia coli* O157:H7 and competition by *Enterobacter asburiae*. *Applied and Environmental Microbiology*, 69: 4915–4926.

Costa, W.A., Vanetti, M.C.D. and Puschmann, R. 2009. Biocontrol of *Listeria monocytogenes* by *Pediococcus acidilactici* in fresh-cut kale. *Ciencia E Tecnologia De Alimentos*, 29: 785–792.

Deegan, L.H., Cotter, P.D., Hill, C. and Ross, P. 2006. Bacteriocins: biological tools for bio-preservation and shelf-life extension. *International Dairy Journal*, 16: 1058–1071.

Deresinski, S. 2009. Bacteriophage therapy: exploiting smaller fleas. *Clin. Infectious Diseases*, 48: 1096–1101.

Engelhardt, T., Albano, H., Kiskó, G., Mohácsi-Farkas, C. and Teixeira, P. 2015. Antilisterial activity of bacteriocinogenic *Pediococcus acidilactici* HA6111-2 and *Lactobacillus plantarum* ESB 202 grown under pH and osmotic stress conditions. *Food Microbiology*, 48: 109–115.

FAO/WHO. 2008. Microbiological hazards in fresh leafy vegetables and herbs: meeting report, Microbiological risk assessment series. Food Agriculture Organization of the United Nations/ World Health Organization, Rome, Italy, p. 158.

Ferguson, S., Roberts, C., Handy, E. and Sharma, M. 2013. Lytic bacteriophages reduce *Escherichia coli* O157. *Bacteriophage*, 3: e24323.

Fett, W.F. 2006. Inhibition of *Salmonella enterica* by plant-associated pseudomonads *in vitro* and on sprouting alfalfa seed. *Journal of Food Protection*, 69: 719–728.

Food and Drug Administration, H.H.S. 2008. Irradiation in the production, processing and handling of food. Final rule. *Federal Register*, 73: 49593–49603.

Froder, H., Martins, C.G., de Souza, K.L.O., Landgraf, M., Franco, B. and Destro, M.T. 2007. Minimally processed vegetable salads: Microbial quality evaluation. *Journal of Food Protection*, 70: 1277–1280.

Galvez, A., Abriouel, H., Lopez, R.L. and Ben Omar, N. 2006. Bacteriocin-based Strategies for Food Biopreservation. Elsevier Science Bv, Bologna, ITALY, pp. 51–70.

Gálvez, A., Cobo, A., Abriouel, H. and Pulido, R.P. 2011. Natural Antimicrobials for Biopreservation of Sprouts. INTECH Open Access Publisher.

Gálvez, A., Lucas, A., Hikmate, A., Burgos, M.J.G. and Pulido, R.P. 2012. Bacteriocins Decontamination of Fresh and Minimally Processed Produce. Wiley-Blackwell, pp. 317–332.

Gálvez, A., Pulido, R.P., Hikmate, A., Omar, N. and Burgos, M.J.G. 2012. Protective Cultures, Decontamination of Fresh and Minimally Processed Produce. Wiley-Blackwell, pp. 297–312.

Gálvez, A., López, R.L., Pulido, R.P. and Burgos, M.J.G. 2014. Natural Antimicrobials for Food Biopreservation, Food Biopreservation. Springer New York, pp. 3–14.

Gandhi, M. and Chikindas, M.L. 2007. *Listeria*: a foodborne pathogen that knows how to survive. *International Journal of Food Microbiology*, 113: 1–15.

Gao, Y., Li, D., Liu, S. and Zhang, L. 2015. Garviecin LG34, a novel bacteriocin produced by *Lactococcus garvieae* isolated from traditional Chinese fermented cucumber. *Food Control*, 50: 896–900.

García, P., Rodríguez, L., Rodríguez, A. and Martínez, B. 2010. Food biopreservation: promising strategies using bacteriocins, bacteriophages and endolysins. *Trends in Food Science & Technology*, 21: 373–382.

Goodburn, C. and Wallace, C.A. 2013. The microbiological efficacy of decontamination methodologies for fresh produce: a review. *Food Control*, 32: 418–427.

Gragg, S.E. and Brashears, M.M. 2010. Reduction of *Escherichia coli* O157:H7 in Fresh Spinach, using lactic acid bacteria and chlorine as a multihurdle intervention. *Journal of Food Protection*, 73: 358–361.

Guenther, S., Huwyler, D., Richard, S. and Loessner, M.J. 2009. Virulent bacteriophage for efficient biocontrol of *Listeria monocytogenes* in ready-to-eat foods. *Applied and Environmental Microbiology*, 75: 93–100.

Hagens, S. and Loessner, M.J. 2010. Bacteriophage for biocontrol of foodborne pathogens: calculations and considerations. *Curr. Pharm. Biotechnol.* 11: 58–68.

Heng, N.C.K. and Tagg, J.R. 2006. What's in a name? Class distinction for bacteriocins. *Natural Reviews Microbiology*, 4.

Holzapfel, W.H., Geisen, R. and Schillinger, U. 1995. Biological preservation of foods with reference to protective cultures, bacteriocins and food-grade enzymes. *International Journal of Food Microbiology*, 24: 343–362.

Jamuna, M., Babusha, S.T. and Jeevaratnam, K. 2005. Inhibitory efficacy of nisin and bacteriocins from *Lactobacillus* isolates against food spoilage and pathogenic organisms in model and food systems. *Food Microbiology*, 22: 449–454.

Johnston, M.A., Harrison, M.A. and Morrow, R.A. 2009. Microbial antagonists of *Escherichia coli* O157:H7 on fresh-cut lettuce and spinach. *Journal of Food Protection*, 72: 1569–1575.

Kocharunchitt, C., Ross, T. and McNeil, D.L. 2009. Use of bacteriophages as biocontrol agents to control *Salmonella* associated with seed sprouts. *International Journal of Food Microbiology*, 128: 453–459.

Kostrzynska, M. and Bachand, A. 2006. Use of microbial antagonism to reduce pathogen levels on produce and meat products: a review. *Canadian Journal of Microbiology*, 52: 1017–1026.

Liao, C.H. and Sapers, G.M. 1999. Influence of soft rot bacteria on growth of *Listeria monocytogenes* on potato tuber slices. *Journal of Food Protection*, 62: 343–348.

Liao, C.H. and Fett, W.F. 2001. Analysis of native microflora and selection of strains antagonistic to human pathogens on fresh produce. *Journal of Food Protection*, 64: 1110–1115.

Liao, C.H. 2006. 19 - *Pseudomonas* and related genera. pp. 507–540. *In*: Blackburn, C.d.W. (Ed.). Food Spoilage Microorganisms. Woodhead Publishing.

Liao, C.H. 2008. Growth of *Salmonella* on sprouting alfalfa seeds as affected by the inoculum size, native microbial load and *Pseudomonas fluorescens* 2–79. *Letters in Applied Microbiology*, 46: 232–236.

Liao, C.H. 2009. Control of foodborne pathogens and soft-rot bacteria on bell pepper by three strains of bacterial antagonists. *Journal of Food Protection*, 72: 85–92.

Magnone, J.P., Marek, P.J., Sulakvelidze, A. and Senecal, A.G. 2013. Additive approach for inactivation of *Escherichia coli* O157:H7, *Salmonella* and *Shigella* spp. on contaminated fresh fruits and vegetables using bacteriophage cocktail and produce wash. *Journal of Food Protection*, 76: 1336–1341.

Mahony, J., McAuliffe, O., Ross, R.P. and van Sinderen, D. 2011. Bacteriophages as biocontrol agents of food pathogens. *Current Opinion in Biotechnology*, 22: 157–163.

Matamoros, S., Pilet, M.F., Gigout, F., Prevost, H. and Leroi, F. 2009. Selection and evaluation of seafood-borne psychrotrophic lactic acid bacteria as inhibitors of pathogenic and spoilage bacteria. *Food Microbiology*, 26: 638–644.

Matos, A. and Garland, J.L. 2005. Effects of community versus single strain inoculants on the biocontrol of *Salmonella* and microbial community dynamics in alfalfa sprouts. *Journal of Food Protection*, 68: 40–48.

Molinos, A.C., Abriouel, H., Ben Omar, N., Valdivia, E., Lopez, R.L., Maqueda, M., Canamero, M.M. and Galvez, A. 2005. Effect of immersion solutions containing enterocin AS-48 on *Listeria monocytogenes* in vegetable foods. *Applied and Environmental Microbiology*, 71: 7781–7787.

Molinos, A.C., Abriouel, H., Lopez, R.L., Valdivia, E., Omar, N.B. and Galvez, A. 2008a. Combined physico-chemical treatments based on enterocin AS-48 for inactivation of Gram-negative bacteria in soybean sprouts. *Food and Chemical Toxicology*, 46: 2912–2921.

Molinos, A.C., Abriouel, H., Lucas Lopez, R., Ben Omar, N., Valdivia, E. and Galvez, A. 2008b. Inhibition of *Bacillus cereus* and *Bacillus weihenstephanensis* in raw vegetables by application of washing solutions containing enterocin AS-48 alone and in combination with other antimicrobials. *Food Microbiology*, 25: 762–770.

Naghmouchi, K., Kheadr, E., Lacroix, C. and Fliss, I. 2007. Class I/Class IIa bacteriocin cross-resistance phenomenon in *Listeria monocytogenes*. *Food Microbiology*, 24: 718–727.

64 *Novel Approaches in Biopreservation for Food and Clinical Purposes*

Nandiwada, L.S., Schamberger, G.P., Schafer, H.W. and Diez-Gonzalez, F. 2004. Characterization of an E2-type colicin and its application to treat alfalfa seeds to reduce *Escherichia coli* O157:H7. *International Journal of Food Microbiology*, 93: 267–279.

O'Bryan, C.A., Crandall, P.G., Ricke, S.C. and Ndahetuye, J.B. 2015. 6—Lactic acid bacteria (LAB) as antimicrobials in food products: types and mechanisms of action. pp. 117–136. *In*: Taylor, T.M. (Ed.). Handbook of Natural Antimicrobials for Food Safety and Quality. Woodhead Publishing, Oxford.

Olanya, M.O., Ukuku, D.O., Annous, B.A., Niemira, B.A. and Sommers, C.H. 2013. Efficacy of *Pseudomonas fluorescens* for biocontrol of *Escherichia coli* O157:H7 on spinach. *Journal of Food Agriculture & Environment*, 11: 86–91.

Oliveira, M., Abadias, M., Colás-Medà, P., Usall, J. and Viñas, I. 2015. Biopreservative methods to control the growth of foodborne pathogens on fresh-cut lettuce. *International Journal of Food Microbiology*, 214: 4–11.

Palmai, M. and Buchanan, R.L. 2002. Growth of *Listeria monocytogenes* during germination of alfalfa sprouts. *Food Microbiology*, 19: 195–200.

Pao, S., Rolph, S.P., Westbrook, E.W. and Shen, H. 2004. Use of bacteriophages to control *Salmonella* in experimentally contaminated sprout seeds. *Journal of Food Science*, 69: M127–M130.

Perera, M.N., Abuladze, T., Li, M., Woolston, J. and Sulakvelidze, A. 2015. Bacteriophage cocktail significantly reduces or eliminates *Listeria monocytogenes* contamination on lettuce, apples, cheese, smoked salmon and frozen foods. *Food Microbiology*, 52: 42–48.

Ramos, B., Miller, F.A., Brandão, T.R.S., Teixeira, P. and Silva, C.L.M. 2013. Fresh fruits and vegetables— an overview on applied methodologies to improve its quality and safety. *Innovative Food Science & Emerging Technologies*, 20: 1–15.

Ramos, B., Brandão, T.R.S., Teixeira, P. and Silva, C.L.M. 2014. Balsamic vinegar from Modena: an easy and effective approach to reduce *Listeria monocytogenes* from lettuce. *Food Control*, 42: 38–42.

Ramos, B., Brandão, T.R.S., Teixeira, P. and Silva, C.L.M. 2020. Biopreservation approaches to reduce *Listeria monocytogenes* in fresh vegetables. *Food Microbiology*, 85: 103282.

Ramos, B., Ferreira, V., Brandão, T.R.S., Teixeira, P. and Silva, C.L.M. 2016. Antilisterial active compound from lactic acid bacteria present on fresh iceberg lettuce. *Acta Alimentaria*, 45: 416–426.

Randazzo, C.L., Pitino, I., Scifo, G.O. and Caggia, C. 2009. Biopreservation of minimally processed Iceberg lettuces using a bacteriocin produced by *Lactococcus lactis* wild strain. *Food Control*, 20: 756–763.

Reis, J.A., Paula, A.T., Casarotti, S.N. and Penna, A.L.B. 2012. Lactic acid bacteria antimicrobial compounds: characteristics and applications. *Food Engineering Reviews*, 4: 124–140.

Rodgers, S. 2001. Preserving non-fermented refrigerated foods with microbial cultures—a review. *Trends in Food Science & Technology*, 12: 276–284.

Schuenzel, K.M. and Harrison, M.A. 2002. Microbial antagonists of foodborne pathogens on fresh, minimally processed vegetables. *Journal of Food Protection*, 65: 1909–1915.

Scolari, G. and Vescovo, M. 2004. Microbial antagonism of *Lactobacillus casei* added to fresh vegetables. *Italian Journal of Food Science*, 16: 465–475.

Scolari, G., Vescovo, M., Zacconi, C. and Bonadé, A. 2004. Influence of *Lactobacillus plantarum* on *Staphylococcus aureus* growth in a fresh vegetable model system. *European Food Research and Technology*, 218: 274–277.

Settanni, L. and Corsetti, A. 2008. Application of bacteriocins in vegetable food biopreservation. *International Journal of Food Microbiology*, 121: 123–138.

Sharma, M., Patel, J.R., Conway, W.S., Ferguson, S. and Sulakvelidze, A. 2009. Effectiveness of bacteriophages in reducing *Escherichia coli* O157:H7 on fresh-cut cantaloupes and lettuce. *Journal of Food Protection*, 72: 1481–1485.

Sharma, M. and Sharma, G.C. 2012. Bacteriophages, Decontamination of Fresh and Minimally Processed Produce. Wiley-Blackwell, pp. 283–295.

Siroli, L., Patrignani, F., Salvetti, E., Torriani, S., Gardini, F. and Lanciotti, R. 2014. Use of a nisin-producing *Lactococcus lactis* strain, combined with thyme essential oil, to improve the safety and shelf-life of minimally processed lamb's lettuce, Proceeding of 11th International Symposium on lactic acid bacteria, Egmond aan Zee, the Netherlands.

Siroli, L., Patrignani, F., Serrazanetti, D.I., Gardini, F. and Lanciotti, R. 2015a. Innovative strategies based on the use of bio-control agents to improve the safety, shelf-life and quality of minimally processed fruits and vegetables. *Trends in Food Science & Technology*, 46(2): 302–310.

Siroli, L., Patrignani, F., Serrazanetti, D.I., Tabanelli, G., Montanari, C., Gardini, F. and Lanciotti, R. 2015b. Lactic acid bacteria and natural antimicrobials to improve the safety and shelf-life of minimally processed sliced apples and lamb's lettuce. *Food Microbiology*, 47: 74–84.

Sobrino-López, A. and Martín-Belloso, O. 2008. Use of nisin and other bacteriocins for preservation of dairy products. *International Dairy Journal*, 18: 329–343.

Spricigo, D.A., Bardina, C., Cortés, P. and Llagostera, M. 2013. Use of a bacteriophage cocktail to control *Salmonella* in food and the food industry. *International Journal of Food Microbiology*, 165: 169–174.

Todorov, S.D., Prévost, H., Lebois, M., Dousset, X., LeBlanc, J. and Franco, B.D.G.M. 2011. Bacteriocinogenic *Lactobacillus plantarum* ST16Pa isolated from papaya (*Carica papaya*)— from isolation to application: characterization of a bacteriocin. *Food Research International*, 44: 1351–1363.

Torriani, S., Orsi, C. and Vescovo, M. 1997. Potential of *Lactobacillus casei*, culture permeate, and lactic acid to control microorganisms in ready-to-use vegetables. *Journal of Food Protection*, 60: 1564–1567.

Torriani, S., Scolari, G., Dellaglio, F. and Vescovo, M. 1999. Biocontrol of leuconostocs in ready-to-use shredded carrots. *Annali Di Microbiologia Ed Enzimologia*, 49: 23–31.

Trias, R., Badosa, E., Montesinos, E. and Bañeras, L. 2008a. Bioprotective *Leuconostoc* strains against *Listeria monocytogenes* in fresh fruits and vegetables. *International Journal of Food Microbiology*, 127: 91–98.

Trias, R., Bañeras, L., Badosa, E. and Montesinos, E. 2008b. Bioprotection of Golden Delicious apples and Iceberg lettuce against foodborne bacterial pathogens by lactic acid bacteria. *International Journal of Food Microbiology*, 123: 50–60.

Valerio, F., Lonigro, S.L., Di Biase, M., de Candia, S., Callegari, M.L. and Lavermicocca, P. 2013. Bioprotection of ready-to-eat probiotic artichokes processed with *Lactobacillus paracasei* LMGP22043 against foodborne pathogens. *Journal of Food Science*, 78: M1757–1763.

Vescovo, M., Orsi, C., Scolari, G. and Torriani, S. 1995. Inhibitory effect of selected lactic-acid bacteria on microflora associated with the ready-to-use vegetables. *Letters in Applied Microbiology*, 21: 121–125.

Vescovo, M., Torriani, S., Orsi, C., Macchiarolo, F. and Scolari, G. 1996. Application of antimicrobial-producing lactic acid bacteria to control pathogens in ready-to-use vegetables. *Journal of Applied Bacteriology*, 81: 113–119.

Viazis, S., Akhtar, M., Feirtag, J. and Diez-Gonzalez, F. 2011. Reduction of *Escherichia coli* O157:H7 viability on leafy green vegetables by treatment with a bacteriophage mixture and trans-cinnamaldehyde. *Food Microbiology*, 28: 149–157.

Weiss, A., Hertel, C., Grothe, S., Ha, D. and Hammes, W.P. 2007. Characterization of the cultivable microbiota of sprouts and their potential for application as protective cultures. *Systematic and Applied Microbiology*, 30: 483–493.

Yang, E., Fan, L., Jiang, Y., Doucette, C. and Fillmore, S. 2012. Antimicrobial activity of bacteriocin-producing lactic acid bacteria isolated from cheeses and yogurts. *AMB Express*, 2: 48–48.

Ye, J., Kostrzynska, M., Dunfield, K. and Warrineri, K. 2010. Control of *Salmonella* on sprouting mung bean and alfalfa seeds by using a biocontrol preparation based on antagonistic bacteria and lytic bacteriophages. *Journal of Food Protection*, 73: 9–17.

Zacharof, M.P. and Lovitt, R.W. 2012. Bacteriocins Produced by Lactic Acid Bacteria a Review Article. *APCBEE Procedia*, 2: 50–56.

Żaczek, M., Weber-Dąbrowska, B. and Górski, A. 2015. Phages in the global fruit and vegetable industry. *Journal of Applied Microbiology*, 118: 537–556.

CHAPTER 4

Bipreservation in Meat and Meat Products

Annada Das,[1] Dipanwita Bhattacharya,[2] Pramod Kumar Nanda,[3] Santanu Nath[3] and Arun K. Das[3],*

Introduction

The practice of biopreservation of food has been in use since ancient times, with spontaneous fermentation by autochthonous microorganisms and is often termed as humanity's oldest biotechnological tool. However, the recent outbreaks of foodborne illnesses serve as a grim reminder of the importance of food safety, including proper food processing and preservation. According to the World Health Organization (WHO), an estimated 7.69% of the world's population (600 million people, or almost 1 in every 10) suffer from foodborne illness each year, resulting in 420,000 deaths. These figures are equivalent to 31.1% of the annual deaths due to road accidents worldwide (Lee and Yoon, 2021; WHO, 2021; Yu et al., 2021). Children, mostly from low- and middle-income countries and under the age of five, carry 40% of the global burden of foodborne illness, with an economic loss of 110 billion US dollars (WHO, 2022; Worldbank, 2018). Furthermore, approximately 25% of global food loss has been due to microbial spoilage during the post-harvest period (Gram and Dalgaard, 2002; Hussein, 2022).

The microbiota of meat and meat products consists of harmless, beneficial and harmful microorganisms, including pathogenic and spoilage microorganisms. These include bacteria, viruses, fungi and prions viz., *Aeromonas* spp., *Klebsiella* spp., *Clostridium* spp., *Pseudomonas* spp., *Proteus* spp., *Shigella* spp., *Serratia* spp.,

---

[1] Department of Livestock Products Technology, West Bengal University of Animal and Fishery Sciences, Kolkata 700 037, India.

[2] Department of Livestock Products Technology, Faculty of Veterinary and Animal Sciences, Banaras Hindu University, Varanasi 221005, India.

[3] Eastern Regional Station, ICAR-Indian Veterinary Research Institute, Kolkata 700037, India.

* Corresponding author: arun.das@icar.gov.in

Lactobacillus spp., *Salmonella* spp., *Listeria monocytogenes, Bacillus cereus, Escherichia coli, Staphylococcus aureus, Campylobacter* spp., Norovirus, Hepatitis virus, *Aspergillus* spp., *Penicillium* spp., *Mucor* spp., *Cladosporium* spp., *Torula* spp., *Torulopsis* spp., *Toxoplasma gondii, Trichinella spiralis*, etc. (Das et al., 2019; Singh et al., 2022). These biological hazards in meat, responsible for most foodborne illnesses and spoilage, can be limited and regulated to an acceptable limit by employing different biopreservation strategies instead of physical and chemical preservation methods. In fact, the drawbacks of physical and chemical preservation methods, such as heating, chilling, freezing, canning, curing, and chemical preservatives, have resulted in the search for innovative, safe, natural, and effective food preservation methods (Priyadarshini et al., 2019). The physical preservation methods alter the structural, rheological, nutritional and sensory qualities of foods (Martins et al., 2019: Shishodia et al., 2019). Similarly, the chemical preservatives like sulphites and benzoates are allergic to humans causing asthma. Nitrates and nitrites are carcinogenic, while sorbates cause urticaria and contact dermatitis (Pepper et al., 2020; Rajanikar et al., 2021). Again, consumers' concerns about foodborne illnesses and awareness of the detrimental effects of chemical preservatives have urged the use of more natural, safer, and healthier alternatives in food preservation (Das et al., 2023).

The microbial agents including bioprotective cultures mostly lactic acid bacteria (LAB) belonging to the genera *Lactococcus, Lactobacillus, Carnobacterium, Streptococcus, Leuconostoc, Pediococcus*, etc. (Frétin et al., 2020; Suo et al., 2021) and their metabolites like bacteriocins, hydrogen peroxides, lysozymes, cell-free-supernatants or purified molecules, fermentates, postbiotics and other methods like use of bacteriophages, yeasts, etc., exhibiting antimicrobial and/or probiotic activities represent ideal alternatives to chemical preservatives (Leyva Salas et al., 2017; Das et al., 2021; Bhattacharya et al., 2022). This process is called "biopreservation", and the various agents and compounds used are called "biopreservatives". "Biopreservation" can be defined as a method of preserving food using natural or controlled microbiota or antimicrobials to inhibit foodborne pathogens and spoilage microorganisms. Earlier, the biopreservation process was practiced in a limited way in fermented dairy (curd, yoghurt and cheese), bakery products and sausages. Nevertheless, biopreservation was gradually and successfully extended to a wide-variety of foods such as fresh and processed meat (Bhattacharya et al., 2022; Borges et al., 2022; Lahiri et al., 2022). For instance, traditional fermented meat products like ham, salami and sausages include heritage preparations *viz.*, Bulgarian *lulanka* salami, Spanish *chorizo* and *salchichon*, Turkish *sukuk*, American *pepperoni*, Vietnamese *nemchua*, French *saucisson*, Thai *Naem*, etc. (Toldrá et al., 2008; Miranda et al., 2021) are good examples of biopreservation and constitute important source of dietary protein. These products are popular around the world not only for the biopreservative approach, but also for their nutritional benefits, flavor and superior sensory characteristics.

The preservative role of microbial products has been published in many literatures. De Martinez et al. (2002) reported that combination of lactic acid and nisin swabbing reduced microbial contamination of beef carcass. Pediocin and nisin were able to reduce spoilage causing *Lactobacillus sakei* counts in vacuum packaged sliced ham (Kalschne et al., 2014).

68 *Novel Approaches in Biopreservation for Food and Clinical Purposes*

Many biopreservatives, such as nisin, which are commonly used in meat, have been recognized as generally safe (GRAS) and have Qualified Presumption of Safety (QPS) status (Laulund et al., 2017). Furthermore, biopreservatives have been found to possess excellent antimicrobial activity, reduce lipid oxidation and protein degradation, and improve the taste, appearance, and flavor of food products, all while promoting consumer health (Bhattacharya et al., 2022; Lahiri et al., 2022). In view of the importance of biopreservation process, this chapter discusses about microbial composition of meat and meat products, different biopreservative strategies for enhancing food safety and shelf life, application of biopreservation and regulatory frameworks related to it.

Microbial composition

The muscle tissue of healthy food animals and birds is sterile and free from microorganisms immediately after slaughter. But during post-mortem and processing activities like slaughtering, evisceration, washing, cutting, packaging and storage, the muscle foods gets contaminated by microorganisms present in the hair, gastrointestinal tracts, respiratory tracts of the food animal or by several microbes present in the environment like equipment, water, processing surfaces, handlers, etc. (Sofos, 2013; Pradhan et al., 2018; Das et al., 2019). As a result, the meat foods that ultimately come to the consumers are rarely sterile. The microbial associations of different meat and meat products depend upon several intrinsic factors like low acid pH (5.5–6), high water activity (0.99) and extrinsic factors such as temperature, oxygen concentration, processing, storage and distribution in the supply chain (Guo et al., 2021; Bhattacharya et al., 2022). Most importantly, the muscle-based foods from farm animals and poultry are rich sources of good quality proteins, high biological value amino acids, lipids, essential fatty acids, trace minerals (iron, magnesium, zinc, etc.) and most B-complex vitamins (Bhattacharya et al., 2022; Borges et al., 2022). The multitude of nutrients, besides high moisture content of meat foods, makes the atmosphere conducive for the presence and proliferation of a complex ecosystem of microorganisms (Iulietto et al., 2015).

Not all microbes present in meat foods are responsible for spoilage, and only the specific spoilage organisms (SSO), when present above certain standard limits metabolize the available substrates in muscle leading to slime formation, souring, putrefactions, discoloration, off-flavors, off-taste, etc., termed as the indicators of spoilage (Gram and Dalgaard, 2002; Zhu et al., 2022). Most pathogenic microbes, when present above the standard limits cause foodborne illnesses with typical symptoms of abdominal pain, diarrhea, vomiting, nausea, fever, allergies, respiratory distress and in severe case death (Gourama, 2020). The diversity of microbial ecosystem associated with meat and meat products spoilage and illness are discussed below:

Bacteria

An initial microbial load as low as 10^2 and as high as 10^7 colony forming units (CFU) per cm^2 is considered as permissible in different muscle foods as tissue concentration is too low to produce disease or spoilage. The spoilage and pathogenic microorganisms above the permissible limits, i.e., at 10^7 CFU/cm^2 and 10^8 CFU/cm^2 develop off-flavor and sliminess, respectively and contribute significantly to the perishability and safety concerns of muscle foods (Comi, 2017). The bacteria of concern found in meat foods can be classified into three groups *viz.*, (i) bacteria naturally present in the habitat of the consumed species (e.g., *Vibrio* spp., Non-proteolytic *Clostridium* spp., *Aeromonas* spp., *Pseudomonas* spp., *Lactobacillus* spp., etc.), (ii) bacteria present in the processing environment (e.g., *L. monocytogenes*, *E. coli*, *Shigella* spp., *B. cereus*, proteolytic *Clostridium botulinum* type A & B, *Clostridium perfringens*, *Lactobacillus* spp., etc.), (iii) bacteria which are the usual inhabitants of animal body (*Salmonella typhimurium*, *E. coli*, *Campylobacter jejuni*, *Staphylococcus aureus*, etc.) (Iulietto et al., 2015; Das et al., 2019).

The bacterial genera responsible for spoilage (sliminess, discolouration, spots, off-odour, gassiness, flavor defects, blown pack, etc.) of meat (poultry, pork, lamb, mutton, chevon, beef, etc.) and meat products are *Acinetobacter*, *Aeromonas*, *Alteromonas*, *Bacillus*, *Brochothrix*, *Brucella*, *Flavobacterium*, *Hafnia*, *Kurthia*, *Leuconostoc*, *Micrococcus*, *Moraxella*, *Myroides*, *Pantoea*, *Proteus*, *Pseudomonas*, *Raoultella*, *Schwartzella*, *Shewanella*, *Serratia*, *Vibrio*, *Weissella*, different genera under LAB and Enterobacteriaceae family. Similarly, *Campylobacter* spp., *Salmonella* spp. (raw and under-cooked meat and mainly poultry), *L. monocytogenes*, *S. aureus* (chilled, ready to eat meat and meat products), *E. coli*, *Clostridium* spp., etc., are the main pathogenic bacteria often associated with meat and meat products and are responsible for most foodborne outbreaks (Iulietto et al., 2015; Erkmen and Bozoglu, 2016; Anas et al., 2019).

Many autochthonous beneficial bacteria belonging to LAB family, present in fresh and processed meat products are widely used as probiotics and biopreservatives (Borges et al., 2022). The various spoilage bacterial microbiota associated with different meat products and the defects caused by them are given in Tables 1 and 2.

Fungi

The mold genera like *Alternaria, Aspergillus, Cladosporium, Geotrichum, Mucor, Monilia, Penicillum, Sporotrichum, Thaminidium*, etc., and yeast genera like *Candida, Torulopsis, Trichosporon, Debaryomyces, Rhodotorula*, etc., cause most of the spoilage and illness associated with meat and meat products on prolonged storage. Some of the fungal strains and the defects caused by them are given in Table 3.

Parasites

As per WHO, the foodborne parasites in improperly cooked meat include protozoa (e.g., *Cryptosporidium* spp., *Cyclospora cayetanensis*, *Toxoplasma gondii*, *Giardia* spp., *Entamoeba* spp., *Trypanosoma* spp., etc.), helminths (e.g., *Fasciola* spp., *Fasciolopsis buskii*, *Paragonimus* spp., *Echinococcus* spp., *Taenia* spp., *Angiostrongylus* spp., *Ascaris* spp., *Capillaria* spp., *Toxocara* spp., *Trichinella* spp., *Trichostrongylus* spp., *Gnathostoma* spp., *Spirometra* spp., *Hymenolepis nana*, etc.).

70 *Novel Approaches in Biopreservation for Food and Clinical Purposes*

Table 1. Spoilage bacterial microbiota of different types of meat products.

Meat products	Spoilage bacteria	References
Fresh meat (including poultry)	*Acinetobacter, Aeromonas, Alcaligenes, Alteromonas, Brochothrix, Carnobacterium, Escherichia, Enterobacter, Enterococcus, Flavobacterium, Hafnia, Lactobacillus, Leuconostoc, Micrococcus, Moraxella, Proteus, Pseudomonas, Shewanella, Streptococcus*, etc.	(Comi, 2017; Zhu and Oteiza, 2022)
Chilled/refrigerated meat and meat products	*Acinetobacter, Moraxella, Pseudomonas, Aeromonas, Alcaligenes,* lactic acid bacteria (LAB), *Bacillus, Staphylococcus* and *Micrococcus*	(Hou et al., 2021; Li et al., 2021; Shao et al., 2022)
Frozen meat and meat products	*Pseudomonas, Aeromonas, Bacillus, Serratia, Klebsiella, Flavobacterium, Enterococcus, Staphylococcus, Clostridium*	(Jay et al., 2005; Comi, 2017; Zhu et al., 2022a)
Aerobically packed raw meat	*Pseudomonas, Brochothrix*	(Cenci-Goga et al., 2020; Hoa et al., 2022)
Vacuum packaged (VP) and modified atmosphere packaged (MAP) meat and meat products	*Brochothrix thermosphacta*, lactic acid bacteria (LAB), *Photobacterium*, etc. Mostly by *Lactobacillus* (including homofermentative *L. curvatus, L. sakei*, and heterofermentative *L. viridescens*) and *Leuconostoc* (such as *L. carnosum, L. gelidium*, and *L. mesenteroides*)	(Cenci-Goga et al., 2020; Zagorec and Champomier-Vergès, 2023)
Minced meat	*Pseudomonas, Aeromonas*, Acinetobacter, *Brochothrix*, LAB	(Aït-Kaddour et al., 2011; Xu et al., 2023)
Cured meat and meat products	*Clostridium, Vibrio, Micrococcus, Bacillus, Proteus, Halobacterium*, LAB, *Enterobacteriaceae*, etc.	(Liao et al., 2022; Zhou et al., 2022)
Canned meat and meat products	*Clostridium* spp., *Bacillus* spp.	(Jay et al., 2005; Zagorec and Champomier-Vergès, 2023)
Dried, salted, smoked meat	*Halobacterium salinarum. Bacillus* spp., *Micrococcus* spp., *Pseudomonas* spp.	(Oderinwale et al., 2020; Zhu et al., 2022a)
Fermented meat	*Pseudomonas, Acinetobacter, Moraxella, Lactobacillus, Clostridium, Bacillus*, etc.	(Zagorec and Champomier-Vergès, 2023)

These parasites cause serious diseases like gastroenteritis, fever, nausea, allergies, fatigue, arthritis, pancreatitis, cholangiocarcinoma, etc. (Das et al., 2019; Pozio, 2020). Among these, the top ten foodborne parasites as listed by FAO and WHO are given as follows: *Taenia solium* (pork tapeworm) in pork, *Echinococcus granulosus* (hydatid worm) and *Echinococcus multilocularis* in meat contaminated with dog faeces, *T. gondii* in pork, beef, mutton and other meats, *Cryptosporidium* spp. and *Entamoeba histolytica* in uncooked meat and meat products contaminated with soil and water, *Trichinella spiralis* (pork tapeworm) in pork, *Ascaris lumbricoides* and *Trypanosoma cruzi* in contaminated meat products with other foods, water and environment (Zolfaghari Emameh et al., 2018).

Biopreservation in Meat and Meat Products 71

Table 2. Defects in meat products due to bacterial spoilage.

Defects	Muscle foods	Bacteria	References
Gassiness	Cured ham	*Clostridium* spp.	(Lebret and Čandek-Potokar, 2022)
H_2S production/ Sulphury odour	Cured meat	*Vibrio* spp., *Enterobacteriaceae*	(Liao et al., 2022; Zhou et al., 2022)
Sulfide odour	Vacuum packaged meat	*Clostridium* spp., *Hafnia* spp.	(Palevich et al., 2021)
H_2O_2 greening	Meat	*Weisella* spp., *Leuconostoc* spp., *Enterococcus* spp., *Lactobacillus* spp.	(Comi, 2017; Pellissery et al., 2019)
H_2S greening	Vacuum packaged meat	*Pseudomonas* spp., *Lactobacillus sakei*	(Iulietto et al., 2015; Palevich et al., 2022)
Surface slime production	Aerobically packed meat	*Pseudomonas* spp., *Aeromonas* spp., *Lactobacillus* spp., *Leuconostoc* spp., *Enterococcus* spp., *Weissella* spp., *Brochothrix* spp.	(Shao et al., 2021; Wang et al., 2022)
Blown pack	Vacuum packaged meat	*Clostridium* spp., lactic acid bacteria	(Wambui and Stephan, 2019; Xu et al., 2023)
Putrefaction	Ham	*Enterobacteriaceae*, *Proteus* spp., *Pseudomonas* spp., *Clostridium* spp., *Alcaligenes* spp.	(Zhou et al., 2021; Zhu et al., 2022b)
Bone taint/deep decay	Meats, Ham	*Clostridium* spp., *Enterococcus* spp.	(Zhou et al., 2022, Shedleur-Bourguignon et al., 2023)
Souring	Ham	Lactic acid bacteria (LAB), *Enterococcus* spp., *Micrococcus* spp.	(Taormina, 2021; Shedleur-Bourguignon et al., 2023)
Sulphur stinker	Canned meat	*Clostridium nigrificans*	(Sharma, 2011)
Flat sour	Canned meat	*Bacillus* spp.	(Jay et al., 2005)
Thermophilic acid	Canned meat	*Clostridium thermosachharolyticum*	(Jay et al., 2005)
Blackening	Canned and cured meat	*C. nigrificans*	(Jay et al., 2005)
Yellowing	Vacuum packaged luncheon meat	*Enterococcus* spp. (mostly *Enterococcus casseliflavus*) *Micrococcus* spp.	(Jay et al., 2005)
Lipolysis	Meat	*Pseudomonas* spp., *Achromobacter* spp.	(Jay et al., 2005; Comi, 2017)
Phosphorescence and greenish rot	Meat	*Photobacterium* spp., *Pseudomonas* spp.	(Jay et al., 2005; Comi, 2017)
Bloom	Fresh meat	*Lactobacillus* spp., *Leuconostoc* spp.	(Jay et al. 2005)

Table 2 contd. ...

72   *Novel Approaches in Biopreservation for Food and Clinical Purposes*

...Table 2 contd.

Defects	Muscle foods	Bacteria	References
Surface discolouration	Meat	*Serratia marcescens, Flavobacterium* spp., *Pseudomonas* spp., *Chromobacterium* spp., *Micrococcus* spp.	(Sharma, 2011; Bharti et al., 2020)
Colourless rot	Meat	*Pseudomonas* spp., *Alcaligenes* spp., *Acinetobacter* spp. *Achromobacter* spp.	(Jay et al., 2005)
Pink rot and oniony odour	Meat	*Pseudomonas* spp.	(Keymanesh et al., 2009; Sharma, 2011)
Blue rot	Meat	*Pseudomonas syncyanea*	(Jay et al., 2005; Comi, 2017)
Red rot	Meat	*Serratia marcescens*	(Jay et al., 2005; Comi, 2017)
Red discolouration	Salted meat	*Halobacterium salinarum*	(Jay et al., 2005; Comi, 2017)
Yellow discolouration	Fresh meat	*Flavobacterium* spp.	(Bekhit et al., 2021)
Musty or earthy flavour	Meat	*Actinomycetes* spp., *Achromobacter* spp.	(Bekhit et al., 2021)
Fishy odour	Meat	*Escherichia coli*	(Jay et al., 2005; Comi, 2017)
Cabbage water odour	Meat	*Pseudomonas* spp.	(Jay et al., 2005; Comi, 2017)

Table 3. Defects caused by spoilage fungi in meat and meat products.

Defect	Fungal strains
Black spot	*Cladosporium herbarum, Rhizopus* spp.
White spot	*Sporotrichum* spp.
Green and blue spot	*Penicillium* spp.
Yellow spot	*Penicillium* spp.
Whiskers/Fuzz	*Thamnidium elegans, Mucor miehei, Rhizopus* spp.

Adopted from (Jay et al., 2005; Comi, 2017; Zagorec and Champomier-Vergès, 2023)

Viruses

The most important viruses associated with fresh meat are Norovirus (NoV), Hepatitis A virus (HAV) and Hepatitis E virus (HEV). Norovirus infection causes maximum foodborne viral illness characterized by nausea, explosive vomiting, watery diarrhea and abdominal pain due to consumption of undercooked and contaminated meat products (Price-Hayward and Hartnell, 2016; Bosch et al., 2018). But Adinoviruses, Rotavirus, Rabies virus, FMD virus and Avian influenza viruses are rarely isolated from raw meat foods (Soares et al., 2022).

Prions

Classical Bovine Spongiform Encephalopathy (BSE) agents *viz*. H-type and L-type prions when ingested through bovine specified risk materials (SRM) like brain and spinal cord along with beef cause variant Creutzfeldt-Jakob Disease (vCJD), a neuro-degenerative disease in humans (Das et al., 2019; WHO, 2022).

Biopreservation strategies for enhanced food safety and extended shelf life

Biopreservation strategies are gaining popularity as an effective means of enhancing food safety and extending the shelf life of various food products. These strategies involve using natural or controlled microbiota to inhibit the growth of different pathogenic and specific spoilage microorganisms in food. Various methods such as bacteriocins, protective cultures, organic acids, and other antimicrobial compounds, as well as fermentation, are used to achieve biopreservation. The use of different biopreservation strategies can not only increase the safety and shelf life of food products but also reduce the need for synthetic preservatives and additives, making them more appealing to consumers looking for natural and healthy options.

Protective cultures

As meat production and nutritional demands continue to grow, there is a need for improved quality and safety management systems in meat processing facilities. Spoilage and pathogenic microorganisms in food can have devastating effects on both product shelf life and human health, resulting in significant economic losses and foodborne outbreaks. Traditional preservation techniques such as refrigeration, freezing, thermal processing, and chemical preservatives have proven insufficient in meeting safety challenges due to issues such as cold chain breakups, faulty packaging, and the health hazards associated. The use of biopreservation techniques has, therefore, become increasingly important in the meat industry to combat spoilage and contamination issues. Biopreservation techniques involve the use of beneficial microbes, specifically LAB and their metabolites, to combat pathogenic microorganisms. Common food-grade LAB species used for biopreservation include *Lactococcus, Lactobacillus, Pediococcus, Leuconostoc, Staphylococcus, Carnobacterium, Enterococcus, Streptococcus*, and *Bifidobacterium* (Katiyar and Jain, 2018). These protective cultures dominate pathogenic microorganisms for available nutrients and space and produce antagonistic bioprotective metabolites such as bacteriocins, organic acids, hydrogen peroxide, and enzymes to ensure safety (Barcenilla et al., 2022). By using an adequate amount of these protective cultures or their biologically active metabolites, food waste and food insecurity can be minimized on a global level. These protective cultures are considered safe and can extend the shelf life of meat products, inhibiting the growth of foodborne pathogens and other spoilage organisms (Hernández-Aquino et al., 2019; Xu et al., 2022).

The protective approach of benign groups of microbiota, especially against different pathogenic microorganisms, such as the *Enterobacteriaceae* group, *S. aureus*, *L. monocytogenes*, and *C. botulinum*, in raw and minimally processed meat has been reported. The bioprotective activity of protective cultures can be enhanced with the application of hurdle concept associated with temperature regulation and proper packaging technology during storage without altering any sensory attributes (Xu et al., 2022).

In a recent study, it was found that the application of two mixed protective cultures, *Latilactobacillus sakei* and *Staphylococcus carnosus*, to low-fat minced beef under vacuum packed refrigerated condition for 12 days resulted in a positive outcome. These two commercially available cultures were able to suppress the growth of spoilage bacteria such as *Enterobacteriaceae* and *Pseudomonas* spp. (Xu et al., 2022). A similar experiment was conducted by Xu et al. (2021) using a group of protective cultures containing *L. sakei*, *Pediococcus pentosaceus*, *Staphylococcus xylosus*, and *S. carnosus* in different combinations. These cultures were applied to raw lamb meat under vacuum packaging and modified atmosphere packaging at 4°C for 15 days and 7 days, respectively. Vacuum packed lamb meat showed a favorable reduction of spoilage organisms such as LAB, *Brochothrix thermosphacta*, total viable count, *Pseudomonas* spp., and *Enterobacteriaceae*, as well as an extension of shelf life. Furthermore, the natural antimicrobial cultures did not affect the freshness, retaining a balanced pH and favorable sensorial characteristics of raw lamb meat. The combination of *S. carnosus* and *L. sakei* reinforced the vacuum-packed storage life by inhibiting 2.7 $\log_{10}$ CFU/cm^2 of *B. thermosphacta* and 2.8 $\log_{10}$ CFU/cm^2 *Enterobacteriaceae* after the 15th day (Trabelsi et al., 2019). Similar studies have also been reported with an inhibitory effect of *L. sakei*, *L. curvatus*, or *Lactiplantibacillus plantarum* in vacuum-packaged beef for *Enterobacteriaceae* (Zhang et al., 2018). Therefore, these research studies corroborate the novelty of protective culture's remarkable antimicrobial activity intertwined with additional hurdles, like appropriate packaging technology.

Another protective bacterium of LAB group, *Lactococcus pisicum*, has been reported in many literatures as biopreservation in meat. Adequate application of *L. pisicum* in meat foods markedly curbed the growth of wide spectrum of spoilage as well as pathogenic microorganisms without undesirable sensory aromas. *L. pisicum* was found to reduce the growth of notorious spoilage bacterium *B. thermosphacta* up to 3–4 log CFU/g under refrigerated vacuum packaging and MAP for meat (Saraoui et al., 2016). Application of protective cultures *L. carnosum* 4010, *L. curvatus* CRL705, and CWBI-B28-wt are noteworthy in the race of antimicrobial properties of different bioprotective cultures without organoleptic manipulations (Castellano et al., 2008). All these living inoculums proved their efficacy as protective cultures against *L. monocytogenes*, *B. thermosphacta* and other spoilers in different species of meat under fermentation or vacuum-packed refrigerated storage conditions. It was apparently reported that appositeness of these protective cultures instead of their purified or semi purified metabolites before heat treatment responded well in fermented meat products (Favaro and Todorov, 2017).

Microbial by-products (primary and secondary metabolites, acids, cell parts, etc.)

A large number of antagonistic metabolites are produced as by-products of microbial fermentation (Fig. 1). The various metabolites include organic acids (lactic, acetic, propionic, butyric, formic acid, etc.), bacteriocins, hydrogen peroxide, fatty acids, diacetyl, reuterin, reutericyclin, carbon dioxide, acetoin, cyclic dipeptides, etc. (Katiyar and Jain, 2018; Bhattacharya et al., 2022; Singh et al., 2022). Some of the important microbial metabolites are discussed hereafter.

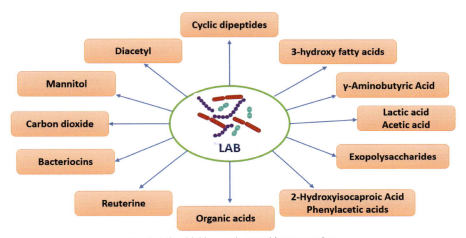

Fig. 1. Microbial by-products as biopreservatives.

Bacteriocins

Bacteriocins are natural antimicrobial compounds that are produced by certain types of bacteria, including LAB. These proteinaceous molecules are synthesized by ribosomes and are non-toxic to humans. Bacteriocins are effective against a broad range of bacteria, including pathogenic and spoilage bacteria, and are known to extend the shelf life of meat products. Unlike traditional chemical preservatives, bacteriocins are naturally occurring and are inherently stable in adverse conditions, such as changes in temperature and pH. They are also easily broken down by proteolytic enzymes, making them suitable for use as natural preservatives in food products (Daba and Elkhateeb, 2020; Bhattacharya et al., 2022). Bacteriocins are similar to antibiotics in their antimicrobial properties, but they are generally more effective and have a lower risk of resistance development (Daba and Elkhateeb, 2020). Due to their proteinaceous nature and small size, bacteriocins act quickly and at very low concentrations, making it difficult for bacteria to develop resistance (Favaro and Todorov, 2017). Additionally, bacteriocins have a simple biosynthetic mechanism that can be easily bioengineered to target specific pathogens with greater accuracy (O'Connor et al., 2015). Several commercially available bacteriocins have been granted Generally Recognized As Safe (GRAS) status by the US Food and Drug Administration (FDA) and Health Canada for use as food additives and biopreservatives. These include pediocin from *Pediococcus acidilactici*, nisin from *Lactococcus lactis, and carnobacteriocin BM1, carnocyclin A, and piscicolin 126* from *Carnobacterium maltaromaticum* (Bhattacharya et al. 2022). Overall, bacteriocins represent a promising alternative to traditional chemical preservatives in meat preservation due to their natural origin and effectiveness in preventing bacterial growth.

Bacteriocins are a heterogeneous group of compounds that exhibit various functional traits, leading to their classification based on different factors such as their bacterial source, molecular size, heat stability, chemical structure, and mechanism of action. Although can be obtained from both Gram-positive and Gram-negative bacteria, a new group of bacteriocins called 'archaeocins' have also been reported from representatives of the Archaea domain. Archaeocins are halocins and sulfolobicins produced by Halobacteriales and Sulfolobales, respectively, and have unique characteristics that inhibit the growth of pathogens instead of killing them (Zimina et al., 2020). Bacteriocins produced by Gram-positive LAB can be classified into four categories: class I, class II, class III, and class IV. Class I bacteriocins are small, thermostable, ribosomally synthesized peptides with non-proteinogenic thioether amino acids, which are known as lantibiotics (Cotter et al. 2005). Nisin A produced by *L. lactis* is the first commercially available most experimented bacteriocin under this group.

Class II bacteriocins are small, hydrophobic, thermostable peptides (< 10 kDa) with 30–60 amino acids, known as non-lantibiotics. Due to their structural complexity and diverse action of antigenicity, these bacteriocins are further subdivided into four subclasses: pediocin-like bacteriocins (class IIa), two-peptide bacteriocins (class IIb), circular bacteriocins (class IIc), and unmodified, linear, nonpediocin-like bacteriocins (class IId). Class IIa bacteriocins are known for their strong inhibitory activity against certain bacteria and are called pediocin-like. This group is very much selective and specific for antilisterial activity. Class IIb bacteriocins require both peptides to be fully active and they penetrate the target pathogen's cellular membrane followed by lowering the ATP concentration, the source of energy for survival of pathogen. Lactocin 705 secreted from *Lactobacillus curvatus* CRL705, plantaricin from *L. plantarum*, enterocin from *Enterococcus faecalis*, and lactococcin from *L. lactis* are few examples of class IIb bacteriocins. Class IIc bacteriocins have a circular configuration due to post translational modification and covalent interlinking between C and N termini and are known for their stability. The discovery of circular bacteriocins is becoming more common and could potentially be used as biopreservatives. Garvicin ML, gassericinA, lactocyclin Q, and leucocyclin Q, circularin A, reutericin 6 are different heat resistant class IIc bacteriocins with an outstanding pathogen and spoilage bacteria suppressing properties. The last subclass of class II bacteriocins, called as class IId are considered as miscellaneous or one-peptide, nonpediocin linear group. They include sec-dependent bacteriocins and an increasing number of bacteriocins without leader peptides. Out of total 31 discovered class IId bacteriocins, majority are contributed by Lactococcin. Both the class III and IV bacteriocins are still under research. Class III bacteriocins are heat sensitive macromolecules and they show the pathogenicity by cell wall lysis. Lysostaphin and enterolysin A are examples of class III bacteriolysins. Class IV bacteriocins are complex in structural as well as antimicrobial activity, therefore it is difficult to call them as true bacteriocins (Favaro and Todorov, 2017; Bhattacharya et al., 2022). The classification of bacteriocins is important, as it helps understand their mechanism of action, target range and structure, which can aid in the development of new antimicrobial agent.

Mechanism of action of bacteriocins: Bacteriocins are peptide molecules that are synthesized in the ribosomes of LAB. The genes responsible for the synthesis and secretion of bacteriocins can be found in various genetic elements like conjugative transposable elements, plasmids, and mobile genetic materials in the genome of LAB. Initially, bacteriocins are synthesized as precursor forms which undergo further processing and modifications within the cells. The production of bacteriocins is dependent on the availability of synthesized peptides or pheromones and typically occurs during the exponential growth phase of LAB. The transportation and modification of preformed bacteriocins are regulated by different genes to get a functioning molecule. ABC transporters and sec-dependent exporters play important roles in the secretion of bacteriocins (Mokoena et al., 2021). Bacteriocins can bind to specific receptors on the cell surface of target pathogenic and non-pathogenic bacteria, leading to their death or reduced pathogenicity. They typically have positively charged peptides with hydrophobic regions that interact electrostatically with the negatively charged bacterial cell surface. Bacteriocins attack phosphatidylethanolamine (PE), phosphatidylglycerol (PG), lipopolysaccharide (LPS), lipoteichoic acid (LTA), and other components of the target bacteria's cell surface, disrupting the integrity of the cell membrane, and inhibiting or killing the pathogen (Montville et al., 1995; Ríos Colombo et al., 2018). In Gram-positive bacteria, there are two different mechanisms by which bacteriocins can work. Class I models inhibit the synthesis of components related to the bacterial cell wall and lipid II in the cell membrane, while class II models form ion-selective pores in the cell membrane that cause dissipation of the proton motive force and depletion of intracellular ATP, leading to eventual death (Kumariya et al., 2019). However, the effectiveness of bacteriocins can depend on various factors, including their concentration, purity, growth phase, pH, temperature, and the presence of other antimicrobial substances. The mode of action of bacteriocins also varies based on their versatile physico-chemical properties. For instance, lantibiotics form pores in the cell membrane, while class II bacteriocins increase membrane permeability. Some bacteriocins require a docking molecule to interact with the membrane, while others do not. Research has also suggested that the external environment can play a role in the efficacy of bacteriocins. Chemical or physical stresses, such as exposure to organic acids, essential oils, EDTA, temperature, pulsed electric field, pH, and high hydrostatic pressure, can alleviate the antimicrobial action of bacteriocins, particularly for Gram-negative bacteria (Da Costa et al., 2019).

78 *Novel Approaches in Biopreservation for Food and Clinical Purposes*

Organic acids

Organic acids are a type of metabolites mainly produced as end products of the fermentation process, and sometimes during oxidation by microorganisms (Baptista et al., 2020). These acids have been used as food preservatives for centuries due to their ability to act as antimicrobials, antioxidants, food stabilizers, and flavor enhancers (Papadochristopoulos et al., 2021). Since organic acids are of natural origin, they are classified as "Generally Recognized As Safe" (GRAS). When using organic acids in meat foods, it is essential to follow the Acceptable Daily Intake (ADI) guidelines and not exceed the No Observable Adverse Effect Level (NOAEL) (Hauser et al., 2016). Several organic acids, including lactic acid, acetic acid, propionic acid, citric acid, benzoic acid, sorbic acid, formic acid, butyric acid, and phenyl lactic acid, are synthesized as byproducts of microbial fermentation and have excellent antimicrobial properties (Hochreutener et al., 2017; Kure et al., 2020; Barcenilla et al., 2022). Heterofermentative LABs produce acetic acid in addition to lactic acid, and *Acetobacter aceti* is responsible for its synthesis. Acetic acid is more effective against Gram-negative bacteria and is bacteriostatic up to a 0.2% concentration, becoming bactericidal above 0.3% concentration in meat foods. Propionic acid, on the other hand, is produced by *Propionibacterium* spp. and is effective against both Gram-positive and Gram-negative bacteria, yeasts, and molds (Erginkaya et al., 2011a; Sigh, 2018; Mahajan et al., 2020).

Organic acids cause acidification of cytoplasm impacting acid-base equilibrium by lowering pH, hamper homeostasis, interfere with gene expression, disrupt cytoplasmic membrane, thus causing cell lysis of the target microorganism in muscle foods (Geng et al., 2017; Taylor et al., 2019; Rathod et al., 2022). Organic acids in lower concentrations, usually below 1% and in combination with other antimicrobials are economical, and can achieve the desired antimicrobial effect without influencing the sensory qualities of meat foods (Kure et al., 2020). The sour effect of the organic acids can be nullified by the use of encapsulation technique (Rathod et al., 2022). Despite the GRAS status, food application of organic acids is sometimes questionable due to the undesirable toxicological effects on human health after ingestion of meat products treated with organic acids and their salts (Ben Braïek and Smaoui, 2021).

Application of organic acids in meat foods: It has been reported that lactic acid solution (1.25%) along with modified atmosphere packaging significantly reduced the total aerobic count, preserved the meat color and reduced lipid oxidation in chicken meat under refrigerated storage (Jaspal et al., 2021). Similarly, as per the findings of Carpenter et al. (2011), 2% lactic acid displayed strong antimicrobial activity against *E. coli* O157:H7, *Salmonella* spp. and *L. monocytogenes* during decontamination of beef plate surface, chicken and pork skin and turkey rolls, respectively. A recent study reported that lactic, acetic and citric acids at 1, 2 and 3% concentrations reduced *Salmonella enteritidis, E. coli* and *L. monocytogenes* counts in beef, where best results were observed with 3% concentration (Dan et al., 2017).

Hydrogen Peroxide (H_2O_2)

Hydrogen peroxide (H_2O_2) is an oxidizing compound produced by LABs in presence of O_2 during aerobic growth by the action of flavoprotein oxidase or NADH peroxidase enzyme (Ammor et al., 2006). The antimicrobial effect of H_2O_2 against most bacterial vegetative cells and spores is through oxidative damage of proteins and increase in membrane permeability of the microbial cell. It also alters the chemical structure and biochemical enzyme characteristics of the target microbe (Barberis et al., 2018).

Fatty acids

Fatty acids like capric acid, caproic acid, lauric acid, etc., are produced by the action of lipolytic LABs and their associated lipase enzymes during fermentation of fat-rich food products (Schnürer and Magnusson, 2005). Although the fatty acid production affects the sensory properties, the antibacterial and antifungal activity of fatty acids depends upon their chain length (Leyva Salas et al., 2017).

Diacetyl

The diacetyl (2,3-butanedione) is a flavor compound produced during citrate fermentation by some LABs. The activity of diacetyl is more evident at low pH (< 5.0) against Gram-negative bacteria. The antimicrobial activity of diacetyl is due to its reaction with arginine binding sites of Gram-negative bacterial protein, hence antagonizing arginine utilization (Schnürer and Magnusson, 2005; Katiyar and Jain, 2018; Rathod et al., 2022).

Reuterin

Reuterin is a low molecular weight antibacterial compound produced by *Lactobacillus reuteri* and some other LABs (Mahajan et al., 2020; Rathod et al., 2022). Also known as 3-hydroxypropionaldehyde (3-HPA), it is formed as an intermediate during glycerol metabolism to 1,3-propanediol under anaerobic conditions (Katiyar and Jain, 2018). The antimicrobial activity of reuterin is attributed to its ability to inhibit DNA synthesis (Sigh, 2018). It has abroad spectrum of activity against Gram-positive and Gram-negative bacteria, fungi and protozoa. It has shown bacteriostatic activity against *L. monocytogenes* and bactericidal activity against *S. aureus, E. coli* O157:H7, *Salmonella cholerasuis, Y. enterocolitica, A. hydrophilla* and *C. jejuni* (Arqués et al., 2004; Katiyar and Jain, 2018).

Reutericyclin

This unique tetramic acid is a negatively charged, highly hydrophobic bacterial antagonist and is produced by some LABs (Erginkaya et al., 2011b). Reutericyclin acts as a proton ionophore, dissipating the proton motive force of Gram-positive bacteria including *Bacillus* spp., *S. aureus, Listeria* spp., *Enterococcus* spp., etc. This is less effective towards Gram-negative bacteria and ineffective for fungi (Jay, 1992; Katiyar and Jain, 2018).

Carbon dioxide (CO_2)

CO_2 is produced by heterofermentative LABs during fermentation of sugar. CO_2 dissolves in water producing carbonic acid, which is also antagonistic. The CO_2 production makes the environment anaerobic, thereby inhibiting the growth and survival of aerobic bacteria (Rathod et al., 2022).

Starters/Fermentation

Fermentation is a traditional food preservation method and is widely used for improving food safety, shelf life, organoleptic properties, nutritional value and functionality of meat foods since time immemorial. Fermented meat products are popular in different parts of the world and are an integral part of many traditional foods (Hassoun et al., 2020; Zang et al., 2020). Traditionally fermentation is carried out using autochthonous microflora present in foods, but now a days commercialized starter cultures are employed in meat food fermentation, thereby controlling the process and ensuring safety (Laranjo et al., 2019). Fermentation typically involves introduction of starter cultures that convert certain nutrients in muscle foods into mixture of carbon dioxide, alcohol, organic acids, bacteriocins and other beneficial metabolites (Zang et al., 2020; Bourdichon et al., 2021). Fermentation may be of lactic acid fermentation (homofermentative and heterofermentative), acetic acid fermentation and propionic acid fermentation based on the type of microorganisms used in the starter culture and their metabolism (Bourdichon et al., 2021).

Starters can be defined as preparations with a large number of cells that include a single type or a mixture of two or more microbial cultures added to foods in order to take advantage of their metabolites or enzymatic activity to carry out the fermentation process (Laranjo et al., 2019; García-Díez and Saraiva, 2021). Microorganisms used as starter cultures are bacteria, moulds and yeasts. Lactic acid bacteria (LAB) is the most representative group of bacteria deciphering fermentation of meat foods, with *Lactobacillus* spp., *Streptococcus* spp. and *Bifidobacterium* spp. being the most commonly used starter cultures (Hassoun et al., 2020; García-Díez, and Saraiva, 2021). Additionally, other bacteria like coagulase-negative Staphylococcus (CNS) (e.g., *Staphylococcus carnosus* and *S. xylosus*) and Micrococcaceae (e.g., *Kocuria* spp.) are also used as starter cultures in meat food fermentation (Semedo-Lemsaddek et al., 2016; Altieri et al., 2017; Correia Santos et al., 2017; Stavropoulou et al., 2018). While different moulds are used for fermentation of animal origin foods, yeasts have their major use in baking and beverages industry (García-Díez and Saraiva, 2021). However, the yeasts belonging to the genera Debaryomyces and Candida and moulds belonging to the species *Penicillium nalgiovense* and *Penicillium gladioli* are mainly inoculated as starter cultures in meat foods (Laranjo et al., 2019).

Starter cultures are the good microbes that work by competitive exclusion of target spoilage and pathogenic microorganisms by competing for nutrients, oxygen and changing the environment through quorum sensing (Young and O'Sullivan, 2011; Laranjo et al., 2019).

The meat food fermentation starter cultures have a long history of use and have gained the GRAS (Generally Recognized As Safe) and QPS (Qualified Presumption of Safety) status (Bhattacharya et al., 2022; Singh et al., 2022). The various beneficial effects of fermentation and starter cultures in meat foods, safeguarding consumer health and ensuring food safety, are given below (Semedo-Lemsaddek et al., 2016; Laranjo et al., 2019; Bavisetty et al., 2021; Bourdichon et al., 2021; García-Díez and Saraiva, 2021):

1. Preservation of meat foods by rapid matrix acidification (fast pH decrease), production of antimicrobial substances against foodborne spoilage organisms.

2. Killing or retarding the growth of several pathogenic foodborne bacteria (*Salmonella* spp., *Listeria* spp., *Staphylococcus* spp., etc.) by competitive exclusion principle.
3. Reducing the formation of chemical hazards like biogenic amines (BAs), polycyclic aromatic hydrocarbons (PAHs), mycotoxins etc. during processing.
4. Improving the organoleptic qualities of fermented meat products.
5. Enhancing the functionality and nutritive value of fermented meat foods by production of bioactive compounds.

In conclusion, the fermentation of meat and meat products is an important process in the production of many popular food items. It involves the use of selected strains of LAB or other microorganisms, which can help to ensure a unique flavor, texture, and preservation characteristics with consistent and desirable results. Biopreservative effect of lactic acid bacteria and bacteriocins on quality and safety of meat and meat products is presented in Fig. 2. However, the process requires careful control and can be affected by various factors, such as the quality of the meat, temperature, humidity, pH, the presence of other microorganisms, and conditions in the production environment.

Application in meat foods: Traditional fermented meat products like ham, salami and sausages include heritage preparations *viz.*, Bulgarian *lulanka* salami, Spanish *chorizo* and *salchichon*, Turkish *sukuk*, American *pepperoni*, Vietnamese *nemchua*, French *saucisson*, Thai *Naem*, etc. (Toldrá et al., 2008; Miranda et al., 2021). The application of fermentation starter cultures for biopreservation of meat foods by different researchers is given in Table 4.

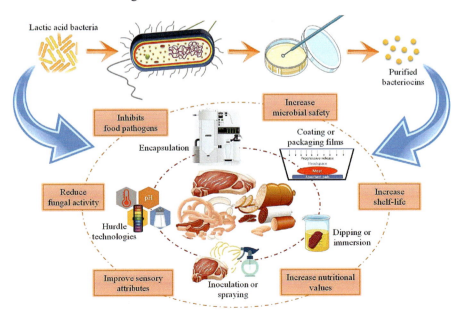

Fig. 2. Effect of LAB and bacteriocin on various quality aspects of muscle foods (with permission from Bhattacharya et al., 2022).

82 *Novel Approaches in Biopreservation for Food and Clinical Purposes*

Table 4. Application of fermentation starter cultures for biopreservation of meat foods.

Starter culture	Meat food	Biopreservative action/effect	References
L. sakei, L. plantarum and *L. curvatus*	Low acid fermented sausage	Decrease in enterobacteria count (3.5 log cfu/g to < 1.0 log cfu/g) in 16 days	(Baka et al., 2011)
P. acidilactici and *S. vitulinus*	*Salchichon* – A traditional Iberian dry sausage	Inhibitory effect on enterobacteria and coliforms	(Casquete et al., 2012)
Lactococcus lactis and *Lactococcus casei*	Salami	Inhibits the growth of coliforms and *S. aureus*	(Cenci-Goga et al., 2012)
L. plantarum	Pork salami	Inhibits the growth of artificially inoculated *Clostridium perfringens* and other *Clostridium* spp.	(Di Gioia et al., 2016)
Debaryomyces hansenii, D. marasmus and *Hyphopichia* spp.	Dry cured hams	Antifungal activity against toxigenic *Penicillium nordicum* mould	(Simoncini et al., 2015)
CNS Staphylococci (*S. xylosus*)	Fermented meat	Preserving meat by synthesizing nitric oxide (NO) via nitric oxide synthase (NOS)	(Ras et al., 2017, 2018)

Bacteriophages

Bacteriophages and their endolysins are emerging as a new approach for the biocontrol of foodborne bacteria. Bacteriophages are specific viruses that infect bacterial cells and have a large genome size ranging from 3.4 kb to 500 kb (Veesler and Cambillau, 2011). The principal virulent bacteriophages against foodborne bacteria belong to three families: *Myoviridae*, *Siphoviridae*, and *Podoviridae* (Garvey, 2022). Bacteriophages outnumber bacterial cells by 10-fold and hence, are the most abundant species on earth (Romero-Calle et al., 2019) and are categorized into two types based on their life cycle: lytic and lysogenic. Lytic phages cause lysis of bacterial cells, while lysogenic phages do not result in cell lysis and only the lysogenic phage DNA is incorporated into the bacterial cell (Garvey, 2022; Rendueles et al., 2022). Bacteriophages used in the food industry should be strictly lytic, as they only target specific food-borne bacteria, without causing much harm to the beneficial microflora. Use of lysogenic phages is undesirable due to horizontal gene transfer (Fernández et al., 2018).

Lytic phases have several potential applications as food biocontrol and biopreservation agents (Deka et al., 2022). Moreover, phage-encoded lytic proteins, called "endolysins", are specific peptidoglycan hydrolases that extend the antibacterial activity of the bacteriophages (Rendueles et al., 2022).

Phages are mainly used in three sectors of the food industry: primary production, bio-sanitization, and biopreservation (Połaska and Sokołowska, 2019). Phages have several unique features and advantages that make them ideal biopreservatives, especially in meat foods. These features include high host specificity, safety (harmless to mammalian cells), preservation of the natural microbiota of food and gut, effective killing of antibiotic-resistant pathogenic bacteria, ubiquitous nature (present in a wide variety of food, soil, water, etc.), no alteration of sensory quality of food, resistance to different processing and cooking methods, GRAS status, and approval by food regulating agencies as "edible viruses" (Połaska and Sokołowska, 2019; Endersen and Coffey, 2020; Rathod et al., 2022).

As stated, endolysins derived from bacteriophages are useful for biopreservation of muscle foods due to their broad host specificity. They can be used alone or in combination with other antimicrobials to effectively control the growth of foodborne bacteria such as *L. monocytogenes, S. enterica, C. jejuni, E. coli* O157:H7, *Shigella* spp., *S. aureus, Carnobacter sakazkii*, etc., in different types of meat foods (Fernández et al., 2018; Endersen and Coffey, 2020). Commercial preparations of phages such as ListShield, SalmoFresh, SalmoLyse, and BacWash targeting specific bacteria *viz.*, *L. monocytogenes, S. enterica, Salmonella omnilytics*, etc., have been approved by the FDA as GRAS for use in food products. Phage therapy has also been successfully used for surface decontamination of poultry and beef in the food industry (Połaska and Sokołowska, 2019). Since the FDA's approval of the first phage-based product ListShield in 2006, many phage-based cocktails have been used in poultry, meat and ready to eat (RTE) products (Rasolofo et al., 2011). Table 5 provides a summary of some of the phage-based products and their applications in the food industry.

Furthermore, bacteriophages along with other antimicrobial hurdles have shown synergistic effects in biopreservation of meat foods. In recent studies, the antimicrobial effect of phages was improved by using in combination with bacteriocins like nisin, cinnamaldehyde oil, sodium diacetate, potassium lactate, modified atmosphere packaging, etc. (Połaska and Sokołowska, 2019; Garvey, 2022). However, the main disadvantage associated with phages is their problematic contamination of fermented foods causing "the dead vat condition", where they negatively impact the fermentation process by lysis of some bacterial starter cultures (Garvey, 2022).

Others

Other biopreservatives used in meat foods include some compounds and substances obtained from microbial sources other than LAB, animal and plant sources. Some methods or processes individually or in combination, using different biological substances are also being used as unconventional biopreservation strategies (Villalobos-Delgado et al., 2019; Ren et al., 2021). The other important biopreservation strategies for enhanced food safety and extended shelf life are summarized in Table 6.

Applications/market products

Some commonly used market products of purified or partially purified bacteriocins are nisin A or E234 (under the trademarks Nisaplin™), pediocin PA1 (under the trademarks Microgard™), Sakacin (Bactoferm™) and Leucocin A (Bactoferm™ B-SF-43) with their ability to control Gram positive bacteria (*Bacillus cereus,*

84   Novel Approaches in Biopreservation for Food and Clinical Purposes

Table 5. Application of bacteriophages in biopreservation of meat foods against foodborne bacteria.

Type of food	Phage used	Target Pathogen	Findings	References
Chicken skin, raw poultry	NCTC 12675; NCTC 12672	*Campylobacter* spp.	2 and 1 log reductions in Campylobacter count in frozen-thawed and fresh chicken samples, respectively.	(Atterbury et al., 2003; Goode et al., 2003)
Raw and cooked beef	Cj6	*Campylobacter* spp.	*Campylobacter* spp. levels significantly reduced in both raw and cooked beef (51°C) over a period of 8 days.	(Bigwood et al., 2009)
Beef surface	Ecoshield™ (contains 3 phage cocktail *viz.*, e11/2, pp01 & e4/1c) {GRAS status in 2011}	*Escherichia coli* O157:H7	5 log unit reduction in *Escherichia coli* O157:H7 number in 1 hour at 37°C.	(O'Flynn et al., 2004; Lewis and Hill, 2020)
Ready to eat (RTE) meat products	PhageGuardListex™ (formerly Listex™; P100) {GRAS status in 2007}	*Listeria monocytogenes*	Reduction in colony forming units (CFU) by 1.4–2 log units at 4°C, 1.7–2.1 log at 10°C and 1.6–2.3 log at 22°C.	(Bolocan et al., 2019)
Mixed meat, RTE meat products, Raw poultry	ListShield (formerly LMP-102) {GRAS status in 2006}	*L. monocytogenes*	95% reduction in *L. monocytogenes* viability when applied at 1 ml/500 cm² of food before packaging.	(Zhang, 2018; EFSA; ECDC, 2021; Garvey, 2022)
Poultry	SalmoPro, SalmoFresh™	*Salmonella enterica*	Significant reduction in Salmonella count.	(Tang et al., 2019; Garvey, 2022)

S. aureus and *Listeria monocytogenes*) or Gram negative pathogens (*E. coli, Pseudomonas* spp., etc.) and sometimes bacterial spores in meat food products (Silva et al., 2018; Daba and Elkhateeb, 2020). An approved market product of organic acids (OA) with the tradename 'Optimizer' manufactured by Pacific Vet Group, USA is a combination of lactic, acetic, tannic, propionic and caprylic acids and being used in broiler industry at the rate of 4L OA /1000 L of water for the control of Salmonella (Menconi et al., 2014). Commercial phage preparations for specific bacteria *viz.*, *Listeria monocytogenes* are available in the tradename of ListShield & PhageGuard Listex™ (formerly Listex™ P100). SalmoFresh, Salmopro, SalmoLyse & BacWash are approved commercial phage preparations against *Salmonella enterica*, *Salmonella omnilytics* (Połaska and Sokołowska, 2019; Mahajan et al., 2020; Tang et al., 2019; Garvey, 2022). Ecoshield™ and ShigaShield™ are the market phage products against *Escherichia coli* O157:H7 and *Shigella* spp., respectively (O'Flynn et al., 2004; Lewis and Hill, 2020). Some commercial protective cultures in a combination of either *Pediococcus acidilactici*, *Latilactobacillus curvatus*, and *Staphylococcus xylos* are being marketed for biopreservation of meat against *L. monocytogenes*

Table 6. Other biopreservation strategies for enhanced food safety and extended shelf life.

Classification	Antimicrobial substances/compounds	Source/Origin	Biopreservative action	References
Microbial-derived	Natamycin (Primaricin)	A class of polyene macrolides, produced from the eubacterium *Streptomyces natalensis*. (the only approved broad spectrum antifungal biopreservative in food).	Binds irreversibly to fungal cell membrane due to its high affinity for ergosterol resulting in membrane hyperpermeability ultimately causing cell lysis.	(Choo et al., 2021; Wang et al., 2021)
	ε-Polylysine	A class of antimicrobial peptide obtained from fermentation of glucose by *Streptomyces albicans*.	Interacts with the cell wall, cell membrane, genetic materials and enzymes of the target bacteria.	(Alirezalu et al., 2021a; Alirezalu et al., 2021b)
Animal derived	Chitosan	Obtained from insect, shellfish, fungal cell wall, algae, protozoa and yeast with chitin as the active component obtained by the deproteinization and demineralization.	Acts on the surface of the bacterial and fungal cell wall thus leads to increase in permeability of the microbial cell and disrupts the cell membrane. Causes leakage of cellular materials and deformation of the cell.	(Baptista et al., 2020; Li and Zhuang, 2020)
	Propolis/Bee glue	Resin like material made by bees from the buds of poplar and cone-bearing trees.	Inhibits the growth of microbes by retarding the total volatile basic nitrogen or by protein degradation.	(Mehdizadeh and Mojaddar Langroodi, 2019; Cunha et al. 2021)
	Lactoperoxidase system	Antimicrobial system originating from milk and consists of lactoperoxidase, thiocyanate and H_2O_2.	Effective against Gram-positive bacteria.	(Molayi et al., 2018; Yousefi et al., 2022)
	Ovotransferrin	Second most abundant protein naturally present in egg white.	Retards microbial growth by iron depletion.	(Legros et al., 2021; Wang et al., 2021)
	Protamine	An arginine rich polycationic protein, naturally obtained from spermatic cells of fish, birds and mammals.	Broad spectrum antimicrobial activity against Gram-positive and Gram-negative bacteria, fungi. More effective against Gram-positive bacteria. Cause disruption of cell membrane.	(Fujiki et al., 2019; Saeed et al., 2019; Marrez et al., 2022)
	Pleurocidin (Ple)	Novel heat stable, salt tolerant antimicrobial peptide with 25 amino acids, derived from skin mucous membrane of winter flounder (*Pleuronectes americanus*).	Exhibits broad spectrum antimicrobial activity against Gram-positive and Gram-negative foodborne bacteria. Causes membrane translocation and pore formation in bacterial phospholipid cell membrane.	(Mogoşanu et al., 2017; Yusuf, 2018; Villalobos-Delgado et al., 2019)
	Lysozyme	Low molecular weight (15000 Da) cationic enzyme, naturally found as part of defense system of living organisms. It can be obtained from egg white, milk, etc.	Causes hydrolysis of β-1,4-glycosidic bond between N-acetyl muramic acid and N-acetylglucosamine in peptidoglycan layer of Gram-positive bacterial cell wall.	(Singh, 2018)

Table 6 contd. ...

Classification	Antimicrobial substances/compounds	Source/Origin	Biopreservative action	References
Plant derived	Plant derivatives (e.g., Bamboo, rosemary, clove, thyme essential oils, moringa flower, dragon fruit, litchi, pomegranate peel, etc.	Source of polyphenols, terpenes, terpenoids, aldehydes, ketones, aliphatic alcohols, isoflavonoids, etc.	Acts against several foodborne bacteria by rupture of the cell membrane, affects nucleic acid metabolism, decay of proton motive force, depletion of adenosine triphosphate (ATP), etc.	(Das et al., 2016, 2023; Madane et al., 2019, 2020; Das et al., 2021)
Bio-based packaging films	Antimicrobial packaging films, edible coatings	Polysachharides (starch, carageenan, agar agar, carboxymethyl cellulose, alginate), chitosan, gelatin, enzymes, plant extracts-based packaging films/coatings.	Maintains low pH of muscle foods and decrease the microbial load under storage conditions.	(Molayi et al., 2018; Bhattacharya et al., 2019; Alirezalu et al., 2021)
Hurdle technology (Composite biopreservation methods)	Combination of different biopreservation methods and/or agents	Packaging films with EOs, combination of antimicrobial substances, blend of EOs, etc. (e.g., Chitosan film and spice extract, CMC coating and turmeric, gelatin and lemon EOs, etc.).	Ruptures cell membrane, affects nucleic acid metabolism of food borne bacteria.	(Dalvandi et al., 2020; Ucak et al., 2021)

and additionally they are preventing slime formation and off odours in meat during storage period (Daba and Elkhateeb, 2020).

Regulatory framework (EFSA and FDA general guidelines)

Biopreservation using LAB and their metabolites is currently the main alternative to chemical and physical methods of preserving meat foods. These biopreservatives have a long history of use and have been granted the GRAS and QPS (Qualified Presumption of Safety) status by the FDA and European Food Safety Agency (EFSA), respectively (Laulund et al., 2017; Mokoena et al., 2021). While food cultures have been used as source of microorganisms in food for centuries, the EFSA does not currently define them as a food ingredient (Laulund et al., 2017). The QPS list was introduced by the EFSA scientific panel in 2007 to provide a regularly updated pre-evaluation of the safety of microorganisms intended for use in food or feed chains (Herman et al., 2019). The QPS list consists of taxonomic units (TU) with specific qualifications, including the lack of virulence factors, toxic metabolites, and resistance to some antifungals, transmissible antibiotic resistance, and "for production purpose only" on a strain-by-strain basis (Ricci et al., 2017). The last published QPS status list, updated in 2017, includes 95 TU, including 60 species of Gram-positive non-sporulating bacteria (mainly LAB), 13 bacillus species, 2 Gram-negative bacteria (*Gluconobactero xydans* and *Xanthomonas campestris*), 15 yeast species, and 3 viruses (plant viruses *Alphaflexiviridae*, *Potyviridae*, insect virus *Baculoviridae*) only for use as food additives (Ricci et al., 2017; Herman et al., 2019; Koutsoumanis et al., 2022). The Panel on Biohazard (BIOHAZ) delivers scientific opinions on maintenance of the QPS list of microorganisms and the QPS panel statement includes microbes and their metabolites like organic acids, bacteriocins, enzymes, etc. The QPS list is revised every six months, with an overall total review of previously accessed microorganisms done only every three years (Laulund et al., 2017; Herman et al., 2019). Some TUs, such as filamentous fungi, bacteriophages, *Enterococcus faecium*, *E. coli*, *Streptomyces* spp., and Oomycetes, have been excluded from the QPS list after revisions (Herman et al., 2019).

Similar to the EFSA, the FDA has also published lists of microorganisms and microbial-derived ingredients that are deemed safe and suitable for human consumption as food additives (Gad and Sullivan, 2014; de Lacerda et al., 2016). Some of these have received GRAS status, which means they are considered safe for use in food based on scientific procedures or past experience of their common use in food. However, not all LABs can be described as GRAS from a regulatory point of view, although many non-genetically modified LABs are often claimed to have this status. The GRAS status differs from that of a food additive in that FDA itself determines the safety of the ingredient for the latter, while for a GRAS substance, the evaluation is conducted by outside experts assigned by the FDA. The GRAS status can be found in an inventory by the FDA, which is updated monthly and listed in title 21 of the Code of Federal Regulation (21 CFR) (Gad and Sullivan, 2014; FDA 2022a, 2022b).

In contrast, the QPS is a status that applies to microorganisms only and not all ingredients, and it applies to the taxonomic unit (TU) of a species of microorganism

88 *Novel Approaches in Biopreservation for Food and Clinical Purposes*

rather than at the strain level or for a particular food product (Laulund et al., 2017). Although both QPS and GRAS lists are not exhaustive and only include the microorganisms notified to the standard agencies, the absence of a microbe or its metabolite from these lists does not necessarily imply that it is unsafe.

Conclusion

The quality and safety of food products are becoming a growing concern globally. To address this, there is an increasing interest in using LAB (lactic acid bacteria) and antimicrobial metabolites to inhibit the growth of various pathogenic and specific spoilage microorganisms in meat and meat products. Bacteriocins, the active metabolites of LAB, have great potential for preserving meat-based food products. However, their use in the commercial sector is limited due to factors such as a narrow antimicrobial spectrum and high production costs. To overcome these hurdles, more advanced research is needed to optimize production and purification processes. Several strategies are being employed, including combining bacteriocins of two classes through genetic bioengineering, conjugating existing bacteriocin-producing strains with nanoparticles, and using non-thermal techniques like high hydrostatic pressure, irradiation, pulsed electric fields, and ultrasound. It is essential to conduct toxicological testing and evaluate the potential effects of these additives before seeking regulatory approval for their commercial use. Advancements in these areas will pave the way for suitable commercial applications of LAB and their various metabolites, particularly bacteriocins, in different processed foods, including the muscle food sector.

Funding

This research received no external funding.

References

Aït-Kaddour, A., Boubellouta, T. and Chevallier, I. 2011. Development of a portable spectrofluorimeter for measuring the microbial spoilage of minced beef. *Meat Science*, 88(4): 675–681.

Alirezalu, K., Movlan, H.S., Yaghoubi, M., Pateiro, M. and Lorenzo, J.M. 2021a. ε-Polylysine coating with stinging nettle extract for fresh beef preservation. *Meat Science*, 176: 474.

Alirezalu, K., Pirouzi, S., Yaghoubi, M., Karimi-Dehkordi, M., Jafarzadeh, S. and Mousavi Khaneghah, A. 2021b. Packaging of beef fillet with active chitosan film incorporated with ε-polylysine: an assessment of quality indices and shelf life. *Meat Science*, 176: 475.

Altieri, C., Ciuffreda, E., Di Maggio, B. and Sinigaglia, M. 2017. Lactic acid bacteria as starter cultures. *Starter Cultures in Food Production*, 1–15.

Ammor, S., Tauveron, G., Dufour, E. and Chevallier, I. 2006. Antibacterial activity of lactic acid bacteria against spoilage and pathogenic bacteria isolated from the same meat small-scale facility - 1 - Screening and characterization of the antibacterial compounds. *Food Control*, 17(6): 454–461.

Anas, M., Ahmad, S. and Malik, A. 2019. Microbial escalation in meat and meat products and its consequences. *In: Health and Safety Aspects of Food Processing Technologies*, 29–49.

Arqués, J.L., Fernández, J., Gaya, P., Nuñez, M., Rodríguez, E. and Medina, M. 2004. Antimicrobial activity of reuterin in combination with nisin against food-borne pathogens. *International Journal of Food Microbiology*, 95(2): 225–229.

Atterbury, R.J., Connerton, P.L., Dodd, C.E.R., Rees, C.E.D. and Connerton, I.F. 2003. Application of host-specific bacteriophages to the surface of chicken skin leads to a reduction in recovery of Campylobacter jejuni. *Applied and Environmental Microbiology*, 69(10): 6302–6306.

Baka, A.M., Papavergou, E.J., Pragalaki, T., Bloukas, J.G. and Kotzekidou, P. 2011. Effect of selected autochthonous starter cultures on processing and quality characteristics of Greek fermented sausages. *LWT-Food Science and Technology*, 44(1): 54–61.

Baptista, R.C., Horita, C.N. and Sant'Ana, A.S. 2020. Natural products with preservative properties for enhancing the microbiological safety and extending the shelf-life of seafood: a review. *Food Research International*, 127, 762.

Barberis, S., Quiroga, H.G., Barcia, C., Talia, J.M. and Debattista, N. 2018. Natural food preservatives against microorganisms. *In*: *Food Safety and Preservation*. Elsevier, 621–658.

Barcenilla, C., Ducic, M., López, M., Prieto, M. and Álvarez-Ordóñez, A. 2022. Application of lactic acid bacteria for the biopreservation of meat products: a systematic review. *Meat Science*, 183, 661.

Bavisetty, S.C.B., Benjakul, S., Olatunde, O.O. and Ali, A.M.M. 2021. Bioactive compounds in fermented fish and meat products: health aspects. *Bioactive Compounds in Fermented Foods*, 325–362.

Bekhit, A.E.D.A., Holman, B.W.B., Giteru, S.G. and Hopkins, D.L. 2021. Total volatile basic nitrogen (TVB-N) and its role in meat spoilage: a review. *Trends in Food Science and Technology*, 109: 280–302.

Bharti, S.K., Pathak, V., Alam, T., Arya, A., Basak, G. and Awasthi, M.G. 2020. Materiality of edible film packaging in muscle foods: a worthwhile Conception. *Journal of Packaging Technology and Research*, 4(1): 117–132.

Bhattacharya, D., Gurunathan, K., Mendiratta, S., Vishnuraj, M. and Soni, A. 2019. Estimation of microbial load in anthocyanin-CMC based edible film coated chicken meat slices in refrigeration storage using Fluorescein Di-acetate (FDA) hydrolysis assay. *International Journal of Livestock Research*, 9(6): 99–108.

Bhattacharya, D., Nanda, P.K., Pateiro, M., Lorenzo, J.M., Dhar, P. and Das, A.K. 2022. Lactic acid bacteria and bacteriocins: novel biotechnological approach for biopreservation of meat and meat products. *Microorganisms*, 10(10): 2058.

Bigwood, T., Hudson, J.A. and Billington, C. (2009). Influence of host and bacteriophage concentrations on the inactivation of food-borne pathogenic bacteria by two phages: research letter. *FEMS Microbiology Letters*, 291(1): 59–64.

Bintsis, T. 2017. Foodborne pathogens. *AIMS Microbiology*, 3(3): 529–563.

Bolocan, A.S., Upadrasta, A., De Almeida Bettio, P.H., Clooney, A.G., Draper, L.A., Ross, R.P. and Hill, C. 2019. Evaluation of phage therapy in the context of *Enterococcus faecalis* and its associated diseases. *Viruses*, 11(4): 366.

Borges, F., Briandet, R., Callon, C., Champomier-Vergès, M.C., Christieans, S., Chuzeville, S., Denis, C., Desmasures, N., Desmonts, M.H., Feurer, C., Leroi, F., Leroy, S., Mounier, J., Passerini, D., Pilet, M.F., Schlusselhuber, M., Stahl, V., Strub, C., Talon, R. and Zagorec, M. 2022. Contribution of omics to biopreservation: toward food microbiome engineering. *Frontiers in Microbiology*, 13: 182.

Bosch, A., Gkogka, E., Le Guyader, F.S., Loisy-Hamon, F., Lee, A., van Lieshout, L., Marthi, B., Myrmel, M., Sansom, A., Schultz, A.C., Winkler, A., Zuber, S. and Phister, T. 2018. Foodborne viruses: detection, risk assessment and control options in food processing. *International Journal of Food Microbiology*, 285: 110–128.

Bourdichon, F., Arias, E., Babuchowski, A., Bückle, A., Dal Bello, F., Dubois, A., Fontana, A., Fritz, D., Kemperman, R., Laulund, S., McAuliffe, O., Hanna Miks, M., Papademas, P., Patrone, V., Sharma, D.K., Sliwinski, E., Stanton, C., Von Ah, U., Yao, S. and Morelli, L. 2021. The forgotten role of food cultures. *FEMS Microbiology Letters*, 368(14), 85.

Ben Braïek, O. and Smaoui, S. 2021. Chemistry, safety and challenges of the use of organic acids and their derivative salts in meat preservation. *Journal of Food Quality*, 2021, 1–20.

Carpenter, C.E., Smith, J.V. and Broadbent, J.R. 2011. Efficacy of washing meat surfaces with 2% levulinic, acetic, or lactic acid for pathogen decontamination and residual growth inhibition. *Meat Science*, 88(2): 256–260.

90 *Novel Approaches in Biopreservation for Food and Clinical Purposes*

Casquete, R., Benito, M.J., Martín, A., Ruiz-Moyano, S., Pérez-Nevado, F. and Córdoba, M.G. 2012. Comparison of the effects of a commercial and an autochthonous *Pediococcus acidilactici* and *Staphylococcus vitulus* starter culture on the sensory and safety properties of a traditional Iberian dry-fermented sausage 'salchichón'. *International Journal of Food Science and Technology*, 47(5): 1011–1019.

Castellano, P., Belfiore, C., Fadda, S. and Vignolo, G. 2008. A review of bacteriocinogenic lactic acid bacteria used as bioprotective cultures in fresh meat produced in Argentina. *Meat Science*, 79(3): 483–499.

Cenci-Goga, B., Iulietto, M., Sechi, P., Borgogni, E., Karama, M. and Grispoldi, L. 2020. New trends in meat packaging. *Microbiology Research*, 11(2): 56–67.

Cenci-Goga, B.T., Rossitto, P.V., Sechi, P., Parmegiani, S., Cambiotti, V. and Cullor, J.S. 2012. Effect of selected dairy starter cultures on microbiological, chemical and sensory characteristics of swine and venison (Dama dama) nitrite-free dry-cured sausages. *Meat Science*, 90(3): 599–606.

Cenci-Goga, B.T., Sechi, P., Iulietto, M.F., Amirjalali, S., Barbera, S., Karama, M., Aly, S.S. and Grispoldi, L. 2020. Characterization and growth under different storage temperatures of ropy slime-producing leuconostoc mesenteroides isolated from cooked meat products. *Journal of Food Protection*, 83(6): 1043–1049.

Choo, K.S.O., Bollen, M., Ravensdale, J.T., Dykes, G.A. and Coorey, R. 2021. Effect of chitosan and gum Arabic with natamycin on the aroma profile and bacterial community of Australian grown black Périgord truffles (Tuber melansoporum) during storage. *Food Microbiology*, 97: 743.

Comi, G. 2017. Spoilage of meat and fish. *The Microbiological Quality of Food: Foodborne Spoilers*, 179–210.

Correia Santos, S., Fraqueza, M.J., Elias, M., Salvador Barreto, A. and Semedo-Lemsaddek, T. 2017. Traditional dry smoked fermented meat sausages: characterization of *Autochthonous enterococci*. *LWT-Food Science and Technology*, 79: 410–415.

Da Costa, R.J., Voloski, F.L.S., Mondadori, R.G., Duval, E.H. and Fiorentini, Â.M. 2019. Preservation of meat products with bacteriocins produced by lactic acid bacteria isolated from meat. *Journal of Food Quality*, 2019, 510.

Cotter, P.D., Hill, C. and Ross, P.R. 2005. Bacteriocins: developing innate immunity for food. *Nature Reviews Microbiology*, 3(10): 777–788.

Cunha, G.F., Soares, J.C., de Sousa, T.L., Egea, M.B., de Alencar, S.M., Belisário, C.M. and Plácido, G.R. 2021. Cassava-starch-based films supplemented with propolis extract: physical, chemical and microstructure characterization. *Biointerface Research in Applied Chemistry*, 11(4): 12149–12158.

Daba, G.M. and Elkhateeb, W.A. 2020. Bacteriocins of lactic acid bacteria as biotechnological tools in food and pharmaceuticals: current applications and future prospects. *Biocatalysis and Agricultural Biotechnology*, 28: 750.

Dalvandi, F., Almasi, H., Ghanbarzadeh, B., Hosseini, H. and Khosroshahi, K. 2020. Effect of vacuum packaging and edible coating containing black pepper seeds and turmeric extracts on shelf life extension of chicken breast fillets. *Journal of Food and Bioprocess Engineering*, 3(1): 69–78.

Dan, S.D., Mihaiu, M., Reget, O., Oltean, D. and Tăbăran, A. 2017. Pathogens contamination level reduction on beef using organic acids decontamination methods. Bulletin of University of Agricultural Sciences and Veterinary Medicine Cluj-Napoca. *Veterinary Medicine*, 74(2): 212.

Das, A., Chauhan, G., Agrawal, R.K., Das, A.K., Tomar, S., Uddin, S., Satyaprakash, K., Pateiro, M. and Lorenzo, J.M. 2021. Characterization of crude extract prepared from Indian curd and its potential as a biopreservative. *Food Science and Technology International*, 27(4): 313–325.

Das, A.K., Rajkumar, V., Nanda, P.K., Chauhan, P., Pradhan, S.R. and Biswas, S. 2016. Antioxidant efficacy of litchi (Litchi chinensis Sonn.) pericarp extract in sheep meat nuggets. *Antioxidants*, 5(2): 16.

Das, A.K., Nanda, P.K., Das, A. and Biswas, S. 2019. Hazards and safety issues of meat and meat products. *Food Safety and Human Health*, 145–168.

Das, A.K., Nanda, P.K., Chowdhury, N.R., Dandapat, P., Gagaoua, M., Chauhan, P., Pateiro, M. and Lorenzo, J.M. 2021. Application of pomegranate by-products in muscle foods: oxidative indices, colour stability, shelf life and health benefits. *Molecules (Basel, Switzerland)*, 26(2): 467.

Das, J.K., Chatterjee, N., Pal, S., Nanda, P.K., Das, A., Das, L., Dhar, P. and Das, A.K. 2023. Effect of bamboo essential oil on the oxidative stability, microbial attributes and sensory quality of chicken meatballs. *Foods*, 12(1): 218.

Deka, D., Annapure, U.S., Shirkole, S.S. and Thorat, B.N. 2022. Bacteriophages: an organic approach to food decontamination. *Journal of Food Processing and Preservation*, 46(10): 101.

EFSA; ECDC. 2021. The European Union One Health 2020 Zoonoses Report. *EFSA Journal*, 19(12): 6971.

Endersen, L. and Coffey, A. 2020. The use of bacteriophages for food safety. *Current Opinion in Food Science*, 36: 1–8.

Erginkaya, Z., Ünal, E. and Kalkan, S. 2011a. Importance of microbial antagonisms about food attribution. *Science against Microbial Pathogens: Communicating Current Research and Technological Advances*, 1342–1348.

Erginkaya, Z., Ünal, E. and Kalkan, S. 2011b. Importance of microbial antagonisms about food attribution. *Science against Microbial Pathogens: Communicating Current Research and Technological Advances*, 1342–1348.

Erkmen, O. and Bozoglu, T.F. 2016. Spoilage of meat and meat products. pp. 279–295. *In*: Erkmen, O. and Bozoglu, T.F. (eds.). *Food Microbiology: Principles into Practice*. Wiley.

Favaro, L. and Todorov, S.D. 2017. Bacteriocinogenic LAB strains for fermented meat preservation: perspectives, challenges and limitations. *Probiotics and Antimicrobial Proteins*, 9(4): 444–458.

FDA. 2022a. Generally Recognized as Safe (GRAS) [online]. Available from: https://www.fda.gov/food/food-ingredients-packaging/generally-recognized-safe-gras#:~:text=%22GRAS%22 is an acronym for,phrase Generally Recognized As Safe.

FDA. 2022b. Microorganisms & Microbial-Derived Ingredients Used in Food (Partial List) [online]. Available from: https://www.fda.gov/food/generally-recognized-safe-gras/microorganisms-microbial-derived-ingredients-used-food-partial-list.

Fernández, L., Gutiérrez, D., Rodríguez, A. and García, P. 2018. Application of bacteriophages in the agro-food sector: a long way toward approval. *Frontiers in Cellular and Infection Microbiology*, 8(AUG): 296.

Frétin, M., Chassard, C., Delbès, C., Lavigne, R., Rifa, E., Theil, S., Fernandez, B., Laforce, P. and Callon, C. 2020. Robustness and efficacy of an inhibitory consortium against *E. coli* O26:H11 in raw milk cheeses. *Food Control*, 115: 282.

Fujiki, M., Abe, K., Hayakawa, T., Yamamoto, T., Torii, M., Iohara, K., Koizumi, D., Togawa, R., Aizawa, M. and Honda, M. 2019. Antimicrobial activity of protamine-loaded calcium phosphates against oral bacteria. *Materials*, 12(7): 2816.

Gad, S.E. and Sullivan, D.W. 2014. Generally Recognized as Safe (GRAS). *Encyclopedia of Toxicology: Third Edition*, 706–709.

García-Díez, J. and Saraiva, C. 2021. Use of starter cultures in foods from animal origin to improve their safety. *International Journal of Environmental Research and Public Health*, 18(5): 1–25.

Garvey, M. 2022. Bacteriophages and food production: biocontrol and bio-preservation options for food safety. *Antibiotics*, 11(10), :1324.

Geng, P., Zhang, L. and Shi, G.Y. 2017. Omics analysis of acetic acid tolerance in *Saccharomyces cerevisiae*. *World Journal of Microbiology and Biotechnology*, 33(5): 599.

Di Gioia, D., Mazzola, G., Nikodinoska, I., Aloisio, I., Langerholc, T., Rossi, M., Raimondi, S., Melero, B. and Rovira, J. 2016. Lactic acid bacteria as protective cultures in fermented pork meat to prevent *Clostridium* spp. growth. *International Journal of Food Microbiology*, 235: 53–59.

Goode, D., Allen, V.M. and Barrow, P.A. 2003. Reduction of experimental Salmonella and Campylobacter contamination of chicken skin by application of lytic bacteriophages. *Applied and Environmental Microbiology*, 69(8): 5032–5036.

Gourama, H. 2020. Foodborne pathogens. pp. 25–49. *In*: Demirci, A., Feng, H. and Krishnamurthy, K. (eds.). *Food Engineering Series*.

Guo, X., Wang, Y., Lu, S., Wang, J., Fu, H., Gu, B., Lyu, B. and Wang, Q. 2021. Changes in proteolysis, protein oxidation, flavor, color and texture of dry-cured mutton ham during storage. *LWT-Food Science and Technology*, 149: 860.

Hassoun, A., Gudjónsdóttir, M., Prieto, M.A., Garcia-Oliveira, P., Simal-Gandara, J., Marini, F., Di Donato, F., D'Archivio, A.A. and Biancolillo, A. 2020. Application of novel techniques for

92 Novel Approaches in Biopreservation for Food and Clinical Purposes

monitoring quality changes in meat and fish products during traditional processing processes: Reconciling novelty and tradition. *Processes*, 8(8): 988.

Hauser, C., Thielmann, J. and Muranyi, P. 2016. Organic acids: usage and potential in antimicrobial packaging. *Antimicrobial Food Packaging*, 563–580.

Herman, L., Chemaly, M., Cocconcelli, P.S., Fernandez, P., Klein, G., Peixe, L., Prieto, M., Querol, A., Suarez, J.E., Sundh, I., Vlak, J. and Correia, S. 2019. The qualified presumption of safety assessment and its role in EFSA risk evaluations: 15 years past. *FEMS Microbiology Letters*, 366(1): 260.

Hernández-Aquino, S., Miranda-Romero, L.A., Fujikawa, H., de Jesús Maldonado-Simán, E. and Alarcón-Zuñiga, B. 2019. Antibacterial activity of lactic acid bacteria to improve shelf life of raw meat. *Biocontrol Science*, 24(4): 185–192.

Hoa, V.B., Song, D.H., Seol, K.H., Kang, S.M., Kim, H.W., Kim, J.H. and Cho, S.H. 2022. Coating with chitosan containing lauric acid (C12:0) significantly extends the shelf-life of aerobically— Packaged beef steaks during refrigerated storage. *Meat Science*, 184: 696.

Hochreutener, M., Zweifel, C., Corti, S. and Stephan, R. 2017. Effect of a commercial steam-vacuuming treatment implemented after slaughtering for the decontamination of cattle carcasses. *Italian Journal of Food Safety*, 6(3): 120–124.

Hou, W., Yue, Q., Liu, W., Wu, J., Yi, Y. and Wang, H. 2021. Characterization of spoilage bacterial communities in chilled duck meat treated by kojic acid. *Food Science and Human Wellness*, 10(1): 72–77.

Hussein, A.R. 2022. Foods Bio-preservation : a review. *International Journal for Research in Applied Sciences and Biotechnology*, 9(3): 212–217.

Iulietto, M.F., Sechi, P., Borgogni, E. and Cenci-Goga, B.T. 2015. Meat spoilage: a critical review of a neglected alteration due to ropy slime producing bacteria. *Italian Journal of Animal Science*, 14(3): 316–326.

Jaspal, M.H., Ijaz, M., Haq, H.A. ul, Yar, M.K., Asghar, B., Manzoor, A., Badar, I.H., Ullah, S., Islamefussain, S. and Hussain, J. 2021. Effect of oregano essential oil or lactic acid treatments combined with air and modified atmosphere packaging on the quality and storage properties of chicken breast meat. *LWT-Food Science and Technology*, 146: 459.

Jay, J.M. 1992. Taxonomy, role and significance of microorganisms in foods. *Modern Food Microbiology*, 13–37.

Jay, J.M., Loessner, M.J. and Golden, D.A. 2005. *Modern Food Microbiology*. 7th ed. Modern Food Microbiology. Boston, MA: Springer US.

Kalschne, D.L., Geitenes, S., Veit, M.R., Sarmento, C.M.P. and Colla, E. 2014. Growth inhibition of lactic acid bacteria in ham by nisin: a model approach. *Meat Science*, 98(4): 744–752.

Katiyar, V. and Jain, A.K. 2018. Biopreservation: novel technique augmenting shelf life. *Microbial Research: An Overview*, 516.

Keymanesh, K., Soltani, S. and Sardari, S. 2009. Application of antimicrobial peptides in agriculture and food industry. *World Journal of Microbiology and Biotechnology*, 25(6): 933–944.

Koutsoumanis, K., Allende, A., Alvarez-Ordóñez, A., Bolton, D., Bover-Cid, S., Chemaly, M., Davies, R., De Cesare, A., Hilbert, F., Lindqvist, R., Nauta, M., Peixe, L., Ru, G., Simmons, M., Skandamis, P., Suffredini, E., Cocconcelli, P.S., Fernández Escámez, P.S., Maradona, M.P., Querol, A., Sijtsma, L., Suarez, J.E., Sundh, I., Vlak, J., Barizzone, F., Hempen, M., Correia, S. and Herman, L. 2022. Update of the list of QPS-recommended microbiological agents intentionally added to food or feed as notified to EFSA 16: suitability of taxonomic units notified to EFSA until March 2022. *EFSA Journal*, 20(7): 408.

Kumariya, R., Garsa, A.K., Rajput, Y.S., Sood, S.K., Akhtar, N. and Patel, S. 2019. Bacteriocins: classification, synthesis, mechanism of action and resistance development in food spoilage causing bacteria. *Microbial Pathogenesis*, 128: 171–177.

Kure, C.F., Axelsson, L., Carlehög, M., Måge, I., Jensen, M.R. and Holck, A. 2020. The effects of a pilot-scale steam decontamination system on the hygiene and sensory quality of chicken carcasses. *Food Control*, 109: 948.

de Lacerda, J.R.M., da Silva, T.F., Vollú, R.E., Marques, J.M. and Seldin, L. 2016. Generally recognized as safe (GRAS) *Lactococcus lactis* strains associated with Lippia sidoides Cham. are able to solubilize/mineralize phosphate. *SpringerPlus*, 5(1): 828.

Lahiri, D., Nag, M., Sarkar, T., Ray, R.R., Shariati, M.A., Rebezov, M., Bangar, S.P., Lorenzo, J.M. and Domínguez, R. 2022. Lactic Acid Bacteria (LAB): autochthonous and probiotic microbes for meat preservation and fortification. *Foods*, 11(18): 2792.

Laranjo, M., Potes, M.E. and Elias, M. 2019. Role of starter cultures on the safety of fermented meat products. *Frontiers in Microbiology*, 10(APR): 853.

Laulund, S., Wind, A., Derkx, P. and Zuliani, V. 2017. Regulatory and safety requirements for food cultures. *Microorganisms*, 5(2): 28.

Lebret, B. and Čandek-Potokar, M. 2022. Review: pork quality attributes from farm to fork. Part II. Processed pork products. *Animal*, 16: 100383.

Lee, H. and Yoon, Y. 2021. Etiological agents implicated in foodborne illness world wide. *Food Science of Animal Resources*, 41(1): 1–7.

Legros, J., Jan, S., Bonnassie, S., Gautier, M., Croguennec, T., Pezennec, S., Cochet, M.F., Nau, F., Andrews, S.C. and Baron, F. 2021. The role of ovotransferrin in egg-white antimicrobial activity: a review. *Foods*, 10(4): 823.

Lewis, R. and Hill, C. 2020. Overcoming barriers to phage application in food and feed. *Current Opinion in Biotechnology*, 61: 38–44.

Leyva Salas, M., Mounier, J., Valence, F., Coton, M., Thierry, A. and Coton, E. 2017. Antifungal microbial agents for food biopreservation—a review. *Microorganisms*, 5(3): 37.

Li, H.W., Xiang, Y.Z., Zhang, M., Jiang, Y.H., Zhang, Y., Liu, Y.Y., Lin, L.B. and Zhang, Q.L. 2021. A novel bacteriocin from *Lactobacillus salivarius* against *Staphylococcus aureus*: isolation, purification, identification, antibacterial and antibiofilm activity. *LWT-Food Science and Technology*, 140: 826.

Li, J. and Zhuang, S. 2020. Antibacterial activity of chitosan and its derivatives and their interaction mechanism with bacteria: current state and perspectives. *European Polymer Journal*, 138: 984.

Liao, R., Xia, Q., Zhou, C., Geng, F., Wang, Y., Sun, Y., He, J., Pan, D. and Cao, J. 2022. LC-MS/MS-based metabolomics and sensory evaluation characterize metabolites and texture of normal and spoiled dry-cured hams. *Food Chemistry*, 371: 1156.

Madane, P., Das, A.K., Pateiro, M., Nanda, P.K., Bandyopadhyay, S., Jagtap, P., Barba, F.J., Shewalkar, A., Maity, B. and Lorenzo, J.M. 2019. Drumstick (*Moringa oleifera*) flower as an antioxidant dietary fibre in chicken meat nuggets. *Foods*, 8(8): 307.

Madane, P., Das, A.K., Nanda, P.K., Bandyopadhyay, S., Jagtap, P., Shewalkar, A. and Maity, B. 2020. Dragon fruit (*Hylocereus undatus*) peel as antioxidant dietary fibre on quality and lipid oxidation of chicken nuggets. *Journal of Food Science and Technology*, 57(4): 1449–1461.

Mahajan, K., Chatli, M.K. and Kumar, S. 2020. Bacteriophages and competitive cultures: a new vista for food preservation. *Journal of Entomology and Zoology Studies*, 8(3): 1228–1233.

Marrez, D., Shaker, A., Ali, M. and Fathy, H. 2022. Food preservation: comprehensive overview of techniques, applications and hazards. *Egyptian Journal of Chemistry*, 65(8): 345–353.

De Martinez, Y.B., Ferrer, K. and Salas, E.M. 2002. Combined effects of lactic acid and nisin solution in reducing levels of microbiological contamination in red meat carcasses. *Journal of Food Protection*, 65(11): 1780–1783.

Martins, F.C.O.L., Sentanin, M.A. and De Souza, D. 2019. Analytical methods in food additives determination: compounds with functional applications. *Food Chemistry*, 272: 732–750.

Mehdizadeh, T. and Mojaddar Langroodi, A. 2019. Chitosan coatings incorporated with propolis extract and *Zataria multiflora* Boiss oil for active packaging of chicken breast meat. *International Journal of Biological Macromolecules*, 141: 401–409.

Menconi, A., Kuttappan, V.A., Hernandez-Velasco, X., Urbano, T., Matté, F., Layton, S., Kallapura, G., Latorre, J., Morales, B.E., Prado, O., Vicente, J.L., Barton, J., Filho, R.L.A., Lovato, M., Hargis, B.M. and Tellez, G. 2014. Evaluation of a commercially available organic acid product on body weight loss, carcass yield and meat quality during preslaughter feed withdrawal in broiler chickens: a poultry welfare and economic perspective. *Poultry Science*, 93(2): 448–455.

Miranda, C., Contente, D., Igrejas, G., Câmara, S.P.A., Dapkevicius, M. de L.E. and Poeta, P. 2021. Role of Exposure to Lactic Acid Bacteria from Foods of Animal Origin in Human Health. *Foods*, 10(9): 2092.

Mogoşanu, G.D., Grumezescu, A.M., Bejenaru, C. and Bejenaru, L.E. 2017. Natural products used for food preservation. *Food Preservation*, 365–411.

Mokoena, M.P., Omatola, C.A. and Olaniran, A.O. 2021. Applications of lactic acid bacteria and their bacteriocins against food spoilage microorganisms and foodborne pathogens. *Molecules*, 26(22): 7055.

Molayi, R., Ehsani, A. and Yousefi, M. 2018. The antibacterial effect of whey protein–alginate coating incorporated with the lactoperoxidase system on chicken thigh meat. *Food Science and Nutrition*, 6(4): 878–883.

Montville, T.J., Winkowski, K. and Ludescher, R.D. 1995. Models and mechanisms for bacteriocin action and application. *International Dairy Journal*, 5(8): 797–814.

O'Connor, P.M., Ross, R.P., Hill, C. and Cotter, P.D. 2015. Antimicrobial antagonists against food pathogens: a bacteriocin perspective. *Current Opinion in Food Science*, 2: 51–57.

Oderinwale, O.A., Oluwatosin, B.O., Onagbesan, M.O., Akinsoyinu, A.O. and Amosu, S.D. 2020. Performance of kids produced by three breeds of goat fed diets supplemented with graded levels of turmeric powder. *Tropical Animal Health and Production*, 52(3): 1239–1248.

O'Flynn, G., Ross, R.P., Fitzgerald, G.F. and Coffey, A. 2004. Evaluation of a cocktail of three bacteriophages for biocontrol of *Escherichia coli* O157:H7. *Applied and Environmental Microbiology*, 70(6): 3417–3424.

Palevich, N., Palevich, F.P., Maclean, P.H., Altermann, E., Gardner, A., Burgess, S., Mills, J. and Brightwell, G. 2021. Comparative genomics of *Clostridium* species associated with vacuum-packed meat spoilage. *Food Microbiology*, 95: 687.

Palevich, N., Palevich, F.P., Gardner, A., Brightwell, G. and Mills, J. 2022. Genome collection of *Shewanella* spp. isolated from spoiled lamb. *Frontiers in Microbiology*, 13: 152.

Papadochristopoulos, A., Kerry, J.P., Fegan, N., Burgess, C.M. and Duffy, G. 2021. Natural anti-microbials for enhanced microbial safety and shelf-life of processed packaged meat. *Foods*, 10(7): 1598.

Pellissery, A.J., Vinayamohan, P.G., Amalaradjou, M.A.R. and Venkitanarayanan, K. 2019. Spoilage bacteria and meat quality. *Meat Quality Analysis: Advanced Evaluation Methods, Techniques and Technologies*, 307–334.

Pepper, A.N., Sriaroon, P. and Glaum, M.C. 2020. Additives and preservatives: role in food allergy. *Journal of Food Allergy*, 2(1): 119–123.

Połaska, M. and Sokołowska, B. 2019. Bacteriophages—a new hope or a huge problem in the food industry. *AIMS Microbiology*, 5(4): 324–346.

Pozio, E. 2020. How globalization and climate change could affect foodborne parasites. *Experimental Parasitology*, 208: 807.

Pradhan, S.R., Patra, G., Nanda, P.K., Dandapat, P., Bandyopadhyay, S. and Das, A.K. 2018. Comparative microbial load assessment of meat, contact surfaces and water samples in retail chevon meat shops and abattoirs of Kolkata, W.B, India. *International Journal of Current Microbiology and Applied Sciences*, 7(05): 158–164.

Price-Hayward, M. and Hartnell, R. 2016. Summary report of joint scientific workshop on foodborne viruses. *EFSA Supporting Publications*, 13(10): 1103.

Priyadarshini, A., Rajauria, G., O'Donnell, C.P. and Tiwari, B.K. 2019. Emerging food processing technologies and factors impacting their industrial adoption. *Critical Reviews in Food Science and Nutrition*, 59(19): 3082–3101.

Rajanikar, R.V., Nataraj, B.H., Naithani, H., Ali, S.A., Panjagari, N.R. and Behare, P. V. 2021. Phenyllactic acid: a green compound for food biopreservation. *Food Control*, 128: 184.

Ras, G., Zuliani, V., Derkx, P., Seibert, T.M., Leroy, S. and Talon, R. 2017. Evidence for nitric oxide synthase activity in *Staphylococcus xylosus* mediating nitrosoheme formation. *Frontiers in Microbiology*, 8(APR): 598.

Ras, G., Leroy, S. and Talon, R. 2018. Nitric oxide synthase: what is its potential role in the physiology of staphylococci in meat products? *International Journal of Food Microbiology*, 282: 28–34.

Rasolofo, E.A., St-Gelais, D., LaPointe, G., Roy, D., Boulares, M., Mélika, M., Hassouna, M., Lara-aguilar, S., Alcaine, S.D., Munsch-alatossava, P., Gursoy, O., Alatossava, T., Endersen, L., O'Mahony, J., Hill, C., Ross, R.P., McAuliffe, O., Coffey, A., O'Sullivan, L., Bolton, D., McAuliffe, O. and Coffey, A. 2011. Phage therapy in the food industry. *Annual Review of Food Science and Technology*, 5(1): 151–172.

Rathod, N.B., Nirmal, N.P., Pagarkar, A., Özogul, F. and Rocha, J.M. 2022. Antimicrobial impacts of microbial metabolites on the preservation of fish and fishery products: a review with current knowledge. *Microorganisms*, 10(4): 773.

Remenant, B., Jaffrès, E., Dousset, X., Pilet, M.F. and Zagorec, M. 2015. Bacterial spoilers of food: behavior, fitness and functional properties. *Food Microbiology*, 45(PA): 45–53.

Ren, B., Wu, W., Soladoye, O.P., Bak, K.H., Fu, Y. and Zhang, Y. 2021. Application of biopreservatives in meat preservation: a review. *International Journal of Food Science & Technology*, 56(12): 6124–6141.

Rendueles, C., Duarte, A.C., Escobedo, S., Fernández, L., Rodríguez, A., García, P. and Martínez, B. 2022. Combined use of bacteriocins and bacteriophages as food biopreservatives. A review. *International Journal of Food Microbiology*, 368: 109611.

Ricci, A., Allende, A., Bolton, D., Chemaly, M., Davies, R., Girones, R., Herman, L., Koutsoumanis, K., Lindqvist, R., Norrung, B., Robertson, L., Ru, G., Sanaa, M., Simmons, M., Skandamis, P., Snary, E., Speybroek, N., Ter Kuile, B., Threlfall, J., Wahlström, H., Cocconcelli, P., Klein, G. (deceased): Prieto Maradona, M., Querol, A., Peixe, L., Suarez, J., Sundh, I., Vlak, J., Aguilera-Gómez, M., Barizzone, F., Brozzi, R., Correia, S., Heng, L., Istace, F., Lythgo, C. and Fernández Escámez, P. 2017. Scientific opinion on the update of the list of QPS-recommended biological agents intentionally added to food or feed as notified to EFSA. *EFSA Journal*, 15(3): 1–178.

Ríos Colombo, N.S., Chalón, M.C., Navarro, S.A. and Bellomio, A. 2018. Pediocin-like bacteriocins: new perspectives on mechanism of action and immunity. *Current Genetics*, 64(2): 345–351.

Romero-Calle, D., Benevides, R.G., Góes-Neto, A. and Billington, C. 2019. Bacteriophages as alternatives to antibiotics in clinical care. *Antibiotics*, 8(3): 138.

Saeed, F., Afzaal, M., Tufail, T. and Ahmad, A. 2019. Use of natural antimicrobial agents: a safe preservation approach. *In: Active Antimicrobial Food Packaging*, 869.

Saraoui, T., Leroi, F., Björkroth, J. and Pilet, M.F. 2016. *Lactococcus piscium*: a psychrotrophic lactic acid bacterium with bioprotective or spoilage activity in food—a review. *Journal of Applied Microbiology*, 121(4): 907–918.

Schnürer, J. and Magnusson, J. 2005. Antifungal lactic acid bacteria as biopreservatives. *Trends in Food Science and Technology*, 16(1–3): 70–78.

Semedo-Lemsaddek, T., Carvalho, L., Tempera, C., Fernandes, M.H., Fernandes, M.J., Elias, M., Barreto, A.S. and Fraqueza, M.J. 2016. Characterization and technological features of Autochthonous coagulase-negative Staphylococci as potential starters for portuguese dry fermented sausages. *Journal of Food Science*, 81(5): M1197–M1202.

Shao, L., Tian, Y., Chen, S., Xu, X. and Wang, H. 2022. Characterization of the spoilage heterogeneity of *Aeromonas* isolated from chilled chicken meat: *in vitro* and *in situ*. *LWT-Food Science and Technology*, 162: 470.

Shao, P., Liu, L., Yu, J., Lin, Y., Gao, H., Chen, H. and Sun, P. 2021. An overview of intelligent freshness indicator packaging for food quality and safety monitoring. *Trends in Food Science and Technology*, 118: 285–296.

Sharma, D.B. 2011. *Outlines of Meat Science and Technology*. New Delhi: Jypee Publishers.

Shedleur-Bourguignon, F., Duchemin, T., P. Thériault, W., Longpré, J., Thibodeau, A., Hocine, M.N. and Fravalo, P. 2023. Distinct microbiotas are associated with different production lines in the cutting room of a swine slaughterhouse. *Microorganisms*, 11(1): 133.

Shishodia, S.K., Tiwari, S. and Shankar, J. 2019. Resistance mechanism and proteins in *Aspergillus* species against antifungal agents. *Mycology*, 10(3): 151–165.

Singh, V.P. 2018. Recent approaches in food bio-preservation—a review. *Open Veterinary Journal*, 8.

Silva, C.C.G., Silva, S.P.M. and Ribeiro, S.C. 2018. Application of bacteriocins and protective cultures in dairy food preservation. *Frontiers in Microbiology*, 9(APR): 594.

Simoncini, N., Pinna, A., Toscani, T. and Virgili, R. 2015. Effect of added autochthonous yeasts on the volatile compounds of dry-cured hams. *International Journal of Food Microbiology*, 212: 25–33.

Singh, M.P., Nikhanj, P., Singh, P., Singh, R.K. and Sharma, A. 2022. Biopreservation: an alluring method to safeguard food from spoilage. *In: New and Future Developments in Microbial Biotechnology and Bioengineering*. Elsevier, 449–476.

Soares, V.M., dos Santos, E.A.R., Tadielo, L.E., Cerqueira-Cézar, C.K., da Cruz Encide Sampaio, A.N., Eisen, A.K.A., de Oliveira, K.G., Padilha, M.B., de Moraes Guerra, M.E., Gasparetto, R., Brum,

96 *Novel Approaches in Biopreservation for Food and Clinical Purposes*

M.C.S., Traesel, C.K., Henzel, A., Spilki, F.R. and Pereira, J.G. 2022. Detection of adenovirus, rotavirus and hepatitis E virus in meat cuts marketed in Uruguaiana, Rio Grande do Sul, Brazil. *One Health*, 14: 100377.

Sofos, J.N. 2013. Food safety management: Chapter 6. *Meat and Meat Products*, 119–162.

Stavropoulou, D.A., De Vuyst, L. and Leroy, F. 2018. Nonconventional starter cultures of coagulase-negative staphylococci to produce animal-derived fermented foods, a SWOT analysis. *Journal of Applied Microbiology*, 125(6): 1570–1586.

Suo, B., Chen, X. and Wang, Y. 2021. Recent research advances of lactic acid bacteria in sourdough: Origin, diversity and function. *Current Opinion in Food Science*, 37: 66–75.

Tang, S.S., Biswas, S.K., Tan, W.S., Saha, A.K. and Leo, B.F. 2019. Efficacy and potential of phage therapy against multidrug resistant *Shigella* spp. *PeerJ*, 2019(4): 225.

Taormina, P.J. 2021. Microbial Growth and Spoilage. *Food Safety and Quality-Based Shelf Life of Perishable Foods*, 41–69.

Taylor, T.M., Ravishankar, S., Bhargava, K. and Juneja, V.K. 2019. Chemical preservatives and natural food antimicrobials. *Food Microbiology: Fundamentals and Frontiers*, 705–731.

Toldrá, F., Hui, Y.H., Astiasarán, I., Nip, W.K., Sebranek, J.G., Silveira, E.T.F., Stahnke, L.H. and Talon, R. 2008. *Handbook of Fermented Meat and Poultry*. Handbook of Fermented Meat and Poultry.

Trabelsi, I., Ben Slima, S., Ktari, N., Triki, M., Abdehedi, R., Abaza, W., Moussa, H., Abdeslam, A. and Ben Salah, R. 2019. Incorporation of probiotic strain in raw minced beef meat: study of textural modification, lipid and protein oxidation and color parameters during refrigerated storage. *Meat Science*, 154" 29–36.

Ucak, I., Abuibaid, A.K.M., Aldawoud, T.M.S., Galanakis, C.M. and Montesano, D. 2021. Antioxidant and antimicrobial effects of gelatin films incorporated with citrus seed extract on the shelf life of sea bass (*Dicentrarchus labrax*) fillets. *Journal of Food Processing and Preservation*, 45(4): 304.

Veesler, D. and Cambillau, C. 2011. A common evolutionary origin for tailed-bacteriophage functional modules and bacterial machineries. *Microbiology and Molecular Biology Reviews*, 75(3): 423–433.

Villalobos-Delgado, L.H., Nevárez-Moorillon, G.V., Caro, I., Quinto, E.J. and Mateo, J. 2019. Natural antimicrobial agents to improve foods shelf life. *Food Quality and Shelf Life*, 125–157.

Wambui, J. and Stephan, R. 2019. Relevant aspects of *Clostridium estertheticum* as a specific spoilage organism of vacuum-packed meat. *Microorganisms*, 7(5): 142.

Wang, F., Saito, S., Michailides, T.J. and Xiao, C.L. 2021. Postharvest use of natamycin to control Alternaria rot on blueberry fruit caused by *Alternaria alternata* and *A. arborescens*. *Postharvest Biology and Technology*, 172: 383.

Wang, Q., Chen, Q., Xu, J., Sun, F., Liu, H. and Kong, B. 2022. Effects of modified atmosphere packaging with various CO2 concentrations on the bacterial community and shelf-life of smoked chicken legs. *Foods*, 11(4): 559.

Wang, X., Wei, Z. and Xue, C. 2021. The past and future of ovotransferrin: physicochemical properties, assembly and applications. *Trends in Food Science and Technology*, 116: 47–62.

WHO (World Health Organization): Geneva, S. 2021. Estimating the burden of foodborne diseases [online]. Available from: https://www.who.int/activities/estimating-the-burden-of-foodborne-diseases.

WHO (World Health Organization): Geneva, S. 2022. Food Safety [online]. Available from: https://www.who.int/news-room/fact-sheets/detail/food-safety#:~:text=An estimated 600 million – almost,healthy life years (DALYs).

Worldbank. 2018. Food-borne Illnesses Cost US$ 110 Billion Per Year in Low- and Middle-Income Countries [online]. Available from: https://www.worldbank.org/en/news/press-release/2018/10/23/food-borne-illnesses-cost-us-110-billion-per-year-in-low-and-middle-income-countries.

Xu, M.M., Kaur, M., Pillidge, C.J. and Torley, P.J. 2021. Evaluation of the potential of protective cultures to extend the microbial shelf-life of chilled lamb meat. *Meat science*, 181: 108613.

Xu, M.M., Kaur, M., Pillidge, C.J. and Torley, P.J. 2022. Effect of protective cultures on spoilage bacteria and the quality of vacuum-packaged lamb meat. *Food Bioscience*, 50: 148.

Xu, M.M., Kaur, M., Pillidge, C.J. and Torley, P.J. 2023. Culture-dependent and culture-independent evaluation of the effect of protective cultures on spoilage-related bacteria in vacuum-packaged beef mince. *Food and Bioprocess Technology*, 16(2): 382–394.

Young, N.W.G. and O'Sullivan, G.R. 2011. The influence of ingredients on product stability and shelf life. *In*: *Food and Beverage Stability and Shelf Life*. Elsevier, 132–183.

Yousefi, M., Nematollahi, A., Shadnoush, M., Mortazavian, A.M. and Khorshidian, N. 2022. Antimicrobial activity of films and coatings containing Lactoperoxidase system: a review. *Frontiers in Nutrition*, 9: 65.

Yu, H.H., Chin, Y.W. and Paik, H.D. 2021. Application of natural preservatives for meat and meat products against food-borne pathogens and spoilage bacteria: a review. *Foods*, 10(10).

Yusuf, M. 2018. Natural Antimicrobial Agents for Food Biopreservation. pp. 409–438. *In*: Grumezescu, A.M. and Holban, A.M. (eds.). *Food Packaging and Preservation*.

Zagorec, M. and Champomier-Vergès, M.-C. 2023. Meat microbiology and spoilage. pp. 195–218. *In*: Toldrá, F. (ed.). *Lawrie's Meat Science*. Elsevier.

Zang, J., Xu, Y., Xia, W. and Regenstein, J.M. 2020. Quality, functionality and microbiology of fermented fish: a review. *Critical Reviews in Food Science and Nutrition*, 60(7): 1228–1242.

Zhang, L. 2018. The use of biological agents in processing. *Packaging for Nonthermal Processing of Food*, 83–94.

Zhang, Y., Zhu, L., Dong, P., Liang, R., Mao, Y., Qiu, S. and Luo, X. 2018. Bio-protective potential of lactic acid bacteria: effect of *Lactobacillus sakei* and *Lactobacillus curvatus* on changes of the microbial community in vacuum-packaged chilled beef. *Asian-Australasian Journal of Animal Sciences*, 31(4): 585–594.

Zhou, C., Xia, Q., Du, L., He, J., Sun, Y., Dang, Y., Geng, F., Pan, D., Cao, J. and Zhou, G. 2022. Recent developments in off-odor formation mechanism and the potential regulation by starter cultures in dry-cured ham. *Critical Reviews in Food Science and Nutrition*, 62: 7418.

Zhou, C., Zhan, G., Pan, D., Zhou, G., Wang, Y., He, J. and Cao, J. 2022. Charactering the spoilage mechanism of "three sticks" of Jinhua ham. *Food Science and Human Wellness*, 11(5): 1322–1330.

Zhou, C.Y., Pan, D.D., Cao, J.X. and Zhou, G.H. 2021. A comprehensive review on molecular mechanism of defective dry-cured ham with excessive pastiness, adhesiveness and bitterness by proteomics insights. *Comprehensive Reviews in Food Science and Food Safety*, 20(4): 3838–3857.

Zhu, W. and Oteiza, P.I. 2022. Proanthocyanidins at the gastrointestinal tract: mechanisms involved in their capacity to mitigate obesity-associated metabolic disorders. *Critical Reviews in Food Science and Nutrition*, 1–21.

Zhu, Y., Wang, W., Li, M., Zhang, J., Ji, L., Zhao, Z., Zhang, R., Cai, D. and Chen, L. 2022a. Microbial diversity of meat products under spoilage and its controlling approaches. *Frontiers in Nutrition*, 9: 8201.

Zhu, Y., Wang, W., Zhang, Y., Li, M., Zhang, J., Ji, L., Zhao, Z., Zhang, R. and Chen, L. 2022b. Characterization of quality properties in spoiled mianning Ham. *Foods*, 11(12): 1713.

Zimina, M., Babich, O., Prosekov, A., Sukhikh, S., Ivanova, S., Shevchenko, M. and Noskova, S. 2020. Overview of global trends in classification, methods of preparation and application of bacteriocins. *Antibiotics*, 9(9): 1–21.

Zolfaghari Emameh, R., Purmonen, S., Sukura, A. and Parkkila, S. 2018. Surveillance and diagnosis of zoonotic foodborne parasites. *Food Science & Nutrition*, 6(1): 3–17.

CHAPTER 5

Biopreservation of Fish

Vida Šimat,[1] Federica Barbieri,[2] Chiara Montanari,[2] Fausto Gardini,[2] Danijela Skroza,[3] Ivana Generalić Mekinić,[3] Fatih Ozogul,[4] Yesim Ozogul[4] and Giulia Tabanelli[2,]*

Introduction

Biopreservatives are substances obtained from natural sources or formed in foods able to prolong shelf life, by retarding or preventing spoilage and controlling safety risks both of chemical and biological origin (Mani-López et al., 2018). According to this definition, biopreservation has been used by mankind for millennia through the production of fermented foods. In the last decades, research focused on a different vision of the addition of a selected culture, i.e., as antagonistic to contrast autochthonous spoilage or pathogenic microorganisms. In addition, the above definition comprehends the bioprotective strategies also the use of "natural" antimicrobial substances, which can replace "chemical" preservatives.

In this framework, fish and fishery products are food products in which bioprotective approaches have been extensively exploited. One of the most valued Roman products, garum, can be considered the progenitor of several fish-fermented sauces or pastes, obtained through the fermentation of several salty fishes (anchovies, sardines, mackerel, menhaden and even shrimps), nowadays particularly diffused in Southeast Asia (Han et al., 2022). Their traditional production often relies on spontaneous fermentation processes, mainly driven by the addition of NaCl, but in industrial production, the use of starter cultures, in particular, lactic acid bacteria (LAB), staphylococci and sometimes yeasts, is increasing. On the other hand, fermentation is also applied for the fermentation of whole fish and fish pieces (slices), such as herring, skipjack, mackerel and carp (Zang et al., 2020; Xu et al., 2021).

---

[1] University Department of Marine Studies, University of Split, Split, Croatia.
[2] Department of Agricultural and Food Sciences, University of Bologna, Bologna, Italy.
[3] Department of Food Technology and Biotechnology, Faculty of Chemistry and Technology, University of Split, Split, Croatia.
[4] Department of Seafood Processing Technology, Faculty of Fisheries, Cukurova University, Adana 01330, Turkey.
* Corresponding author: giulia.tabanelli2@unibo.it

Fishery products were among the first examples in which bioprotective cultures were used as antagonistic microorganisms. At the end of the last century, Wessels and Huss (1996) proposed the use of a nisin producer strain of *Lactococcus lactis* ssp. *lactis* for the control of *Listeria monocytogenes* in lightly preserved fish. In the following years, several LAB species with bioprotective characteristics were proposed with the same objective. These include *Latilactobacillus sakei* (previously known as *Lactobacillus sakei*), *Carnobacterium maltoaromaticum*, *Carnobacterium divergens* and *Lactococcus piscium*, due to their ability to produce antimicrobial compounds (such as organic acids, hydrogen peroxide, diacetyl and bacteriocins) and their Generally Recognised As Safe (GRAS) status (Leroi, 2010; Aymerich et al., 2019).

The last aspect of biopreservation concerns the use of natural antimicrobial substances derived from plants, algae and animals. In the last decades, a great number of scientific papers described the antimicrobial potential of vegetable derivatives, such as essential oils and phenolic extracts which, in many cases, possess also relevant antioxidant properties. The use of these molecules, which can be often compatible with traditional food preparation such as marinade, could satisfy the need of industry to respond to consumer requests for safer and healthier foods without chemical preservatives and with reduced treatments (Hussain et al., 2021). Another antimicrobial compound with interesting properties is chitosan, mainly obtained by fishery by-products from crustaceans and molluscs, as well as insects and fungi, which can also act as a binding agent, texturizer, antioxidant and edible coating (Inanli et al., 2020).

This chapter reviews these aspects of fishery products and gives an overview of the international legislative framework which regulates the traditional and innovative approaches characterising biopreservation.

Fish spoilage and associated hazards

Seafood safety is about optimum capture, handling, processing and storage, and prevention of foodborne illness due to chemical, physical and biological contaminants. Although novel technologies have been developed to provide seafood safety and quality, fish poisoning outbreaks are still inevitable. Consumers expect fish products to have superior sensorial quality, guaranteed safety and increased nutritional and functional properties as well as extended shelf life. Fish processing involves all activities from harvest time to delivery of the final product, thus, all processing stages must be carried out properly and hygienically since fish products are highly perishable. When fish are not properly stored, processed, packaged, and distributed, they spoil fast, being unsafe due to oxidation and microbial growth (Mukherjee et al., 2020; Farag et al., 2022).

Spoilage of fish occurs as a result of degradation of the tissue caused by both endogenous enzymes and microorganisms present on the surface of the skin, gills and viscera. The process of deterioration implies changes in smell, taste, appearance, and texture of fish, becoming unpleasant or unsafe for consumption. The proximate composition of fish, i.e., the percentages of moisture, ash, lipid, protein and carbohydrate contents, depends on species, even within the same species and different parts of the fish (Sofoulaki et al., 2018; El Oudiani et al., 2019; Ahmed et al., 2020). The main constituents of fish are 20–22% protein, 75–85% water, 0.2%–25% lipid, 1.2%–1.5% mineral and 0%–0.5% carbohydrate (Özogul et al., 2011; Tavares et al., 2021). It is indicated that fish with inferior moisture level has a higher protein and lipid levels. Fish can be categorized into four groups according to fat levels: high-fat (> 8%), medium-fat (4–8%), low-fat (2–4%), and lean (< 2%) (Kolakowska et al., 2010). The reason for variation of proximate profile of fish is attributed to geographical location, season, feed, size and age of fish, metabolic rate, sex, genetic features and migration, etc. (Durmuş et al., 2018). The rate of spoilage can be influenced by some factors which include the composition of fish, type of species, harvesting season, catching method, handling, processing method as well as storage temperature. During spoilage, degradation of different constituents and production of new compounds cause off-odour, off-flavour and changes in colour and texture of the fish (Baird-Parker, 2000).

Oxidative spoilage

Fish are well known as perishable food products due to their superior unsaturated fatty acids content. The main fish spoilage process is lipid degradation, which mainly takes place by means of oxidation or hydrolysis. There are various types of oxidations, including photooxidation, enzymatic oxidation, thermal oxidation and autooxidation. These processes can be enhanced by the presence of prooxidants (such as haemoglobin, myoglobin, cytochrome). Unsaturated fatty acids of triglycerides react with atmospheric oxygen, generating unstable primary compounds such as free fatty acids (FFAs), peroxides and also secondary compounds which contain ketones, aldehydes, alcohols, volatile organic acids, hydrocarbons, epoxy compounds, trienes and carbonyls (Ghaly et al., 2010). Moreover, lipid hydrolysis (known as lipolysis) results from the degradation of triglycerides into free fatty acids (FFAs) by activity of lipase enzyme. Increase in the level of FFAs causes protein denaturation, change of texture, and enhanced drip loss by protein-lipid cross-linkages. In general, protein denaturation occurs in fish by proteolytic enzyme activity (cathepsins) in the muscle and the intestinal tract (trypsins), causing solubility of texture and muscle damage. The subsequent compounds such as amino acids, amines, peptides, hydrogen sulphide (H_2S), ammonia, etc., are accumulated and they play a role as a substrate for microbial growth (Gram and Dalgaard, 2002).

Microbial contamination and spoilage

Fish can be contaminated with chemicals such as heavy metals (lead (Pb), cadmium (Cd), chromium (Cr), arsenic (As) and mercury (Hg)) or organochlorine contaminants (such as PCBc-polychlorinated biphenyls, DDT-Dichloro-Diphenyl-Trichloroethane, dioxins, radioactive substances, etc.) (Maldonado-Simán et al., 2018). These pollutants can affect the gut microbiota composition, resulting in shifts of the microbial population (Rognes et al., 2016). The fish gut is a long digestive organ (such as foregut, midgut and hindgut), having different digestion roles as well as different microbiota niches. Fish gastrointestinal microbiota and spoilage of fish product have extensively been studied (Adeoye et al., 2016; Ramírez and Romero, 2017; Zotta et al., 2019; Syropoulou et al., 2020). The composition, density and function of the microbiota vary along the fish gut (Egerton et al., 2018; Li et al., 2019).

Microbial contamination mainly refers to the pathogenic bacteria, viruses and parasites that naturally occur in aquatic environments. The presence of pathogenic bacteria species, which can cause foodborne diseases in humans, can be associated with direct contact with contaminated environment where fish lives, ingestion of bacteria present in sediments/feed as well as post-capture contamination. When the fish are alive, the fish muscles are regarded as sterile, but after death or harvest, the immune system collapses and the enzymes produced by microorganisms allow them to diffuse into the flesh, stomach and skin surface, where they react with natural substances. Fish meat provides favourable conditions for the growth of microorganisms due to low acidity (pH > 6), high water activity, as well as high protein and free amino acid content and non-protein nitrogenous compounds amount (Gram and Huss, 1996). This spoilage results in alterations in appearance (production of slime), development of off-odours and off-flavours odour and damage of texture, reducing fish shelf life and quality. With the increase in storage time, only a few taxa of bacteria become dominant and cause the final spoilage of fish (Zhuang et al., 2021). These taxa of bacteria generally have strong spoilage capacity and are called specific spoilage organisms (SSOs) in the deterioration of fish products. Zhuang et al. (2020) reported *Aeromonas*, followed by *Pseudomonas*, as the dominant spoilage bacteria when largemouth bass fillets were rejected by sensory analyses. Liu et al. (2018) observed that *Aeromonas*, *Pseudomonas* and *Shewanella* were lower in fresh samples and become dominant in spoiled fillets of bighead carp at the end of storage time. Thus, spoilage of fish products can be evaluated considering the composition of the microbiota, especially the percentage of SSOs in total microflora.

Spoilage caused by microorganisms can form biogenic amines (BAs), volatile amines, sulphides, organic acids, alcohols, ketones and aldehydes with safety risks and unacceptable off-flavours and off-odours (Visciano et al., 2020). During storage, BAs such as histamine, tyramine, cadaverine and putrescine are generated by the decarboxylation of specific free amino acids by microorganisms and their accumulation in seafood can be regarded as a good indicator of spoilage. Among BAs, histamine is the most toxic and it can occur as a poisoning, also known as scombroid fish poisoning, that causes headache, hypotension, vomiting, diarrhoea, flushing of the face and neck, asthma attacks, etc. (Barbieri et al., 2019). *Morganella, Hafnia, Enterobacter, Photobacterium, Proteus, Pseudomonas* spp., former *Lactobacillus* and *Enterococcus* genera can produce BAs (Comas-Basté et al., 2019; Barbieri et al., 2019). Breaking down of nucleotides, especially ATP-related compounds is also correlated with flavour deterioration of fishery products. The potential of microbial spoilage is generally mainly assessed by employing quality indices such as total volatile basic nitrogen (TVB-N) content, BAs, volatile compounds and muscle degradation (K-value).

Biopreservation strategies for enhanced food safety and extended shelf life

The increasing consumer demand for high quality, minimally processed and safe seafood, without synthetic preservatives, led the research towards the study of alternative strategies of biopreservation. One of them is the exploitation of specific microbial groups, mainly Lactic Acid Bacteria (LAB), fungi and *Bacillus* spp., that can develop in raw materials triggering fermentation processes. In addition, some LAB strains, able to produce metabolites with antimicrobial activity, can be used as bio-protective cultures to increase safety and shelf life of fish products. These approaches can be also applied in combination with other techniques, in the framework of hurdle technologies. Among them, the use of plant derivatives such as essential oil or extracts, showed promising applications in the fishery industry, thanks to the strong antimicrobial and antioxidant activity of their constituents against pathogenic and spoilage microorganisms.

Fish fermentation and use of selected starters

Fermentation is one of the most ancient strategies for food preservation. Many products deriving from animal and vegetable raw materials are spread worldwide and represent a valuable cultural heritage (Marco et al., 2017). Among them, fish fermentation is particularly relevant in Southeast Asia, even if some products are still popular in Northern Europe (garum, Hákarl, Rakfisk) and Africa (e.g., Lanhouin and Momoni). The main characteristics of the different products available around the world, which differ in raw material, salt and/or spices addition, temperature and duration of fermentation process, have been recently reviewed (Zang et al., 2020). Since fish fermentation originally took place at domestic level to preserve a raw material extremely perishable because of its intrinsic feature (high nutrient and water content, high pH value), a wide array of recipes and procedures are reported, and the quality of the final products in some cases still relies on artisanal practices and spontaneous processes. Three different types of fermented fish products can be identified: whole fish (or its parts), pastes or sauces. Although significant differences in production methods are present, they generally include salting and drying and, sometimes, also marinating and smoking (Xu et al., 2021). The same products can be also classified according to other parameters (Zang et al., 2020): in relation to the process, fermented fishes can be divided into spontaneously fermented products or products that have undergone guided fermentation through the addition of proper starter cultures. In the first case, strict control of the final product quality is not possible, and safety can also be affected if foodborne pathogens or toxicogenic microorganisms wildly multiply. Despite these issues, the use of starter cultures in such products is not still widely applied, since the variety of fish species used as raw material, as well differences in salt concentration and fermentation conditions, make it difficult to find good candidates able to adapt to these environmental niches and lead the process (Han et al., 2022). Considering the type of substrate, fermented fish derivatives can be divided into two categories, i.e., those obtained only with fish and salts and those in which also carbohydrates (rice, millet, flour) are added.

This allows to speed up the fermentation process, being used as primary energy sources by microflora, and play also a role in the texture and sensorial features of the final product (Zang et al., 2020). Another classification proposed and that can significantly affect the microbiota during fermentation is according to the amount of salt added, whose range can be from 0 up to 30%, therefore also influencing the duration of fermentation process (Xu et al., 2021). It is known that increasing salt concentration results in lowering of a_w value, thus inhibiting spoilage microflora and selecting halophilic or salt-tolerant microorganisms, with consequent modifications of the enzymatic reactions and the final product characteristics.

The variables described above can strongly affect the microbial dynamics during fermentation and the type and amounts of microbial species detected in the different products. In general, LAB and yeast are dominant. Among LAB, the most common are members of lactobacilli or belong to the genera *Leuconostoc*, *Lactococcus*, *Weissella*, *Pediococcus* and, in the products with significant amounts of salt, also *Vagooccoccus* and *Tetragenococcus* can be found. The main contributor of this microbial group relies on pH reduction through the accumulation of lactic and acetic acid. Indeed, the drop of pH results in the gelation of muscle proteins, as well as protein denaturation, contributing to the texture of the fish products (Xu et al., 2021). In addition, some LAB strains are able to produce molecules exerting antagonistic activity against foodborne pathogens or spoilage microflora (i.e., bacteriocins), that will be discussed in details in the next chapter section. Concerning flavor development, LAB isolated from such products generally have low proteolytic and lipolytic activity, and their contribution to the aroma profile is often limited to the accumulation of organic acids or alcohol esterification, the latter being reported for *Lactiplantibacillus plantarum* in Suan yu, a Chinese traditional fermented fish obtained from *Cyprinus carpio* (Gao et al., 2018). Some studies investigated the use of mixed cultures of strains belonging to *Lactobacillus*, *Pediococcus*, *Lactococcus* as starter cultures in different fish products, particularly in those in which also a carbon source (mainly rice) is added, in order to give them a competitive advantage with respect to the microflora naturally occurring. Moreover, the selection of LAB to be used in these products has recently focused also on their ability to release amino acids, improving the aroma profile (Zang et al., 2020) and probiotic features (Speranza et al., 2017). Concerning possible side effects, it is noteworthy that some LAB can also be responsible for the accumulation of considerable amounts of BAs, and therefore the use of selected strains can be also a strategy to limit the accumulation of these toxic compounds. Besides LAB, fungi can dominate the microbiota of some products: for example, species of the genera *Aspergillus* and *Actinomucor* have been used to ferment surimi (Zhao et al., 2017) or silver carp fish paste (Kasankala et al., 2012), with an increase of aromatic notes, while many yeast species belonging to *Saccharomyces*, *Candida*, *Pichia*, *Hanseniaspora*, *Debaryomyces* are frequently found in fermented fish products. Because of their proteolytic and lipolytic activity, they give a significant contribution to the final characteristics of the product, influencing both texture and taste. For example, some strains of *Saccharomyces cerevisiae* have been reported to produce in Suan yu alcohols with peculiar aromatic notes, such as 3-methyl-1-butanol, 2-methyl-butanol, 2-methyl-propanol and phenethyl alcohol, starting for the corresponding amino acids valine, leucine, isoleucine and phenylalanine (Wang et al., 2017).

Staphylococci also play important role during fish fermentation. Indeed, even if they are not the dominant population, their lipolytic activity favours the release of free fatty acids from lipids, significantly contributing to flavour development (Xu et al., 2021). In addition, recent studies showed that strains belonging to the species *Staphylococcus carnosus* and *Staphylococcus xylosus* are able to reduce histamine accumulation during fermentation of fish sauce and in salted and fermented anchovies (Mah and Hwang 2009; Zaman et al., 2014).

Bacillus spp. is another microbial group that contributes to the final characteristics of fermented fish products, thanks to its high proteolytic activity. Indeed, strains belonging to the species *Bacillus megaterium* and *Bacillus subtilis* have been reported to produce proteases that, combined with the activity of endogenous proteases occurring in fish muscles, allow protein degradation and accumulation of small peptides and amino acids, that can be further metabolized into aromatic impact compounds (Xu et al., 2021). Other enzymes reported for *Bacillus* spp. are lipases, which contribute to the lipid fraction degradation, and amine-oxidases, which can have a positive effect on food quality by reducing biogenic amine accumulation (Zaman et al., 2011).

Other microorganisms that can be isolated from fermented fish products belong to the genera *Micrococcus*, whose lipase activity can improve sensory features of fish sauces, *Halomonas*, *Enterobacter*, *Photobacterium*, etc. (Han et al., 2022).

It is noteworthy that in many cases the taste and flavour of the final product is obtained through the succession of these microbial groups that dominate different stages of fermentation. For example, in fish sauces, yeast prevail in the first weeks of fermentation but, when the pH drops, their number decrease while LAB increase. Then, other species can develop, and this evolution in microbiota favours the formation of a more complex aroma profile, since they produce a wide array of metabolites including alcohols, aldehydes, esters, acids, sulfur compounds, etc.

Metabolites and by-products of bioprotective cultures for fish biopreservation

The spoilage and safety of fishery products are a crucial issue, and much research has been focused on the development of new strategies to extend the shelf life of these products. To reduce the proliferation and presence of spoilage or pathogenic microorganisms, the use of LAB strains as bioprotective cultures has been proposed, due to their competition with spontaneous microbiota and their ability to produce specific metabolites, characterised by antimicrobial activity (Rathod et al., 2021a). In particular, LAB strains can accumulate different bioactive compounds, such as organic acids, antimicrobial peptides (bacteriocins), antifungals and other low molecular weight compounds, that can exert activity against both Gram positive and Gram negative bacteria, increasing their importance for natural preservation (Tabanelli et al., 2019; Šimat et al., 2021).

Organic acids are the principal metabolites obtained from LAB fermentations, in which they can accumulate only lactic acid (homofermentative process) or lactic and acetic acid (heterofermentative process). Moreover, LAB can use other pathways to produce these compounds, i.e., citrate and pyruvate metabolisms that can lead to the accumulation also of succinic and/or formic acid (Gänzle, 2015).

Organic acids cause an environmental pH decrease that can affect microbial cell activities and membrane permeability. These compounds, due to their lipophilic characteristics, can solubilise in the cell membrane and can enter the cytoplasm. The resulting intracellular environment acidification inhibits the enzymatic reaction, reducing microorganism proliferation (García-Díez and Saraiva, 2021). Moreover, these cell membrane modifications can enhance the effect of other antimicrobial substances (Coban, 2020). Nevertheless, organic acids can also affect the organoleptic and sensory characteristics of fish products. For this reason, other metabolites with antimicrobial characteristics are studied to be used in industrial food preservation.

In particular, the attention is focused on bacteriocins, which are ribosomally synthesized antimicrobial peptides with bacteriostatic or bactericidal effect on target microorganisms and a minimal impact on the nutritional and sensory properties of food products (Chen et al., 2020). These metabolites are characterised by different targets and can vary in terms of molecular size, chemical structure and mode of action. Generally, they are more active in the inhibition of Gram-positive bacteria, such as *Listeria monocytogenes*, and, despite their differences, common mechanisms of action are hypothesised such as pore formation with consequent cell membrane permeabilization, that lead to cell death or the inactivation of the enzymes required for cell wall synthesis (Kumariya et al., 2019). Currently, nisin is the only bacteriocin approved by the European legislation and by the Food and Drug Administration (FDA) for its direct addition into foods as pure additives (EU Directive 1129/2011/ EC). This metabolite, produced by several LAB such as *Lactococcus lactis*, attacks the cell wall determining the lysis of the target microorganism. There are three different variants (A, Z and Q) differing only in one amino acid and characterised by different antimicrobial potential and chemical characteristics (temperature and pH stability, sensitivity to proteolysis, etc.). Nisin has a relatively broad-spectrum activity against several Gram positive microorganisms compared with other bacteriocins, but in particular it is known to inhibit *List. monocytogenes* and *Clostridium botulinum* (Gharsallaoui et al., 2016). However, the antibacterial activity of this compound can be affected by several factors, including environmental conditions. In fact, once the purified bacteriocin is added into the products, its activity can be limited or influenced by food storage conditions, pH, food constituents, proteolytic enzymes, solubility, or other food additives. Therefore, the use of microorganisms (mainly LAB) able to produce bacteriocins *in situ* can represent a feasible strategy to prevent the growth of undesired microorganisms, by exploiting the antimicrobial activity of these metabolites in the industrial processes, without legal restriction (Castellano et al., 2017). In fish farming, *Lc. lactis* is used to contrast fish diseases: a strain of this species, isolated from water, was able to prevent lactococcosis, a disease caused by *Lactococcus garviae*, producing nisin Z active against the pathogen when added into rainbow trout feed (Araujo et al., 2015). The addition of *Lc. lactis* to feed was also efficient in reducing a disease of farmed sea bass, known as vibriosis, determined by *Vibrio anguillarum* (Touraki et al., 2013). In general, LAB belonging to different genera (i.e., *Lactococcus*, *Leuconostoc*, *Carnobacterium* and *Pediococcus*) have been proposed to prolong shelf life of seafood products decreasing the presence of spoilage or foodborne pathogen microorganisms (Munoz-Atienza et al., 2013).

Moreover, protective cultures added in seafood products can be combined with other treatments (i.e., vacuum packaging, addition of essential oils or their components or other extracts, enzymes, etc.) in order to reduce the growth of undesired microorganisms (Tabanelli et al., 2019). Based on the hurdle technology, nisin is widely applied in combination with preserving technologies to enhance product safety and quality. In particular, this metabolite was used in combination with natural plant derivatives characterised by antioxidant and antimicrobial activities, such as cinnamon, thyme, and rosemary essential oils (EOs), and shallot and turmeric extract, that were able to reduce *List. monocytogenes* growth in cooked minced fish (silver carp) and increase the shelf life of seafood products (Abdollahzadeh et al., 2014). Nisin was also used for pompano biopreservation in combination with rosemary EO (Gao et al., 2014), and to guarantee the quality of snakehead fillets thanks to its application with cinnamon in an alginate coating (Lu et al., 2010).

Other reported bacteriocins are sakacin, enterocin, pediocin and lacticin, characterised by activity against Gram positive microorganisms, including *Listeria* (Baptista et al., 2020; Smid and Gorris, 2020).

Moreover, LAB strains are able to metabolize lipids and proteins during fermentation, producing specific aroma compounds that can act as a biopreservatives. On the other hand, they can influence the food organoleptic profiles. Among these, diacetyl (2,3-butanedione) can affect microbial enzymes (Kontominas et al., 2021). This compound derives from pyruvate through the citrate metabolism. Hydrogen peroxide, another LAB metabolite, is characterised by an antimicrobial activity against both bacteria and fungi: it can have a strong oxidizing effect on cell membranes and can affect the protein molecular structures. This compound can be produced in the presence of oxygen by LAB flavoprotein oxidases (Siedler et al., 2019). Finally, another studied antimicrobial compounds are reuterin and reutericyclin. The first one can induce oxidative stress in cells, modifying thiol groups in proteins and small molecules (Schaefer et al., 2010), while the second one presents an effect against Gram positive microorganisms, based on proton-ionophore activity, but not against Gram negative ones, due to the barrier properties of their membrane (Gänzle, 2004).

Finally, some LAB strains can reduce the accumulation of toxic compounds in foods through the production of degrading enzymes. One example can be the production of biogenic amine-degrading enzymes, which are able to detoxify these molecules and represent an efficient strategy to preserve food safety, avoiding high BAs amounts (Gardini et al., 2016).

Plant extracts and essential oils and their constituents as natural preservatives

Plant-derived compounds have attracted a lot of attention to enhance the microbial and chemical (oxidative) stability of foods and extend the shelf life of the products. Those with notable antioxidant and antimicrobial activity, like polar bioactive compounds (i.e., phenolics) and essential oils show special interest. Therefore, in the last few years the preservative effects of different plant extracts and/or their essential oils in fish, seafood and their products are extensively documented and reviewed by numerous researchers (Hassoun and Çoban, 2017; Šimat and Generalić Mekinić, 2020; Hao et al., 2021; Huang et al., 2021). In most cases their application is usually combined with other preservation methods, the so called 'hurdle technology', and the preservative effects are usually related to their main bioactive compounds.

Over 1300 herbs and spices are proved to possess biological potential and they have been used for hundred of years in culinary purposes for improving the organoleptic properties of foods (Nieto, 2020). Numerous studies reported their preservative properties as potential in delaying lipid oxidation or inhibition of microbial growth, as well as their positive effect on textural properties of food. In seafood industry, the applications of plant extract have been carried out on chilled, frozen and dried whole fish, filet or fish mince where they help to eliminate the risk for food spoilage (Viji et al., 2017; Rathod et al., 2021b).

There are various ways for the application of plant extracts and EOs in fish, seafood and their products. It can be direct (by adding them to the products during the manufacturing and processing, by immersion/dip treatment, pipette dropping method followed by massaging, via spraying or by adding in encapsulated form) or indirect (adding to ice/water during chilling, adding to enrich animal feed, incorporating/ integrating into food coatings, edible/active films, packaging materials, etc.) (Nieto, 2020; Rathod et al., 2021b; Hao et al., 2021). It has been proved that EOs effectively inhibit lipid oxidation and growth of undesirable microorganisms but, due to their strong flavour and odour, they could negatively affect product organoleptic characteristics and cause its sensory rejection due to unpleasant aroma. Due to these limitations, EOs are often encapsulated in order to mask their flavours but also to improve their stability or solubility and control their release. Furthermore, EOs are usually effective at higher concentrations in comparison to the synthetic additives so they can also make foods unsafe (limitations due to allergic reactions). Therefore, special attention should be devoted to the applied dose (Hao et al., 2021).

Preservative effects of plant extracts and EOs as natural antioxidants and antimicrobials for preserving the quality of seafood products and increasing their shelf-life by inhibiting/delaying the bacterial spoilage and oxidative processes was reviewed by numerous studies (Hassoun and Çoban, 2017; Viji et al., 2017, Mei et al., 2019; Moosavi-Nasab et al. 2020; Šimat and Generalić Mekinić, 2020; Hao et al., 2021; Huang et al., 2021; Rathod et al., 2021b; Vijayan et al., 2021). Some of the most investigated plants with application is seafood industry are oregano, rosemary, thyme, laurel, sage, cinnamon, clove, and basil, and the literature overview of their use as natural preservatives for extending the shelf life of seafood products and improving its quality is reported in Table 1.

Applications of bioprotective strategies on fish products

The food production sectors are challenged to produce safe and affordable foods with improved nutritional values and health benefits. The trend of healthy eating and the research on healthy and nutritious foods has increased over the years. With high levels of polyunsaturated fatty acids (PUFA) and high-quality-bioavailable proteins, fish is accepted as a healthy choice that contributes to consumers' overall wellbeing. However, the prevention of fish spoilage and assurance of its safety is a challenge. The distribution chain starts with the finishing and aquaculture activities, followed by the processing (if not marketed chilled/fresh) and marketing of the fish

Table 1. Overview of the studies (from 2015 up to date) on the application of plant extracts (PEs) and essential oils (EOs) used as natural preservatives in fish, seafood and products.

PE/EO	Seafood product	Storage conditions	Mode of EO application and processing	Salient findings	References
Oregano extract	Meagre (*Argyrosomus regius*) fillets cooked with sous-vide	4 ± 1°C/42 days	Immersion in infusion (5%, w/v) for 30 min at 4°C in ratio of 1:1 (w/v); Souse vide (85°C, 36 min); Vacuum packaging	The treatment with extract resulted in lower values of total volatile basic nitrogen and thiobarbituric acid in comparison to the control, and the sensory quality of the product was improved.	Bozova and Izci, 2021
	Rainbow trout (*Oncorhynchus mykiss*) fillets	4°C/6 days	Immersion in extract (470 mg/L)	The use of extracts in marinade improves oxidative stability of fillets and their colour.	Fellenberg et al., 2020
	Hairtail fish balls	4°C /15 days	Addition of extracts (0.1% sage, 0.1% oregano and 0.01% grape seed extract, w/v); Cooking (45°C; 20 min)	The extract addition significantly reduced the concentration of volatile compounds and inhibited fishy odour by inhibiting oxidation and bacteria growth.	Guan et al., 2019
	Frigate tuna (*Auxis thazard*) fillets	3 ± 1°C/18 days	Dipping in extract (0.5%) for 5 min; Vacuum packaging	The combination of oregano extract and vacuum packaging improved texture of the products and extended its shelf-life (from 12 to 18 days).	Lahreche et al., 2019
	Atlantic salmon (*Salmo salar*)	Iced salmon at 18 ± 1°C during day/ at 2 ± 1°C during night/4 days	Ice with extracts (1:100, v/v)	The mesophilic aerobic bacteria count, total volatile basic nitrogen, thiobarbituric acid reactive substances and pH values remained within the limits of acceptability, and extract addition improved the sensory attributes and fish flavour.	Dogruyol et al., 2021

Oregano EO	Grass carp (*Ctenopharyngodon idellus*)	4 ± 1°C/10 days	Immersion in EO emulsions (0.1% v/v) for 30 min at room temperature	EO treatment was effective in inhibiting microbial growth, delaying lipid oxidation, and based on sensory analysis, the product shelf-life was extended for 2 days.	Huang et al., 2018
	European eel (*Anguilla anguilla*)	4°C/18 days	EO (0.3% v/w) alone or in combination with chitosan; Vacuum packaging	Treatment of EO in combination with chitosan significantly reduced bacteria, yeasts and moulds during the storage and led to reduction in concentrations of trimethylamine nitrogen, total volatile basic nitrogen and thiobarbituric acid reactive substances compared with the control samples.	Lambrianidi et al., 2019
	Salmon (*Salmo salar*) fillets	4°C/16 days	Immersion in marinade with EO (1% w/w) for 2 min	The microbial shelf life of fillets was prolonged. Total aerobic psychrotrophs in samples were reduced 3 days.	Van Haute et al., 2016
	Scampi (*Penaeus monodon*)	4°C/16 days	Immersion in marinade with EO (1% w/w) for 2 min	The microbial shelf life of the product was prolonged.	Van Haute et al., 2016
	Sea bream (*Sparus aurata*)	4°C/28 days	Vapour EO treatment and dipping in EO (0.1%); Modified atmosphere packaging	Treatment extended shelf life of the products and resulted in better texture.	Navarro-Segura et al., 2020
Rosemary extract	Meagre (*Argyrosomus regius*) fillets cooked with sous-vide	4 ± 1°C/42 days	Immersion in plant infusion (5%, w/v) for 30 min at 4°C in ratio of 1:1 (w/v); Souse vide (85°C, 36 min); Vacuum packaging	The treatment with extract resulted in lower values of total volatile basic nitrogen and thiobarbituric acid in comparison to the control sous vide group samples, and the sensory quality of the product was improved.	Bozova and Izci., 2021
	Atlantic mackerel (*S. scombrus*), minced fish balls	−18°C/10 months	Addition of extracts (0.05%);	Plant extract inhibited microbiological spoilage and the values of biochemical parameters, and prolonged the shelf-life of the product.	Balıkçı et al., 2022

Table 1 contd. ...

...Table 1 contd.

PE/EO	Seafood product	Storage conditions	Mode of EO application and processing	Salient findings	References
	Mackerel	4 ± 1°C/6 days.	Placing samples between PET films containing extract	Commercial PET films were coated with extract and this treatment preserved fish from oxidation what was tested by thiobarbituric acid test.	Farghal et al., 2017
	Rainbow trout (*Oncorhynchus mykiss*) fillets	4.3 ± 0.6 °C/14 days	Washing with cold tap water with extract (from 0.5 to 2.0%) for 10 min; Vacuum packaging	The rosemary extract (0.5, 1.0 and 2%) could effectively inhibit lipid oxidation and retard the microbial growth. It also improved colour and sensory characteristics of the fillets and extend their shelf life to 10 days.	Linhartová et al., 2019
	Mud shrimp (*Solenocera melantho*)	−20°C/24 weeks	Dipping in glazing solution containing extract (0.2%, w/w) at 0 °C for 10 ~ 15 s	Glazing with rosemary extract was more effective in controlling quality changes in shrimps with lower total volatile basic nitrogen, drip loss, peroxide value, free fatty acid and higher lipid content and sensory scores.	Shi et al., 2019
	Nile tilapia (*Oreochromis niloticus*) fillets	2 ± 1°C/18 days	Dipping in extract solution (1.5%) for 10 min	Rosemary extract had efficient antioxidant activity with clear reduction in the thiobarbituric acid reactive substances and weak antimicrobial activity.	Khalafalla et al., 2015
	Rainbow trout (*Oncorhynchus mykiss*) fillets	4°C/6 days	Immersion in extract (7.2 mL/L)	The use of extracts in marinade improves oxidative stability of fillets and their colour.	Fellenberg et al., 2020
	Atlantic salmon (*Salmo salar*)	Iced salamon at 18 ± 1°C during day/ at 2 ± 1°C during night/4 days	Ice with extracts (1:100, v/v)	The mesophilic aerobic bacteria count, total volatile basic nitrogen, thiobarbituric acid reactive substances and pH values remained within the limits of acceptability. Plant extract addition increased the preference of consumption.	Dogruyol et al., 2021

Rosemary EO	Atlantic salmon (*Salmo salar*)	2 ± 1°C/4 days	EO (0.3%) was applied on fillet surface	Significant reduction of bacterial population was recorded in fish fillets inoculated with *List. monocytogenes* and *Salmonela* Enteridis.	Tosun et al., 2017
	Atlantic mackerel (*Scomber scombrus*)	2°C/15 days	Immersion in EO (1%, w/v) for 30 min at 2°C	EO treatment was effective in inhibition of total volatile basic nitrogen and lipid oxidation products by extending the shelf-life of fillets by 2 days, and also in thiobarbituric acid-reactive substance analysis, by extending the shelf-life of products for 2 days, compared to the control.	Karoui and Hassoun, 2017
	Rainbow trout (*Oncorhynchus mykiss*) fillets cooked with sous-vide	4°C/36 days	Addition of EO (100 μL); Vacuum packaging; Cooking (50/55°C, 7 min)	The sous-vide packaging in association with EO showed good results in product quality preservation and inhibition of *List. monocytogenes*.	Ozturk et al., 2021
	Rainbow trout (*Oncorhynchus mykiss*) fillets	4 ± 1°C/9 days	Nanoemulsions with EO (3.63 and 5.50%) applied onto the surface	Treatment with nanoemulsions showed effective antimicrobial activity on some gram-negative bacteria mostly found in fish fillets.	Meral et al., 2019
	Rainbow trout (*Oncorhynchus mykiss*) fillets	2 ± 2°C/24 days	Nanoemulsions with EO (4%); immersion for 3 min	EO nanoemulsions with EO reduced the values of the biochemical parameters and inhibited bacterial growth, but resulted in bitter taste of the products.	Ozogul et al., 2016
Thyme extract	Atlantic mackerel (*S. scombrus*), minced fish balls	−18°C/10 months	Addition of extracts (0.05%);	The extract inhibited microbiological spoilage and the values of biochemical parameters, and prolonged the shelf-life of the product.	Balıkçı et al., 2022
	Nile tilapia (*Oreochromis niloticus*) fillets	2 ± 1°C/18 days	Dipping in extracts (1.5%) for 10 min	The extract had strong antimicrobial and antioxidant activity and did not affected sensory quality of the fillets. It extended the shelf life of refrigerated fillets for 9 days.	Khalafalla et al., 2015

Table 1 contd. ...

...Table 1 contd.

PE/EO	Seafood product	Storage conditions	Mode of EO application and processing	Salient findings	References
	Silver carp (*Hypophthalmichthys molitrix*) minced meat	4 ± 1°C/15 days	Liposomal encapsulated and unencapsulated (0.3 and 0.5%, w/w) extract	The samples containing 0.5% of encapsulated thyme reduced peroxide value and total volatile nitrogen, inhibited the growth of mesophilic and psychrotropic bacteria, and reduced *E. coli* O157:H7.	Javadian et al., 2016
	Sardine (*Sardinella aurita*) fermented fillets	3 ± 1°C/8 weeks	Immersion in brine solution with extract (0.5%) for 3 h at 25°C; Vacuum packaging	Application of cell-free extract of *Pediococcus acidolactici* in combination with thyme extract had considerable effect on reducing trimethylamine formation in fermented fillets.	Kuley et al., 2018
	Common carp (*Cyprinus carpio*) surimi	4 ± 1°C/15 days	Addition of extracts (2 and 4%, w/v)	Treated samples showed decrease of the amount of thiobarbituric acid. Thyme extract (4%) can be used for enhancing the shelf life of the product.	Farjami et al, 2015
	Mackerel	/	Soaking in brine solution with extract (70 g/L) for 30 min at 4°C at ratio 2:1 (w:w); Smoking	The obtained results showed a decrease in pH values and an increase in protein contents of the product. Thyme extract showed a high reduction rate for *Salmonella* spp., *E-coli* and *Pseudomonas* spp. in comparison to the control samples.	Alsaiqali et al., 2016
Thyme EO	European eel (*Anguilla Anguilla*) smoked fillets	4°C/49 days	EO (0.3%) alone or in combination with chitosan dipping	Thyme EO treated samples and chitosan-thyme EO treated samples had prolonged shelf-life compared to control (7 days and more than 14 days, respectively). Thyme EO was effective in inhibiting lipid oxidation in samples and growth of mesophilic bacteria in smoked eel fillets.	El-Obeid et al., 2018
	Grass carp (*Ctenopharyngodon idellus*)	4 ± 1°C/10 days	Immersion in EO emulsions (0.1% v/v) for 30 min at room temperature	EO treatment was effective in inhibiting microbial growth, delaying lipid oxidation, and based on sensory analysis, the product shelf-life was extended for 2 days.	Huang et al., 2018

	Rainbow trout (*Oncorhynchus mykiss*) fillets	4 ± 2°C	Whey protein isolate coating enriched with EO (3, 5 and 7%)	EO addition enhanced quality, inhibited microbiological spoilage and prolonged the shelf-life of the products during refrigerated storage. The effect was concentration dependent.	Tokur et al., 2015
	Rainbow trout (*Oncorhynchus mykiss*) fillets	4 ± 1°C/9 days	Nanoemulsions with EO (3.63 and 5.50%) applied onto the surface	Treatment with EO nanoemulsions showed effective antimicrobial activity on some gram-negative bacteria mostly found in fish fillets.	Meral et al., 2019
	Rainbow trout (*Oncorhynchus mykiss*) fillets	2 ± 2°C/24 days	Immersions in nanoemulsions with EO (4%) for 3 min	EO nanoemulsions enhanced organoleptic quality of fillets, reduced the values of the biochemical parameters and inhibited bacterial growth.	Ozogul et al., 2016
	Scampi (*Penaeus monodon*)	4°C/16 days	Immersion in marinade containing EO (1% w/w) for 2 min	The microbial shelf life of the product was prolonged.	Van Haute et al., 2016
Laurel extract	Sardine (*Sardinella aurita*) fermented fillets	3 ± 1°C/8 weeks	Immersion in brine solution with extract (0.5%) for 3 h at 25°C; Vacuum packaging	The study suggested that cell-free extracts of *Lactobacillus plantarum* and *Pediococcus acidolactici* with laurel extract could improve the shelf-life of fermented sardines.	Kuley et al., 2018
	Atlantic salmon (*Salmo salar*)	Iced salamon at 18 ± 1°C during day/ at 2 ± 1°C during night/4 days	Ice with extracts (1:100, v/v)	The mesophilic aerobic bacteria count, total volatile basic nitrogen, thiobarbituric acid reactive substances and pH values remained within the limits of acceptability. Plant extract improved the sensory attributes and fish flavour.	Dogruyol et al., 2021
Laurel EO	Rainbow trout (*Oncorhynchus mykiss*) fillets	4°C/14 days	Spraying (1 and 2%); Vacuum packaging	The combination of vacuum packing and laurel EO (2%) enhances sensory quality characteristics of fish, delays microbial spoilage in fillets and extends shelf-life for approximately 4 days.	Aksoy and Sezer, 2019
	Rainbow trout (*Oncorhynchus mykiss*) fillets	4 ± 1°C/9 days	Nanoemulsions with EO (3.63 and 5.50%) applied onto the surface	Treatment with nanoemulsions showed effective antimicrobial activity on some gram-negative bacteria mostly found in fish fillets.	Meral et al., 2019

Table 1 contd. ...

...Table 1 contd.

PE/EO	Seafood product	Storage conditions	Mode of EO application and processing	Salient findings	References
Sage extract	Fish fingers	4±1°C/30 days	Addition of extract mixture (clove:sage:kiwifruit peels = 1:1:1) in different proportions (0.1, 0.25 and 0.5%) to the minced fish meat	Extracts addition significantly reduced the lipid and protein oxidation, inhibited microbial growth while organoleptic and sensory attributes were not affected.	Abdel-Wahab et al., 2020
	Hairtail fish balls	4°C /15 days	Addition of extracts (0.1% sage, 0.1% oregano and 0.01% grape seed extract, w/v); Cooking (45°C; 20 min);	The extract addition significantly reduced the concentration of volatile compounds and inhibited fishy odour by inhibiting oxidation and bacteria growth.	Guan et al., 2019
Sage EO	Rainbow trout (*Oncorhynchus mykiss*) fillets	2 ± 2°C/24 days	Nanoemulsions with EO (4%); Immersion for 3 min	EO nanoemulsions enhanced organoleptic quality of fillets, reduced the values of the biochemical parameters and inhibited bacterial growth.	Ozogul et al., 2016
	Rainbow trout (*Oncorhynchus mykiss*) fillets	4 ± 1°C/9 days	Nanoemulsions with EO (3.63 and 5.50%) applied onto the surface	Treatment with nanoemulsions showed effective antimicrobial activity on some gram-negative bacteria mostly found in fish fillets.	Meral et al., 2019
Cinnamon extract	Catfish (*Clarias gariepinus*)	37 ± 2°C/28 days.	Soaking in cinnamon bark extracts (0.5%, 1% and 1.5%, w/v) for 30 min; Smoking (12 h)	The samples treated with 1% solution of extract had the highest moisture content and lowest protein content and those treated with 5 g and 15 g cinnamon bark extract solution, retained very good score for appearance, colour, flavour, texture and general acceptance after 28 days of storage.	Haruna et al., 2021
	Khashm elbanat (*Mormyrus casahive*)		Addition of extracts (5, 10 and 15%) before drying	The results revealed that flavouring with 10% cinnamon extract improved product flavour, but had negative effect on its colour.	Malik et al., 2020
	Mackerel (*Scomber austriasicus*)	35°C/15 days	Addition of extract during fermentation of fish sauce	The ethanolic extract had inhibitory effect on the aerobic bacteria counts, decreased the pH and the total volatile basic nitrogen content of sauce.	Zhou et al., 2016

Cinnamon EO	Fish fillet	5 ± 1°C/20 days	Coating with tea and cinnamon extract (0.25%) and chitosan, tea and cinnamon extract (0.2%:0.25%:0.25%)	Extracts addition to chitosan coating caused reduced levels of free fatty acids and thiobarbituric acid, and samples treated with tea and cinnamon extracts showed the lowest oxidation.	Haghighi et al., 2020
	Common carp (*Cyprinus carpio*) fillets	4 ± 1°C/14 days	Immersion in EO (0.1%) for 10 min; Vacuum packaging	EO treatment extended the shelf-life of common carp fillets for 2 days as EO inhibited the bacterial growth, increase of total volatile basic nitrogen and accumulation of biogenic amines.	Zhang et al., 2016
	Salmon (*Salmo salar*) fillets	4°C/16 days	Immersion in marinade containing EO (1% w/w) for 2 min	The microbial shelf life of fillets was extended. Yeasts and moulds and total aerobic psychrotrophs in samples were reduced for 6 and 3 days, respectively.	Van Haute et al., 2016
	Tiger shrimps (*Penaeus monodon*)	4°C/10 days	Dipping in EO solution (2.5 mL/L) for 30 min at 25°C	The mixture of cinnamon oil and lactic acid was effective in extending the microbial shelf-life of raw shrimps.	Noordin et al., 2018
Clove extract	Fish fingers	4 ± 1°C/30 days	Addition of extract mixture of (clove, :sage and: kiwifruit peels = (1:1:1) in different proportions (0.1, 0.25 and 0.5%) to minced fish meat	Extracts addition significantly reduced the lipid and protein oxidation, inhibited microbial growth while organoleptic and sensory attributes were not affected.	Abdel-Wahab et al., 2020
	Sardines (*Rastrineobola argentea*), sun-dried	Room temperature	Soaking of blanched *dagaa* in clove extract (1:1, w/w) for 40 min at room temperature	Results showed that pre-treatment of sardine with clove water extracts (in range of 5–20 g/L) significantly reduced peroxidation and resulted in higher retention of omega-3 fatty acids and lower concentrations products of secondary lipid oxidation.	Chaula et al., 2018
	Mackerel	4 ± 1°C/6 days.	Minced fish was placed between two treated/coated PET films with extract	Commercial PET films were coated with extract and this treatment preserved fish from oxidation measured as degree of oxidation using thiobarbituric acid test.	Farghal et al., 2017

Table 1 contd. ...

...*Table 1 contd.*

PE/EO	Seafood product	Storage conditions	Mode of EO application and processing	Salient findings	References
	Shrimp (*Fenneropenaeus indicus*)	4 ± 1° C/10 days	Dipping in coating solution with clove extract (1 mg/mL) at 1:2 ratio (w/v) for 30 min at 4°C	The coating treatment of unpeeled shrimps with clove extract/nanochitosan solutions resulted in decrease in the microbial pathogens (*Escherichia coli, Salmonella typhimurium,* and *Staphylococcus aureus*). The most effective combination contained 1.5% of nanochitosan and 1.0% of clove extract.	Tayel et al., 2020
Clove EO	Bluefin tuna (*Thunnus thynnus*) fillets	2°C/14 days	Nanocomposite films based on soy protein isolate-montmorillonite-EO	Films were effective in inhibiting bacterial growth (especially *Pseudomonas* spp.) and lipid oxidation of fillets during storage.	Echeverria et al., 2018
	Bream (*Megalobrama ambycephala*) fillets	4°C/15 days	Pectin coatings with EO	The treatment reduced the extent of lipid oxidation, and extended the shelf life of bream fillets. It improved the weight loss, water holding capacity, textural and colour attributes of the products and was effective in bacterial growth inhibition.	Nisar et al., 2019
	Flounder (*Paralichthys orbignyanus*) fillets	5°C/15 days	Agar films with EO (0.5 g of EO/g)	Fillets covered with films showed lower total volatile bases and pH values and delayed the growth of H$_2$S-producing microorganisms.	da Rocha et al., 2018
	Rainbow trout (*Oncorhynchus mykiss*) gutted	4°C/22 days	Chitosan film with EO (20 g/kg); High-pressure processing; Vacuum packaging	Chitosan film with EO was effective in the inhibition of microbials (total aerobic mesophilic, lactic bacteria and total coliform), its reduced weight loss and water activity in samples.	Albertos et al., 2015
	Red drum (*Sciaenops ocellatus*) fillets	4 ± 1°C/20 days	Treatment with EO (4 μL/L, evaporation in container) for 2 h at 10°C	The essential oil inhibited microbiological growth throughout the storage period and was effective in texture preservation and retarding fish sensory deterioration.	Cai et al., 2015
	Sutchi catfish (*Pangasius hypophtalamus*)	0 – 2°C/15 days	Dipping (0.25%, w/v); Conventional packaging/Vacuum packaging	Coating with clove bud EO was found to be the most effective in protecting the textural quality and retarding lipid oxidation of the fillets.	Binsi et al., 2016

Additive	Fish product	Storage conditions	Treatment	Results	Reference
Basil extract	Atlantic mackerel (*S. scombrus*), minced fish balls	−18°C/10 months	Addition of extracts (0.05%);	Plant extract inhibited microbiological spoilage and the values of biochemical parameters, and provided an 8-month shelf life for frozen fish ball samples. The extract provided negative organoleptic characteristics (an intense odor and a bitter taste).	Balıkçı et al., 2022
	Atlantic salmon (*Salmo salar*)	Iced salamon at 18 ± 1°C during day/ at 2 ± 1°C during night/4 days	Ice with extracts (1:100, v/v)	The mesophilic aerobic bacteria count, total volatile basic nitrogen, thiobarbituric acid reactive substances and pH values remained within the limits of acceptability. Plant extract resulted in a remarkable change in fish flavour.	Dogruyol et al., 2021
	Common carp (*Cyprinus carpio*) surimi	4 ± 1°C/16 days	Addition of extract (2, 4, 6 and 8%)	Total viable count, psychrophilic count, peroxide value, thiobarbitoric acid, total volatile bases nitrogen and pH were lower in treatments with 8% of basil extract.	Zamani et al., 2021
Basil EO	Atlantic mackerel (*Scomber scombrus*)	2°C/15 days	Immersion in EO (1% w/v) for 30 min at 2°C	EO treatment was effective in inhibition of total volatile basic nitrogen and lipid oxidation products by extending the shelf-life of fillets by 5 days, and also in thiobarbituric acid-reactive substance analysis, by extending the shelf-life of products for 3 days, compared to the control.	Karoui and Hassoun, 2017
	Sea bass (*Lates calcarifer*) slices	4°C/12 days	Packaging film (fish protein isolate/ fish skin gelatin films incorporated with 3% ZnO nanoparticles and EO)	The shelf-life of sea bass slices packed in films with EO had 2-fold longer shelf-life compared to the control, showed lower bacterial growth and reduced increase of other quality parameters (pH, total volatile base, peroxide value, TBARS).	Arfat et al., 2015
	Rainbow trout (*Oncorhynchus mykiss*) fillets cooked with sous-vide	4°C/36 days	Addition of EO (100 μL); Vacuum packaging; Cooking (50/55°C, 7 min)	The sous-vide packaging in association with EO showed good results in product quality preservation and inhibition of *List. monocytogenes*.	Ozturk et al., 2021
	Rainbow trout (*Oncorhynchus mykiss*) fillets	2 ± 2°C/24 days	Immersion in nanoemulsions with EO (4%) for 3 min	EO nanoemulsions enhanced organoleptic quality of fillets, reduced the values of the biochemical parameters and inhibited bacterial growth.	Ozogul et al., 2016

or its parts to consumers. According to the Food and Agriculture Organization of the United Nations (FAO, 2022), total fisheries and aquaculture production reached 177.8 million tonnes in 2020, and an estimation is that 157.4 million tonnes is used for human consummation. It is estimated that half of that catch is marketed in fresh/child form, while the rest is marketed prepared into ready-to-eat products and/or preserved in some way. The use of preservation and processing techniques is essential to ensure food safety and the nutritive quality of fish products. Besides temperature reduction (chilling, subchilling and freezing), fish products are preserved by heat treatments (canning, pasteurization and hot smoking), reduction of available water (drying, salting, cold smoking and curing), or by adopting the packaging material and method (vacuum packaging, modified atmosphere packaging) to a specific product (FAO, 2022). Despite the cold distribution chain and new technologies of chilling/superchilling at sub 0 temperatures and packaging, the fish remains a very perishable commodity (Šimat and Generalić Mekinić, 2020). On the other hand, mildly preserved and ready-to-eat fish products require the addition of preservatives to ensure their safety and shelf life. The suspicion of the adverse effect of synthetic additives on human health is present among consumers, so natural alternatives in the form of bioprotective cultures and their metabolites, plant extracts, or EOs present a possible solution to satisfy all mentioned challenges.

A natural preservative is a substance or a mixture of substances that can be used to improve a product's safety, quality, and sensory attributes such as taste, odour/aroma, appearance, texture, colour and others (Gokoglu, 2019). A natural additive can be in the form of dried plant material, extract, derivative, purified compound or EOs. Given the complexity of the natural compounds/extracts used for this purpose they usually improve more than one function in the food. Besides, the diverse biological activity (antimicrobial, antioxidant, anti-inflammatory, anticancer, etc.) of the natural additives may also contribute to the functionality of the product or even bring health benefits to consumers. However, in fish and fishery products they are mainly investigated and applied as natural antimicrobials and antioxidants (Rathod et al., 2021b). To retain the nutritive value and freshness of the fishery products, natural additives are often investigated when used in combination with non-thermal technologies like high-pressure processing, irradiation, cold marinating and others (Arnaud et al., 2018; Šimat et al., 2019; Al-Kuraieef, 2021; Pinto de Rezende et al.; 2022) and modified atmosphere or vacuum packaging (Hassoun and Çoban, 2017). Combining more than one preservation method might have synergic or additive interaction resulting in better preservation and effect against lipid oxidation and pathogen bacteria. There are many examples of successful use of plant extracts/compounds and EOs in fish and fishery products at the laboratory scale (Table 1), however, their industrial application and marketing still have some limitations. The natural additives can be applied directly to the product, even as an additive in the coating, brine/marinade or ice glaze, or as an active substance in the packaging materials and produce a particular effect, however, the results of their application and bioactive effect may vary and be non-reproducible (Rathod et al., 2021a; Pinto de Rezende et al., 2022). The chemical profile and activity of the plant extracts and EOs change with the season, geographical location, environmental factors, extraction/production method and extraction solvent, final concentration in the product, and product characteristic itself. It is possible to produce standard protocols for the extraction and isolation of necessary compounds, but this process requires expertise and equipment thus increasing the production cost which is not acceptable for the industry. *In vitro* studies confirm the biological activity of natural preservatives, however, when they are applied in food models the *in vitro* effect/results are affected by food component interaction and usually higher concentrations of additives are needed. The application of EOs and pure compounds in seafood products in high concentrations can impart some unwanted changes in the taste, colour and aroma of the product and raise the question of the toxicity of these compounds which are rarely tested (Taylor et al., 2021). In accordance with this issue, natural preservatives were investigated for application in packaging films and edible coatings but also in other ways of applications to ensure their stability in the product, such as nanoemulsions (Özogul et al., 2017, 2020; Khezerlou et al., 2019; Khodanazary, 2019; Kanatt, 2020; Zhang and Rhim, 2022). These methods have been successfully applied in fish and seafood products, enhancing their antioxidant and antimicrobial stability thus prolonging the product's shelf life.

The control of bacterial contamination of fish and seafood can be done by application of bacteriocins, secondary metabolites of LAB, or LAB strains themselves. These peptides have promising biological activity, especially antimicrobial activity against fish spoilage bacteria and food-borne pathogens (Daba and Elkhateeb, 2020; Rathod et al., 2021b). LAB bacteriocins such as pediocin ACCEL, BacALP7, bacALP57, and divercin V41 were found to inhibit or decrease the growth of *List. monocytogenes* in different fresh and smoked seafood products (Yin et al., 2007; Mei et al., 2019; Rathod et al., 2022). Beside the retardation of microbial growth, bacteriocins and LAB cell-free supernatants were found to improve seafood physicochemical quality parameters and prolong the shelf life (Jo et al., 2021). The bacteriocin use is contributing to the reduction of chemical preservatives and high temperatures for food preservation, allowing the production of natural foods with rich organoleptic and nutritional properties (Iseppi et al., 2019). This biopreservation method can be applied directly to the product (addition of strains, cell-free supernatants or metabolites) or integrated into coatings and packaging materials (Contessa et al., 2021; Gumienna and Górna, 2021; Pérez-Arauz et al., 2021). Similarly to natural preservatives, the LAB strains and their bacteriocins are much more efficient in maintaining the quality and sensory properties of fish when they are combined with other preservation methods. Further studies are necessary to precise the levels of the bacteriocins necessary to maintain the stability of the product over shelf life.

Regulatory framework (EFSA and FDA general guidelines)

Some food additives are applied to preserve foods from chemical and microbial deterioration and to maintain their organoleptic quality and safety. Food antimicrobial and antioxidant agents can be divided into two groups: chemical and naturally occurring preservatives. The selection of these agents depends on many factors, including solubility and polarity, pH, lipid content, food processing and storage conditions, sensorial effect, cost, etc. They are included during the production of the food and all stages are firmly controlled. There are three governing organisations being in charge of legislation, enforcement of the law and approvement of food additives in the world. These are the FDA (Food and Drug Administration) in USA and the EFSA (European Food Safety Authority) in the European Union (EU). The other important international organisation, which deals with risk and safety issues of food additives, is the Joint FAO (Food and Agriculture Organization)/WHO (World Health Organization) Expert Committee on Food Additives (JECFA). Once the safety assessment is completed by the JECFA, the Codex Alimentarius Commission (CAC), a joint international organization of FAO and WHO, set the levels for maximum use of preservatives and additives in food (WHO, 2018).

According to EFSA, food additives are categorised by six groups: preservatives, nutritional additives, flavouring, colouring, miscellaneous and texturizing agents. Each food additive is given a number, which it belongs the specific group and the letter "E", referring to Europe. This numbering system has been extended by CAC to identify globally all the additives, regardless of whether they have been approved for use. The amount of each additive is firmly estimated for each foodstuff, thus is not allowed to go beyond the Admissible Daily Intake (ADI) with no harmful impacts on health (WHO, 2018). One of main class of additives are the preservatives and their numbers are in the range of E200 and E399. The preservatives are also sectioned into three categories: anti-browning agents, antioxidants and antimicrobials. In particular, anti-browning agents are employed to prevent browning, ntimicrobials are applied to avoid microbial spoilage, while the antioxidants are used to retard oxidation and rancidity, extending the shelf life of foodstuffs. Carocho et al. (2018) reviewed food applications, legislation and roles of antioxidants and antimicrobial agents as preservatives approved by EFSA and FDA. In the EU, guideline EC No 1129/2011 sets the list of antioxidants with code E in the category of "other food additives". According to this guideline, rosemary (E392), tocopherols (E306-E309) and ascorbic acid (E300) extracts are considered natural antioxidants approved as food additives by the EU (EU, 2011). The other groups of antioxidants as food additives include ascorbates (E300-E304), erythrobates (E315-E316), gallates (E310-E312), butylates (E319-E321), citrates (E330-E380), lactates (E325-E327), tartrates (E334-E354), phosphates (E338-E343), adipates (E355-E357), malates (E350-E352), as well as ethylenediaminetetraacetic acid (EDTA, E385) and succinic acid (E363). There are also various natural food antioxidants, including polyphenols, annatto, norbixin, bixin, carotenoids (E160a-E161g), and lutein (E161b). Some GRAS food preservatives considered as antimicrobial agents by FDA are acetic acid (E260), benzoic acid (E210-E213), lactic acid (E270), lactoferrin, lysozyme, natamycin (E235), nitrite (E249), nitrate (E252), parabens, propionic acid ((E280-E283), sorbic acid (E200-E203), sulphites (E221-E226), and sulphur dioxide (E220). Among bacteriocins, nisin is produced by *Lactococcus lactis* subsp. *lactis*, which was the only one approved as a food additive by FDA and later by EFSA with the code E234.

To conclude, there is a growing interest in natural alternative additives, aiming elimination of synthetic preservatives in food. However, this has not occurred yet and, thus, the natural or synthetic antioxidants applied in food industry have not been quantified in the official list which exhibits the levels and permissions for the usage of each additive in each kind of food. In this respect, there is a need for identification and evaluation of natural alternative additives and also to assess the safety issues and mechanisms of action of these natural preservatives together with proper safety regulations.

Conclusions

Fish and fishery products are highly perishable foods due to microbiological and oxidative activities. The literature reviewed in this chapter demonstrates that natural plant derivatives and bioprotective cultures can play an important role in prolonging their shelf life and assuring safety issues, especially in the frame of a hurdle strategy.

Although some applications based on biopreservation have already been implemented in productive processes, a relevant gap persists between scientific reports and applications in industrial environments. In the case of bioprotective cultures, the main concerns are relative to the optimization of the production of antimicrobial compounds (i.e., bacteriocins, organic acids, volatile compounds) without compromising the overall flavour and aroma profile of the food. This challenge can be faced only through a deeper knowledge of the metabolic activities of the bacteria, also including their ability to reduce oxidative reactions through their oxygen scavenging effects, in relation to the process conditions.

Concerning the use of plant derivatives as antimicrobials and antioxidants many important critical aspects need to be elucidated and deeply studied. First, the effect exerted by the different constituents is strongly affected by the extraction method and by their interactions, which can bring synergistic, additive and, sometimes, protective effects. Further, a better comprehension of the action mechanism against microorganisms is important to optimize their use in industrial process, allowing a greater diffusion of their application.

Acknowledgments

This work is supported by the PRIMA program under project BioProMedFood (Project ID 1467). The PRIMA program is supported by the European Union.

References

Abdel-Wahab, M., El-Sohaimy, S.A., Ibrahim, H.A. and Abo El-Makarem, H.S. 2020. Evaluation the efficacy of clove, sage and kiwifruit peels extracts as natural preservatives for fish fingers. *Annals of Agricultural Sciences*, 65: 98–106.

Abdollahzadeh, E., Rezaei, M. and Hosseini, H. 2014. Antibacterial activity of plant essential oils and extracts: the role of thyme essential oil, nisin, and their combination to control *Listeria monocytogenes* inoculated in minced fish meat. *Food Control*, 35: 177–183.

Adeoye, A.A., Yomla, R., Jaramillo-Torres, A., Rodiles, A., Merrifield, D.L. and Davies, S.J. 2016. Combined effects of exogenous enzymes and probiotic on Nile tilapia (*Oreochromis niloticus*) growth, intestinal morphology and microbiome. Aquaculture, 463: 61–70.

Ahmed, M., Liaquat, M., Shah, A.S., Abdel-Farid, I.B. and Jahangir, M. 2020. Proximate composition and fatty acid profiles of selected fish species from pakistan. *JAPS: Journal of Animal & Plant Sciences*, 30: 869–875.

Aksoy, A. and Sezer, Ç. 2019. Combined use of laurel essential oil and vacuum packing to extend the shelf-life of Rainbow trout (*Oncorhynchus mykiss*) fillets. *Kafkas Üniversitesi Veteriner Fakültesi Dergisi*, 25: 779–786.

122   *Novel Approaches in Biopreservation for Food and Clinical Purposes*

Albertos, I., Rico, D., Diez, A.M., González-Arnáiz, L., García-Casas, M.J. and Jaime, I. 2015. Effect of edible chitosan/clove oil films and high-pressure processing on the microbiological shelf life of trout fillets. *Journal of the Science of Food and Agriculture*, 95: 2858–2865.

Al-Kuraieef, A.N. 2021. Microbiological, chemical and organoleptic evaluation of fresh fish and its products irradiated by gamma rays. *Potravinarstvo Slovak Journal of Food Sciences*, 15: 95–100.

Alsaiqali, M.I. and Ghoneim, S.I. 2016. Effect of thyme extract as a natural preservative on. *SINAI Journal of Applied Sciences*, 5: 1–14.

Araujo, C., Munoz-Atienza, E., Perez-Sanchez, T., Poeta, P., Igrejas, G., Hernandez, P.E., Herranz, C., Ruiz-Zarzuela, I. and Cintas, L.M. 2015. Nisin Z production by *Lactococcus lactis* subsp. *cremoris* WA2–67 of aquatic origin as a defense mechanism to protect rainbow trout (*Oncorhynchus mykiss*, Walbaum) against *Lactococcus garvieae*. Marine Biotechnology, 17: 820–830.

Arfat, Y.A., Benjakul, S., Vongkamjan, K., Sumpavapol, P. and Yarnpakdee, S. 2015. Shelf-life extension of refrigerated sea bass slices wrapped with fish protein isolate/fish skin gelatin-ZnO nanocomposite film incorporated with basil leaf essential oil. *Journal of Food Science and Technology*, 52: 6182–6193.

Arnaud, C., de Lamballerie, M. and Pottier, L. 2018. Effect of high-pressure processing on the preservation of frozen and re-thawed sliced cod (*Gadus morhua*) and salmon (*Salmo salar*) fillets. *High Pressure Research*, 38: 62–79.

Aymerich, T., Rodríguez, M., Garriga, M. and Bover-Cid, S. 2019. Assessment of the bioprotective potential of lactic acid bacteria against *Listeria monocytogenes* on vacuum-packed cold-smoked salmon stored at 8°C. *Food Microbiology*, 83: 64–70.

Baird-Parker, T.C. 2000. The production of microbiologically safe and stable foods. *Microbiological Safety and Quality of Food*, 1: 3–18.

Balıkçı, E., Özoğul, Y., Durmuş, M. and Uçar, Y.S.G.T. 2022. The impact of thyme, rosemary, and basil extracts on the chemical, sensory and microbiological quality of mckerel balls stored at –18°C. *Acta Aquatica Turcica*, 18: 217–235.

Baptista, R.C., Horita, C.N. and Sant'Ana, A.S. 2020. Natural products with preservative properties for enhancing the microbiological safety and extending the shelf life of seafood: a review. *Food Research International*, 127: 108762.

Barbieri, F., Montanari, C., Gardini, F. and Tabanelli, G. 2019. Biogenic amine production by lactic acid bacteria: a review. Foods, 8: 17.

Binsi, P.K., Ninan, G. and Ravishankar, C.N. 2017. Effect of curry leaf and clove bud essential oils on textural and oxidative stability of chill stored sutchi catfish fillets. *Journal of Texture Studies*, 48: 258–266.

Bozova, B. and İzci, L. 2021. Effects of plant extracts on the quality of sous vide meagre (*Argyosomus Regius*) fillets. *Acta Aquatica Turcica*, 17: 255–266.

Cai, L., Cao, A., Li, Y., Song, Z., Leng, L. and Li, J. 2015. The effects of essential oil treatment on the biogenic amines inhibition and quality preservation of red drum (*Sciaenops ocellatus*) fillets. *Food Control*, 56: 1–8.

Carocho, M., Morales, P. and Ferreira, I.C. 2018. Antioxidants: reviewing the chemistry, food applications, legislation and role as preservatives. *Trends in Food Science & Technology*, 71: 107–120.

Castellano, P., Perez Ibarreche, M., Blanco Massani, M., Fontana, C. and Vignolo, G.M. 2017. Strategies for pathogen biocontrol using lactic acid bacteria and their metabolites: a focus on meat ecosystems and industrial environments. *Microorganisms*, 5: 38.

Coban, H.B. 2020. Organic acids as antimicrobial food agents: applications and microbial productions. *Bioprocess and Biosystems Engineering*, 43: 569–591.

Comas-Basté, O., Latorre-Moratalla, M.L., Sánchez-Pérez, S., Veciana-Nogués, M.T. and Vidal-Carou, M.C. 2019. Histamine and other biogenic amines in food. From scombroid poisoning to histamine intolerance. pp. 9–19. *In*: Proestos, C. (Ed.). *Biogenic Amines*. IntechOpen: London, UK.

Contessa, C.R., da Rosa, G.S. and Moraes, C.C. 2021. New active packaging based on biopolymeric mixture added with bacteriocin as active compound. *International Journal of Molecular Sciences*, 22: 10628.

Chaula, D., Laswai, H., Chove, B., Dalsgaard, A., Mdegela, R., Jacobsen, C. and Hyldig, G. 2019. Effect of clove (*Syzygium aromaticum*) and seaweed (*Kappaphycus alvarezii*) water extracts pretreatment on lipid oxidation in sun-dried sardines (*Rastrineobola argentea*) from Lake Victoria, Tanzania. *Food Science and Nutrition*, 7: 1406–1416.

Chen, L., Song, Z., Tan, S.Y., Zhang, H. and Yuk, H.G. 2020. Application of bacteriocins produced from lactic acid bacteria for microbiological food safety. *Current Topics in Lactic Acid Bacteria and Probiotics*, 6: 1–8.

da Rocha, M., Alemán, A., Romani, V.P., López-Caballero, M.E., Gómez-Guillén, C., Montero, P. and Prentice, C. 2018. Effects of agar films incorporated with fish protein hydrolysate or clove essential oil on flounder (*Paralichthys orbignyanus*) fillets shelf-life. *Food Hydrocolloids*, 81: 351–363.

Daba, G.M. and Elkhateeb, W.A. 2020. Bacteriocins of lactic acid bacteria as biotechnological tools in food and pharmaceuticals: current applications and future prospects. *Biocatalysis and Agricultural Biotechnology*, 28: 101750.

Dogruyol, H., Ulusoy, Ş., Mol, S. and Alakavuk, D.Ü. 2021. Effect of different plant extracts added to ice on sensory preference of sliced salmon. *Aquatic Sciences and Engineering*, 36: 159–165.

Durmuş, M., Kosker, A.R., Özogul, Y., Aydin, M., UÇar, Y., Ayas, D. and Özogul, F. 2018. The effects of sex and season on the metal levels and proximate composition of red mullet (*Mullus barbatus* Linnaeus 1758) caught from the Middle Black Sea. *Human and Ecological Risk Assessment: an International Journal*, 24: 731–742.

Echeverría, I., López-Caballero, M.E., Gómez-Guillén, M.C., Mauri, A.N. and Pilar Montero, M. 2018. Active nanocomposite films based on soy proteins-montmorillonite-clove essential oil for the preservation of refrigerated bluefin tuna (*Thunnus thynnus*) fillets. *International Journal of Food Microbiology*, 266: 142–149.

Egerton, S., Culloty, S., Whooley, J., Stanton, C. and Ross, R.P. 2018. The gut microbiota of marine fish. *Frontiers in Microbiology*, 9: 873.

El-Obeid, T., Yehia, H.M., Sakkas, H., Lambrianidi, L., Tsiraki, M.I. and Savvaidis, I.N. 2018. Shelf-life of smoked eel fillets treated with chitosan or thyme oil. *International Journal of Biological Macromolecules*, 114: 578–583.

El Oudiani, S., Chetoui, I., Darej, C. and Moujahed, N. 2019. Sex and seasonal variation in proximate composition and fatty acid profile of *Scomber scombrus* (L. 1758) fillets from the Middle East Coast of Tunisia. *Grasas y aceites*, 70: e285–e285.

European Parliament and Council Commission Regulation (EU) No 1129/2011 of 11 November 2011 AmendingAnnex II to Regulation (EC) No 1333/2008 of the European Parliament and of the Council by Establishing a Union List of Food Additives (Text with EEA Relevance. Available online: https://op.europa.eu/en/publication-detail/-/publication/28cb4a37-b40e-11e3-86f9-01aa75ed71a1.

FAO. 2022. The state of world fisheries and aquaculture 2022. Towards Blue Transformation. Rome, Italy: FAO.

Farag, M.A., Zain, A.E., Hariri, M.L., el Aaasar, R., Khalifa, I. and Elmetwally, F. 2022. Potential food safety hazards in fermented and salted fish in Egypt (Feseekh, Renga, Moloha) as case studies and controlling their manufacture using HACCP system. *Journal of Food Safety*, 42:, e12973.

Farghal, H.H., Karabagias, I., El Sayed, M. and Kontominas, M.G. 2017. Determination of antioxidant activity of surface-treated PET films coated with rosemary and clove extracts. *Packaging Technology and Science*, 30: 799–808.

Farjami, B. and Hosseini, S. 2015. Effect of thyme extract on the chemical quality of raw surimi produced from Common carp (*Cyprinus carpio*) during refrigerator storage. *Journal of Fisheries*, 68: 447–456.

Fellenberg, M.A., Carlos, F., Peña, I., Ibáñez, R.A. and Vargas-Bello-Pérez, E. 2020. Oxidative quality and color variation during refrigeration (4°C) of rainbow trout fillets marinated with different natural antioxidants from oregano, quillaia and rosemary. *Agricultural and Food Science*, 29: 43–54.

Gänzle, M. 2004. Reutericyclin: biological activity, mode of action, and potential applications. *Applied Microbiology and Biotechnology*, 64: 326–332.

Gänzle, M.G. 2015. Lactic metabolism revisited: metabolism of lactic acid bacteria in food fermentations and food spoilage. *Current Opinion in Food Science*, 2: 106–117.

Gao, M.S., Feng, L.F., Jiang, T.J., Zhu, J.L., Fu, L.L., Yuan, D.X. and Li, J.R. 2014. The use of rosemary extract in combination with nisin to extend the shelf life of pompano (*Trachinotus ovatus*) fillet during chilled storage. Food Control, 37: 1–8.

Gao, P., Jiang, Q.X., Xu, Y.S. and Xia, W.S. 2018. Biosynthesis of acetate esters by dominate strains, isolated from Chinese traditional fermented fish (Suan yu). *Food Chemistry*, 244: 44–49.

García-Díez, J. and Saraiva, C. 2021. Use of starter cultures in foods from animal origin to improve their safety. *International Journal of Environmental Research and Public Health*, 18: 2544.

Gardini, F., Özogul, Y., Suzzi, G., Tabanelli, G. and Özogul, F. 2016. Technological factors affecting biogenic amine content in foods: a review. *Frontiers in Microbiology*, 7: 1218.

Ghaly, A.E., Dave, D., Budge, S. and Brooks, M.S. 2010. Fish spoilage mechanisms and preservation techniques. *American Journal of Applied Sciences*, 7: 859–877.

Gharsallaoui, A., Oulahal, N., Joly, C. and Degraeve, P. 2016. Nisin as a food preservative: part 1: physicochemical properties, antimicrobial activity, and main uses. *Critical Reviews in Food Science and Nutrition*, 56: 1262–1273.

Gokoglu, N. 2019. Novel natural food preservatives and applications in seafood preservation: a review. *Journal of the Science of Food and Agriculture*, 99: 2068–2077.

Gram, L. and Huss, H.H. 1996. Microbiological spoilage of fish and fish products. *International Journal of Food Microbiology*, 33: 121–137.

Gram, L. and Dalgaard, P. 2002. Fish spoilage bacteria—problems and solutions. *Current Opinion in Biotechnology*, 13: 262–266.

Guan, W., Ren, X., Li, Y. and Mao, L. 2019. The beneficial effects of grape seed, sage and oregano extracts on the quality and volatile flavor component of hairtail fish balls during cold storage at 4°C. *Lebensmittel-Wissenschaft & Technologie*, 101: 25–31.

Gumienna, M. and Górna, B. 2021. Antimicrobial food packaging with biodegradable polymers and bacteriocins. Molecules, 26: 3735.

Haghighi, M. and Yazdanpanah, S. 2020. Chitosan-based coatings incorporated with cinnamon and tea extracts to extend the fish fillets shelf life: validation by FTIR spectroscopy technique. *Journal of Food Quality*, 2020: 8865234.

Han, J., Kong, T., Wang, Q., Jiang, J., Zhou, Q., Li, P., Zhu, B. and Gu, Q. 2022. Regulation of microbial metabolism on the formation of characteristic flavor and quality formation in the traditional fish sauce during fermentation: a review. Critical Reviews in Food Science and Nutrition, 7: 1–20.

Hao, R., Roy, K., Pan, J., Shah, B.R. and Mraz, J. 2021. Critical review on the use of essential oils against spoilage in chilled stored fish: a quantitative meta-analyses. *Trends in Food Science & Technology*, 111: 175–190.

Haruna, M.Y., Bello, M.M., Dadile, M.A. and Mohammed, A.M. 2021. Assessment of cinnamon (*Cinnamomum verum*) bark extract on proximate composition and sensory qualities of smoked-dried African catfish *Clarias gariepinus* (Burchell, 1822). *Asian Journal of Fisheries and Aquatic Research*, 14: 1–6.

Hassoun, A. and Çoban, Ö.E. 2017. Essential oils for antimicrobial and antioxidant applications in fish and other seafood products. *Trends in Food Science & Technology*, 68: 26–36.

Houicher, A., Bensid, A., Regenstein, J.M. and Özogul, F. 2020. Control of biogenic amine production and bacterial growth in fish and seafood products using phytochemicals as biopreservatives: a review. *Food Biosience*, 39: 100807.

Huang, X., Lao, Y., Pan, Y., Chen, Y., Zhao, H., Gong, L., Xie, N. and Mo, C.H. 2021. Synergistic antimicrobial effectiveness of plant essential oil and its application in seafood preservation: a review. *Molecules*, 26: 307.

Huang, Z., Liu, X., Jia, S., Zhang, L. and Luo, Y. 2018. The effect of essential oils on microbial composition and quality of grass carp (*Ctenopharyngodon idellus*) fillets during chilled storage. *International Journal of Food Microbiology*, 266: 52–59.

Hussain, M.A., Sumon, T.A., Mazumder, S.K., Ali, M.M., Jang, W.J., Abualreesh, M.H., Sharifuzzaman S.M., Brown, C.L., Lee, H.T., Lee, E.W. and Hasan, M.T. 2021. Essential oils and chitosan as alternatives to chemical preservatives for fish and fishery products: a review. *Food Control*, 129: 108244.

Inanlia, A.G., Tümerkanb, E.T.A., El Abedd, N., Regensteine, J.M. and Özogul, F. 2020. The impact of chitosan on seafood quality and human health: a review. *Trends in Food Science & Technology*, 97: 404–416.

Iseppi, R., Stefani, S., de Niederhausern, S., Bondi, M., Sabia, C. and Messi, P. 2019. Characterization of anti-*Listeria monocytogenes* properties of two bacteriocin-producing *Enterococcus mundtii* isolated from fresh fish and seafood. *Current Microbiology*, 76: 1010–1019.

Javadian, S.R., Shahosseini, S.R. and Ariaii, P. 2017. The effects of liposomal encapsulated thyme extract on the quality of fish mince and *Escherichia coli* O157:H7 inhibition during refrigerated storage. *Journal of Aquatic Food Product Technology*, 26: 115–123.

Jo, D.M., Park, S.K., Khan, F., Kang, M.G., Lee, J.H. and Kim, Y.M. 2021. An approach to extend the shelf life of ribbonfish fillet using lactic acid bacteria cell-free culture supernatant. *Food Control*, 123: 107731.

Kanatt, S.R. 2020. Development of active/intelligent food packaging film containing Amaranthus leaf extract for shelf life extension of chicken/fish during chilled storage. *Food Packaging and Shelf Life*, 24: 100506.

Karoui, R. and Hassoun, A. 2017. Efficiency of rosemary and basil essential oils on the shelf-life extension of Atlantic mackerel (*Scomber scombrus*) fillets stored at 2°C. *Journal of AOAC International*, 100: 335–344.

Kasankala, L.M., Xiong, Y.L. and Chen, J. 2012. Enzymatic activity and flavor compound production in fermented silver carp fish paste inoculated with douchi starter culture. *Journal of Agricultural and Food Chemistry*, 60: 226–233.

Katarzyna, P. and Katarzyna, J.K.J. 2020. Properties and use of rosemary (*Rosmarinus officinalis* L.) *Pomeranian Journal of Life Sciences*, 66: 76–82.

Khalafalla, F.A., Ali, F.H.M. and Hassan, A.R.H.A. 2015. Quality improvement and shelf-life extension of refrigerated Nile tilapia (*Oreochromis niloticus*) fillets using natural herbs. *Beni-Suef University Journal of Basic and Applied Sciences*, 4: 33–40.

Khezerlou, A., Azizi-Lalabadi, M., Mousavi, M.M. and Ehsani, A. 2019. Incorporation of essential oils with antibiotic properties in edible packaging films. *Journal of Food and Bioprocess Engineering*, 2: 77–84.

Khodanazary, A. 2019. Quality characteristics of refrigerated mackerel *Scomberomorus commerson* using gelatin-polycaprolactone composite film incorporated with lysozyme and pomegranate peel extract. *International Journal of Food Properties*, 22: 2057–2071.

Kolakowska, A., Olley, J. and Dunstan, G.A. 2010. Fish lipids. pp. 221–265. *In*: Zdzislaw, Z., Sikorski, E. and Kolakowska, A. (Eds.). *Chemical and Functional Properties of Food Lipids*; CRC Press Inc.: Boca Raton, USA.

Kontominas, M.G., Badeka, A.V., Kosma, I.S. and Nathanailides, C.I. 2021. Innovative seafood preservation technologies: recent developments. *Animals*, 11: 92.

Kuley, E., Durmus, M., Ucar, Y., Kosker, A.R., Aksun Tumerkan, E.T., Regenstein, J.M. and Ozogul, F. 2018. Combined effects of plant and cell-free extracts of lactic acid bacteria on biogenic amines and bacterial load of fermented sardine stored at 3 ± 1°C. *Food Bioscience*, 24: 127–136.

Kumariya, R., Garsa, A.K., Rajput, Y.S., Sood, S.K., Akhtar, N. and Patel, S. 2019. Bacteriocins: classification, synthesis, mechanism of action and resistance development in food spoilage causing bacteria. *Microbial Pathogenesis*, 128: 171–177.

Lahreche, T., Uçar, Y., Kosker, A.R., Hamdi, T.M. and Özogul, F. 2019. Combined impacts of oregano extract and vacuum packaging on the quality changes of frigate tuna muscles stored at 3±1°C. *Veterinary World*, 12: 155–164.

Lambrianidi, L., Savvaidis, I.N., Tsiraki, M.I. and El-Obeid, T. 2019. Chitosan and oregano oil treatments, individually or in combination, used to increase the shelf life of vacuum-packaged, refrigerated European eel (*Anguilla anguilla*) fillets. *Journal of Food Protection*, 82: 1369–1376.

Leroi, F. 2010. Occurrence and role of lactic acid bacteria in seafood products. *Françoise Food Microbiology*, 27: 698–709.

Li, X., Ringø, E., Hoseinifar, S.H., Lauzon, H.L., Birkbeck, H. and Yang, D. 2019. The adherence and colonization of microorganisms in fish gastrointestinal tract. *Reviews in Aquaculture*, 11: 603–618.

Linhartová, Z., Lunda, R., Dvořák, P., Bárta, J., Bártová, V., Kadlec, J., Samková, E., Bedrníček, J., Pešek, M., Laknerová, I., Možina, S.S., Smetana, P. and Mráz, J. 2019. Influence of rosemary extract (*Rosmarinus officinalis*) Inolens to extend the shelf life of vacuum-packed rainbow trout (*Oncorhynchus mykiss*) fillets stored under refrigerated conditions. *Aquaculture International*, 27: 833–847.

126 *Novel Approaches in Biopreservation for Food and Clinical Purposes*

Liu, X., Huang, Z., Jia, S., Zhang, J., Li, K. and Luo, Y. 2018. The roles of bacteria in the biochemical changes of chill-stored bighead carp (*Aristichthys nobilis*): proteins degradation, biogenic amines accumulation, volatiles production, and nucleotides catabolism. *Food Chemistry*, 255: 174–181.

Lu, F., Ding, Y.T., Ye, X.Q. and Liu, D.H. 2010. Cinnamon and nisin in alginate-calcium coating maintain quality of fresh northern snakehead fish fillets. *LWT-Food Science and Technology*, 43: 1331–1335.

Mah, J.H. and Hwang, H.J. 2009. Inhibition of biogenic amine for mation in a salted and fermented anchovy by *Staphylococcus xylosus* as a protective culture. *Food Control*, 20: 796–801.

Maldonado-Simán, E., González-Ariceaga, C.C., Rodríguez-de Lara, R. and Fallas-López, M. 2018. Potential hazards and biosecurity aspects associated on food safety. pp. 25–61. *In*: Holban, A.M. and Grumezescu, A.M. (Eds.). *Handbook of Food Bioengeneering*; Academic Press: London, UK.

Malik, I.O.M., Yousif, M.H., Mahmoud, H.A.A., Hamadnalla, H.M.Y. and Ali, A.A.S. 2020. Enhancement of dried fish flavor by cinnamon (*Cinnamon Verum*) aqueous extracts. *Journal of Chemistry*, 4: 14–20.

Mani-López, E., Palou, E. and López-Malo, A. 2018. Biopreservatives as agents to revent Food Spoilage. pp. 235–270. *In*: Olban, A.M. and Grumezescu, A.M. (Eds.). *Microbial Contamination and Food Degradation; Microbial Contamination and Food degradation*; Academic Press: London, UK.

Marco, M.L., Heeney, D., Binda, S., Cifelli, C.J., Cotter, P.D., Foligne, B., Gänzle, M., Kort, R., Pasin, G., Pihlanto, A., Smid, E.J. and Hutkins, R. 2017. Health benefits of fermented foods: microbiota and beyond. *Current Opinion in Biotechnology*, 44: 94–102.

Mei, J., Ma, X. and Xie, J. 2019. Review on natural preservatives for extending fish shelf life. *Foods*, 8: 490.

Meral, R., Ceylan, Z. and Kose, S. 2019. Limitation of microbial spoilage of rainbow trout fillets using characterized thyme oil antibacterial nanoemulsions. *Journal of Food Saftey*, 39: e12644.

Moosavi-Nasab, M., Mirzapour-Kouhdasht, A. and Oliyaei, N. 2019. Application of essential oils for shelf-life extension of seafood products. *In*: El-Shemy, H.A. (Ed.). *Essential Oils - Oils of Nature*, IntechOpen, London, UK.

Mukherjee, S., Nath, S., Chowdhury, S. and Chatterjee, P. 2020. Antimicrobial activity of garlic (*Allium sativum*) and its potential use in fish preservation and disease prevention. *International Journal of Microbiology Research*, 12: 1879–1883.

Munoz-Atienza, E., Gomez-Sala, B., Araujo, C., Campanero, C., del Campo, R., Hernandez, P.E., Herranz, C. and Cintas, L.M. 2013. Antimicrobial activity, antibiotic susceptibility and virulence factors of lactic acid bacteria of aquatic origin intended for use as probiotics in aquaculture. *BMC Microbiology*, 13: 15.

Navarro-Segura, L., Ros-Chumillas, M., Martínez-Hernández, G.B. and López-Gómez, A. 2020. A new advanced packaging system for extending the shelf life of refrigerated farmed fish fillets. *Journal of the Science of Food and Agriculture*, 100: 4601–4611.

Nieto, G. 2020. A review on applications and uses of thymus in the food industry. Plants, 9: 1–29.

Nisar, T., Yang, X., Alim, A., Iqbal, M., Wang, Z.C. and Guo, Y. 2019. Physicochemical responses and microbiological changes of bream (*Megalobrama ambycephala*) to pectin-based coatings enriched with clove essential oil during refrigeration. *International Journal of Biological Macromolecules*, 124: 1156–1166.

Noordin, W.N.M., Shunmugam, N., Huda, N. and Adzitey, F. 2018. The effects of essential oils and organic acids on microbiological and physicochemical properties of whole shrimps at refrigerated storage. *Current Research in Nutrition and Food Science*, 6: 273–283.

Özogul, Y., Polat, A., Uçak, İ. and Özogul, F. 2011. Seasonal fat and fatty acids variations of seven marine fish species from the Mediterranean Sea. *European Journal of Lipid Science and Technology*, 113: 1491–1498.

Özogul, Y., Yuvka, İ., Ucar, Y., Durmus, M., Kösker, A.R., Öz, M. and Özogul, F. 2017. Evaluation of effects of nanoemulsion based on herb essential oils (rosemary, laurel, thyme and sage) on sensory, chemical and microbiological quality of rainbow trout (*Oncorhynchus mykiss*) fillets during ice storage. *LWT - Food Science and Technology*, 75: 677–684.

Özogul, Y., Kuley Boğa, E., Akyol, I., Durmus, M., Ucar, Y., Regenstein, J.M. and Köşker, A.R. 2020. Antimicrobial activity of thyme essential oil nanoemulsions on spoilage bacteria of fish and foodborne pathogens. *Food Bioscience*, 36: 100635.

Öztürk, F., Gündüz, H. and Sürengil, G. 2021. The effects of essential oils on inactivation of *Listeria monocytogenes* in rainbow trout cooked with sous-vide. *Journal of Food Processing and Preservation*, 45: e15878.

Pérez-Arauz, Á., Rodríguez-Hernández, A., Rocío López-Cuellar, M., Martínez-Juárez, V. and Chavarría-Hernández, N. 2021. Films based on Pectin, Gellan, EDTA, and bacteriocin-like compounds produced by *Streptococcus infantarius* for the bacterial control in fish packaging. *Journal of Food Processing and Preservation*, 45: e15006.

Pinto de Rezende, L., Barbosa, J. and Teixeira, P. 2022. Analysis of alternative shelf life-extending protocols and their effect on the preservation of seafood products. *Foods*, 11: 1100.

Ramírez, C. and Romero, J. 2017. The microbiome of *Seriola lalandi* of wild and aquaculture origin reveals differences in composition and potential function. *Frontiers in Microbiology*, 8: 1844.

Rathod, N.B., Phadke, G.G., Tabanelli, G., Mane, A., Ranveer, R.C., Pagarkar, A. and Özogul, F. 2021a. Recent advances in bio-preservatives impacts of lactic acid bacteria and their metabolites on aquatic food products. *Food Bioscience*, 44: 101440.

Rathod, N.B., Ranveer, R.C., Benjakul, S., Kim, S.K., Pagarkar, A.U., Patange, S. and Özogul, F. 2021b. Recent developments of natural antimicrobials and antioxidants on fish and fishery food products. *Comprehensive Reviews in Food Science and Food Safety*, 20: 4182–4210.

Rathod, N.B., Nirmal, N.P., Pagarkar, A., Özogul, F. and Rocha, J.M. 2022. Antimicrobial impacts of microbial metabolites on the preservation of fish and fishery products: a review with current knowledge. *Microorganisms*, 10: 773.

Rognes, T., Flouri, T., Nichols, B., Quince, C. and Mahé, F. 2016. VSEARCH: a versatile open source tool for metagenomics. *PeerJ*, 4: e2584.

Schaefer, L., Auchtung, T.A., Hermans, K.E., Whitehead, D., Borhan, B. and Britton, R.A. 2010. The antimicrobial compound reuterin (3-hydroxypropionaldehyde) induces oxidative stress via interaction with thiol groups. *Microbiology*, 156: 1589–1599.

Shi, J., Lei, Y., Shen, H., Hong, H., Yu, X., Zhu, B. and Luo, Y. 2019. Effect of glazing and rosemary (*Rosmarinus officinalis*) extract on preservation of mud shrimp (*Solenocera melantho*) during frozen storage. *Food Chemistry*, 272: 604–612.

Siedler, S., Balti, R. and Neves, A.R. 2019. Bioprotective mechanisms of lactic acid bacteria against fungal spoilage of food. *Current Opinion in Biotechnology*, 56: 138–146.

Šimat, V., Mićunović, A., Bogdanović, T., Listeš, I., Generalić Mekinić, I., Hamed, I. and Skroza, D. 2019. The impact of lemon juice on the marination of anchovy (*Engraulis encrasicolus*): chemical, microbiological and sensory changes. *Italian Journal of Food Science*, 31: 604–617.

Šimat, V. and Generalić Mekinić, I. 2020. Advances in chilling. pp. 1–25. *In*: Özogul, Y. (Ed.). *Innovative Technologies in Seafood Processing*; CRC Press Inc.: Boca Raton, USA.

Šimat, V., Čagalj, M., Skroza, D., Gardini, F., Tabanelli, G., Montanari, C., Hassoun, A. and Özogul. 2021. Sustainable sources for antioxidant and antimicrobial compounds used in meat and seafood products. pp. 55–118. *In*: Fidel, T. (Ed). *Advances in Food and Nutrition Research*. Elsevier Inc.: Oxford, UK.

Smid, E.J. and Gorris, L.G.M. 2020. Natural antimicrobials for food preservation. pp. 283–298. *In*: Rahman, M.S. (Ed.). *Handbook of Food Preservation* (3rd ed.), CRC Press Inc.: Boca Raton, USA.

Sofoulaki, K., Kalantzi, I., Machias, A., Mastoraki, M., Chatzifotis, S., Mylona, K., Pergantis, S.A. and Tsapakis, M. 2018. Metals and elements in sardine and anchovy: species specific differences and correlations with proximate composition and size. *Science of the Total Environment*, 645: 329–338.

Speranza, B., Racioppo, A., Beneduce, L., Bevilacqua, A., Sinigaglia, M. and Corbo, M.R. 2017. Autochthonous lactic acid bacteria with probiotic aptitudes as starter cultures for fish-based products. *Food Microbiology*, 65: 244–253.

Syropoulou, F., Parlapani, F.F., Bosmali, I., Madesis, P. and Boziaris, I.S. 2020. HRM and 16S rRNA gene sequencing reveal the cultivable microbiota of the European sea bass during ice storage. *International Journal of Food Microbiology*, 327: 108658.

Tabanelli, G., Barbieri, F., Montanari, C. and Gardini, F. 2019. Application of natural antimicrobial strategies in seafood preservation. pp. 243–262. *In*: Özogul, Y. (Ed.). *Innovative Technologies in Seafood Processing*; CRC Press Inc.; Boca Raton, USA.

Tavares, J., Martins, A., Fidalgo, L.G., Lima, V., Amaral, R.A., Pinto, C.A., Silva, A.M. and Saraiva, J.A. 2021. Fresh fish degradation and advances in preservation using physical emerging technologies. *Foods*, 10: 780.

Tayel, A.A., Elzahy, A.F., Moussa, S.H., Al-Saggaf, M.S. and Diab, A.M. 2020. Biopreservation of shrimps using composed edible coatings from chitosan nanoparticles and cloves extract. *Journal of Food Quality*, 2020: 8878452.

Taylor, T.M., Davidson, P.M. and Jairus, R.D.D. 2021. Food antimicrobials—an introduction. pp. 1–12. *In*: Davidson, P.M., Taylor, T.M. and David, J.R.D. (Eds.). *Antimicrobials in Food*. CRC Press Inc.: Boca Raton, USA.

Tokur, B.K., Sert, F., Aksun, E.T. and Özoğul, F. 2016. The effect of whey protein isolate coating enriched with thyme essential oils on trout quality at refrigerated storage (4 ± 2°C). *Journal of Aquatic Food Product Technology*, 25: 585–596.

Tosun, Ş.Y., Üçok Alakavuk, D., Ulusoy, Ş. and Erkan, N. 2018. Effects of essential oils on the survival of *Salmonella* Enteritidis and *Listeria monocytogenes* on fresh Atlantic salmons (*Salmo salar*) during storage at 2 ± 1°C. *Journal of Food Safety*, 38: e12408.

Touraki, M., Karamanlidou, G., Koziotis, M. and Christidis, I. 2013. Antibacterial effect of *Lactococcus lactis* subsp *lactis* on *Artemia franciscana* nauplii and *Dicentrarchus labrax* larvae against the fish pathogen *Vibrio anguillarum*. *Aquaculture International*, 21: 481–495.

Van Haute, S., Raes, K., Van der Meeren, P. and Sampers, I. 2016. The effect of cinnamon, oregano and thyme essential oils in marinade on the microbial shelf life of fish and meat products. *Food Control*, 68: 30–39.

Vijayan, A., Sivaraman, G.K., Visnuvinayagam, S. and Mothadaka, M.P. 2021. Role of natural additives on quality and shelf life extension of fish and fishery products. *In*: Lage, M.A.A.P. and Otero, P. (Eds.). *Food Additives*. IntechOpen, London, UK.

Viji, P., Venkateshwarlu, G., Ravishankar, C.N. and Srinivasa Gopa, T.K. 2017. Role of plant extracts as natural additives in fish and fish products—a review. Fishery Technology, 54: 145–154.

Visciano, P., Schirone, M. and Paparella, A. 2020. An overview of histamine and other biogenic amines in fish and fish products. *Foods*, 9: 1795.

Wang, W., Xia, W., Gao, P., Xu, Y. and Jiang, Q. 2017. Proteolysis during fermentation of Suanyu as a traditional fermented fish product of China. *International Journal of Food Properties*, 20: S166–S176.

Wessels, S. and Huss, H.H. 1996. Suitability of *Lactococcus lactis* subsp. *lactis* ATCC 11454 as a protective culture for lightly preserved fish products. *Food Microbiology*, 13: 323.332.

WHO. 2018. Food additives. Available online: https://www.who.int/news-room/fact-sheets/detail/food-additives.

Xu, Y., Zang, J., Regenstein, J.M. and Xia, W. 2021. Technological roles of microorganisms in fish fermentation: a review. *Critical Reviews in Food Science and Nutrition*, 61: 1000–1012.

Yin, L.J., Wu, C.W. and Jiang, S.T. 2007. Biopreeservative effect of pediocin ACCEL on refrigerated seafood. *Fisheries Science*, 73: 907–912.

Zaman, M.Z., Bakar, F.A., Jinap, S. and Bakar, J. 2011. Novel starter cultures to inhibit biogenic amines accumulation during fish sauce fermentation. *International Journal of Food Microbiology*, 145: 84–91.

Zaman, M.Z., Bakar, F.A., Selamat, J., Bakar, J., Ang, S.S. and Chong, C.Y. 2014. Degradation of histamine by the halotolerant *Staphylococcus carnosus* FS19 isolate obtained from fish sauce. *Food Control*, 40: 58–63.

Zamani, Z. and Abaei, F.A. 2021. Assessment of chemical and bacterial indices of common carp (*Cyprinus carpio*) surimi under various concentrations of basil (*Ocimum basilicum*) extract during storage in refrigerator. *Iranian Journal of Nutrition Sciences and Food Technology*, 16: 95–107.

Zang, J., Xu, Y., Xia, W. and Regenstein, J.M. 2020. Quality, functionality, and microbiology of fermented fish: a review. *Critical Reviews in Food Science and Nutrition*, 60: 1228–1242.

Zhang, Y., Li, L., Lv, J., Li, Q., Kong, C. and Luo, Y. 2017. Effect of cinnamon essential oil on bacterial diversity and shelf-life in vacuum-packaged common carp (*Cyprinus carpio*) during refrigerated storage. *International Journal of Food Microbiology*, 249: 1–8.

Zhang, W. and Rhim, J.W. 2022. Functional edible films/coatings integrated with lactoperoxidase and lysozyme and their application in food preservation. *Food Control*, 133: 108670.

Zhao, D., Lu, F., Gu, S., Ding, Y. and Zhou, X. 2017. Physicochemical characteristics, protein hydrolysis, and textual properties of surimi during fermentation with *Actinomucor elegans*. *International Journal of Food Properties*, 20: 538–548.

Zhou, X., Qiu, M., Zhao, D., Lu, F. and Ding, Y. 2016. Inhibitory effects of spices on biogenic amine accumulation during fish sauce fermentation. *Journal of Food Science*, 81: M913–M920.

Zhuang, S., Li, Y., Hong, H., Liu, Y., Shu, R. and Luo, Y. 2020. Effects of ethyl lauroyl arginate hydrochloride on microbiota, quality and biochemical changes of container-cultured largemouth bass (*Micropterus salmonides*) fillets during storage at 4 C. *Food Chemistry*, 324: 126886.

Zhuang, S., Hong, H., Zhang, L. and Luo, Y. 2021. Spoilage-related microbiota in fish and crustaceans during storage: research progress and future trends. *Comprehensive Reviews in Food Science and Food Safety*, 20: 252–288.

Zotta, T., Parente, E., Ianniello, R.G., De Filippis, F. and Ricciardi, A. 2019. Dynamics of bacterial communities and interaction networks in thawed fish fillets during chilled storage in air. *International Journal of Food Microbiology*, 293: 102–113.

CHAPTER 6

Biopreservation in Flours and Bread

*Biljana Kovacevik,[1] Sanja Kostadinović Veličkovska,[1]
Tuba Esatbeyoglu,[2] Aleksandar Cvetkovski,[3]
Muhammad Qamar[4] and João Miguel Rocha[5,6,7]**

Introduction

The sustainable food security encompasses physical, social and economic access to safe, nutritious and varied diets to everyone towards supporting an active and healthy lifestyle. Therefore, international agencies like the United Nations (UN) are evolving a hierarchy of strategies to reduce food insecurity so that to achieve the goal of sustainable hunger eradication due to an ever-increasing population (Bankefa et al., 2021). One disturbing statistics is, for instance, the continuous significant increase of chronic food shortage since the last decade, affecting approximately

---

[1] Faculty of Agriculture, Goce Delcev University, Krste Misirkov 10 A, 2000 Stip, North Macedonia.
Email: biljana.kovacevik@ugd.edu.mk, ORCID: http://orcid.org/0000-0002-3361-0759.
Email: sanja.kostadinovik@ugd.edu.mk, ORCID: https://orcid.org/0000-0003-2402-3306
[2] Institute of Food Science and Human Nutrition, Gottfried Wilhelm Leibniz University Hannover, Am Kleinen Felde 30, 30167 Hannover, Germany.
Email: esatbeyoglu@lw.uni-hannover.de, ORCID: https://orcid.org/0000-0003-2413-6925.
[3] Faculty of Medical Science, Goce Delcev University, Krste Misirkov 10 A, 2000 Stip, North Macedonia.
Email: aleksandar.cvetkovski@ugd.edu.mk, ORCID: http://orcid.org/0000-0002-8827-0245.
[4] Institute of Food Science and Nutrition, Bahauddin Zakariya University, Multan 60800, Pakistan.
Email: muhammad.qamar44@gmail.com, ORCID: https://orcid.org/0000-0002-6207-553X
[5] Universidade Católica Portuguesa, CBQF - Centro de Biotecnologia e Química Fina – Laboratório Associado, Escola Superior de Biotecnologia, Rua Diogo Botelho 1327, 4169-005 Porto, Portugal.
[6] LEPABE – Laboratory for Process Engineering, Environment, Biotechnology and Energy, Faculty of Engineering, University of Porto, Rua Dr. Roberto Frias, 4200-465 Porto, Portugal.
ORCID: http://orcid.org/0000-0002-0936-2003.
[7] ALiCE – Associate Laboratory in Chemical Engineering, Faculty of Engineering, University of Porto, Rua Dr. Roberto Frias, 4200-465 Porto, Portugal.
* Corresponding author: jmfrocha@fc.up.pt, joao.rocha73@gmail.com

2 billion of people (25.9% of the world's population) (WHO, 2020). Novel solutions for a sustainable food security future, without compromising food safety, consists of hunger eradication, responsible food production with sustainable land and water use, climate change minimization, advanced food preservation methodologies and food wastage minimization.

Biopreservation of bread is supposed to be a promising niche in contributing to the shelf-stability and food security. Bread is a staple food for humans and for ages has been the most dominant source of nutrients in diets. Spontaneous sourdough fermentation [mainly consortia of endogenous acid-tolerant yeasts and lactic acid bacteria (LAB)] is the earliest breadmaking biotechnology, whereas the baker's yeast (*Saccharomyces cerevisiae*) fermentation was industrialized in the early 20th century (Shiekha, 2015). A total 9 billion kg of baking products are annually produced with an average per capita consumption of 41–303 kg (Cho and Peterson, 2010). Almost 68% of North Americans were reported to eat bread or related products six times per two weeks (Collar, 1991). Despite a medium growth rate from 2007 to 2016, the bread shares a global revenue of nearly $358 billion in 2016 (Garcia et al., 2019a). It is also an efficient source of energy and, for instance, a source of iron, calcium, proteins (7.5–7.8%) and various vitamins (Rosell et al., 2011; Khanom et al., 2016; Saranraj et al., 2012).

Most global diets, like the European, lack partially fibers due to cereal adjustment. Thus, commercial bakery products are ideal for fiber addition. Nevertheless, due to the limited shelf-life and staling phenomenon, the quality and palatability of bread deteriorates, resulting in mostly undesirable physical, biochemical, sensorial and microbial changes (Garcia et al., 2019a,b). In packaged bread and related baking items, mold deterioration (which includes their mycotoxins) are the primary factor contributing to significant negative financial losses (Ju et al., 2018; Melini et al., 2018; Pinilla et al., 2019). In wheat-based bakery products, deterioration have been attributed to several fungal genera, chiefly *Penicillium* spp., *Aspergillus* spp., *Cladosporium* spp. and *Neurospora* spp. (Ashiq, 2015). A quantitative study reported that mold infestation caused losses in the production section and in the final product transportation (Goryńska et al., 2020).

Bacillus genus, such as *Bacillus cereus*, *Bacillus subtilis* among other species, are known as foodborne pathogens. Negative impacts in the human health and development of multi-drug resistance of *B. cereus* is suggestive of the need to search for novel antimicrobial compounds (Peng and Yuan, 2018). In baking industry, *B. subtilis* is mainly responsible for the bread spoilage because of the endospores that resist to the employed high baking temperature. *Bacillus subtilis* can develop itself in the bread loaf within 35–48 h and release volatile compounds including acetaldehyde, isovaleric-aldehyde, acetoin and diacetyl, which originates off flavors and color changes (Melini et al., 2017). Bacteria also produce amylases and proteases that degrade bread crumb. In addition to *B. cereus* and *B. subtilis*, other species involved in food spoilage are *Bacillus pumilus*, *Bacillus amyloliquefaciens*, *Bacillus megaterium* and *Bacillus licheniformis* (Lavermicocca et al., 2016). A significant proportion of food is wasted around the world due to these spoilage microorganisms. For instance, 34.7% of total produced bread was wasted in Germany in 2015 (Alpers et al., 2021). An estimated 10% of produced bread was also wasted in Brazil and

132 *Novel Approaches in Biopreservation for Food and Clinical Purposes*

presumably in other tropical countries (Freire et al., 2011). Moreover, mycotoxin infection in these foods is also a global apprehension, affecting almost 25% of all grain yields (Rodrigues et al., 2012a,b; Wagacha et al., 2008).

Nowadays, one of the foremost concerns in food industry, from the consumer's perspective, is the quality, safety, freshness and enhanced shelf-stability of foodstuffs. Using organic preservatives to achieve these requirements is quite viable (Luz et al., 2018). Prolonged exposure to chemical preservatives may pose health threats, so the use of natural antimicrobial preservatives is highly acknowledged (Erickson et al., 2017). Nevertheless, the use of chemical preservatives (as antimicrobial agents), like sorbates and propionates to extend shelf-life of bakery products, is of considerable interest (Belz et al., 2012; da Cruz Cabral et al., 2013). Therefore, the biopreservation of bread and other baking goods is an important tool towards limitation of the economic losses triggered by the spoilage by yeasts, molds and bacteria, along with minimizing health concerns (Ghabraie et al., 2016). The organic biopreservatives may be exploited to inhibit microbial propagation while improving food quality. It is a radical concept for preserving perishable foods with generally recognized as safe (GRAS) status. Several studies report the scope of biopreservation. The beneficial bacteria and fermentation products are generally selected in the process to reduce viable pathogens and control food spoilage (Rahman et al., 2022). This chapter results from a comprehensive revision of the literature to integrate all notable aspects of bread spoilage and methods of biopreservation to extend the shelf-life and safety of bakery products.

Microbial composition of flours and bread

The quality of flours is crucial in breadmaking and depends significantly on the quality of raw cereals. Endogenous cereal grain microbiota extends to the microbiota of flours and, partially, sourdough and is influenced by the type of grain, grain quality as well as geographical region, and method of cultivation, transport and storage conditions of grains (Lhomme et al., 2016; Minervini et al., 2015). Most frequently detected field molds (filamentous fungi) considered as grain contaminants are *Alternaria* spp. (*A. alternata*), *Fusarium* spp. (*F. culmorum* and *F. gramineraum*), *Cladosporium* spp. (*C. cladosporioides*) and *Helminthosporium* spp. (Laca et al., 2006; Bensassi et. al., 2011). Molds with less incidence includes *Eurotium* spp., followed by *Aureobasidium* spp., *Endomyces* spp., *Rhyzopus* spp., *Absidia* spp., *Scopulariopsis* spp., *Trichoderma* spp., *Epicoccum* spp., *Mucor* spp. and *Wallemia* spp. (Berghofer et al., 2003; Weidenbörner et al., 2000; Gashgari et al. 2010; Cabañas et al., 2008; Sperber et al., 2003). During storage of cereal grains, the molds *Aspergillus* spp. (*A. flavus, A. parasiticus, A. candidus* and *A. versicolor*) and *Penicillium* spp. (*P. verrucosum*) become dominant (Berghofer et al., 2003).

The most important microbial risk in the pre-harvest and postharvest period of grain maintenance is associated with the synthesis of their secondary metabolites such as mycotoxins. According to the annual report of the Rapid Alert System for Food and Feed (RASFF) from 2019, mycotoxins together with the pathogenic microorganisms belong to the group of most frequently reported hazards (European Commission, 2020). Worldwide scale investigations conducted by el

Khoury and Atoui (2010) showed that 25 to 40% of cereals are contaminated by mycotoxins. Mycotoxins are thermally stable and persistent and, therefore, it is important to prevent fungus to metabolize them, not only during grain production in the field but also during storage, transportation, processing and post-processing steps.

In general, mycotoxin levels in flours are lower than in the germ fractions. The extent of their presence depends on the cereal variety, penetration degree of the mycotoxin-producing molds, transfer of mycotoxins to the inner parts of the kernel and flour extraction rate. Knowing the incidence of the mycotoxigenic species on cereals and the agri-ecological conditions that favor their production, it is of utmost importance to decrease the impact of mycotoxins and improve food quality and safety. Nevertheless, some of the mycotoxigenic fungi will be suppressed during the ecological niche from some other antagonistic species and by unfavorable environment conditions during storage. Moreover, mycotoxigenic fungi will be inactivated by the low pH and acidic environment in baking sourdough preparation and by high temperatures used during baking. Yet, most of the mycotoxins are thermostable compounds which will persist and, consequently, will contaminate the flours and bread (Milani et al., 2014; Benkerroum, 2020).

Organic flours are more susceptible to fungal contamination and mycotoxins compared to the non-organic ones. Also, flour packed in plastic packages is observed to retain more fungi and mycotoxins than flour packed in paper bags (Sacco et al., 2020). The most relevant mycotoxins for public health include the groups of aflatoxins (AF), ochratoxins (OT), trichothecenes (T), fumonisins (F) and ergot alkaloids (Hussein, 2001). There are around 13 types of aflatoxins structurally belonging to the group of difuranocoumarins, which are synthesized via the polyketide pathway by some species of the genus *Aspergillus* distributed in the section *Flavi* (B- and G-type aflatoxins), section *Ochraceorosei* (aflatoxins B1 and B2) and section *Nidulantes* (aflatoxin B1). Species of section *Flavi*, primarily *A. flavus* and *A. parasiticus*, are most frequently associated with aflatoxins in wheat grains and flour (Benkerroum, 2020). Aflatoxins of greatest importance due to their high incidence and high toxicities include AFB1, AFB2, AFG1 and AFG2. However, the toxicity of other aflatoxins should not be negligible, as is the case of sterigmatocystin (an intermediate metabolite of aflatoxins B1 and G1), because they can readily invert to the most potent AFB1 or act as intermediates for the biosynthesis of more toxic compounds (Yu et al., 2004).

Optimal temperature for mold to produce aflatoxins is 28 to 33°C and a water activity (a_w) of 0.99 (Mannaa and Kim, 2017). The greatest incidence of aflatoxins is observed in developing countries from the tropical and sub-tropical regions such as African countries, South Asia and South America, where the growth of toxigenic filamentous fungi is mostly favored by the climatic conditions (di Stefano, 2019; Benkerroum, 2020). The most toxic AFB1 shows hepatotoxic, genotoxic, carcinogenic, immune-toxic and teratogenic effects. Its oral median lethal dose (LD_{50}) is 7.2–17.9 mg/kg and possess the greatest toxicity, being followed by AFG1. Maximum acceptable levels in grain and flour established by the European Committee Regulations (ECR) are 4 ppb for total AF and 2 ppb of AFB1 (CR 1881/2006). Numerous studies have shown that AF are heat resistant, so the temperature used during breadmaking will not reduce the aflatoxins (Hwang and Lee, 2006; Nazhand et al., 2020; Samarajeewa et al., 1990). Nonetheless, there are studies showing that aflatoxins are sensitive to fermentation and existent activate enzymes during fermentation are found to be effective in their reduction (Giray et al., 2007; Uma et al., 2010), as well as by the rise of moisture content and acid concentration (Méndez-Albores et al., 2009; Plessas et al., 2005). This may explain the importance of sourdough fermentation as an effective natural biopreservation method to reduce or vanish mycotoxins in baked goods.

Some strains from the genus *Aspergillus* and *Penicillium* produce ochratoxins in grains and flours, such as is the cases of *Penicillium verrucosum*, *Aspergillus ochraceus*, *Aspergillus carbonarius*, *Aspergillus niger* and some of their closely related species (WHO, 2006). *Penicillium verrucosum* is the most prevalent species in Europe and Canada. It grows at temperatures below 30°C (optimum at 25°C) and $a_w > 0.80$ (optimum at $a_w = 0.95$) (Nazhand et al., 2020). The most toxic mycotoxin from the group of ochratoxins is ochratoxin A (OTA). Its oral LD_{50} is about 20–25 mg kg^{-1}. The scientific commission of the European Union (EU) has established the maximum levels of OTA for unprocessed cereals of 5 ng/g, and for products derived from cereals of 3 ng/g (EC 1881/2006). OTA has nephrotoxic, immunosuppressive and teratogenic effect in humans and animals, and it is considered as probably carcinogenic. It is considered that 50% of human daily intake of OTA is due to the consumption of different cereal derived products, including rye and wheat breads (di Stefano, 2019). The importance of OTA in flours is due to its high stability and resistance to acidity and high temperatures (el Khoury and Atoui, 2010). OTA is only partially degraded in normal cooking conditions (Milani and Heidari, 2016). Moreover, even at baking temperatures of 250°C its destruction is not complete (Boudra et al., 1995). Thus, once the flours are contaminated, it is very difficult to absolutely remove it from bread and other baking foods. Duarte et al. (2010) investigated the occurrence of OTA in different types of bread worldwide. The mean values of positive samples of wheat bread ranged from 0.07 ng/g in Switzerland to 13 ng/g in Morocco, although they are found mostly below 0.50 ng/g (Osborne, 1980; Zinedine et al., 2007).

Numerous studies have shown the potential of lactic acid bacteria and *Saccharomyces cerevisiae* as aflatoxin decontaminating agents, that is a biopreservation method. According to Shetty and Jespersen (2006), *S. cerevisiae* and LAB such as *Lacticaseibacillus rhamnosus*, *Lactiplantibacillus plantarum*, *Limosilactobacillus fermentum*, possess high mycotoxin binding abilities. They can be used as part of the microbial starter cultures in the fermentation of food and feed or used as additives in small quantities without compromising the characteristics of the final product.

The presence of trichothecenes (T-2, HT-2, DON, DAS, FUS-X, NIV, diacetylnivalenol, neosolaniol and ZEA) and fumonisins (FB1, FB2, FB3, FB4) in flours are associated with *Fusarium* grain contamination. *Fusarium* species are considered the most prevalent toxin producing fungi in the Northern temperate zone such as North America, Europe and Asia (Creppy, 2002). *F. graminearum* and *F. culmorum* produce ZEA, NIV, FUS-X and DON. T-2 is produced by *Fusarium sporotrichoides*, while fumonisins are produced by *Fusarium verticillioides*.

Other fungi such as *Fusarium proliferatum*, *Fusarium anthophilum*, *Fusarium nygamai* as well as *Alternia alternata* f.sp. *lycopersici* may also produce fumonisins (Bennett, 2003). *Fusarium* spp. mycotoxins are synthetized mainly during the pre-harvest, although some toxin synthesis may occur during the storage period. Water activity greater than 0.9 and temperatures of 25–30°C affects colonization of *Fusarium* spp. and production of FB1 and FB2 (Marin et al., 1995). FB1 has been classified as a "2B" carcinogen (IARC, 1993). The chemical structures of fumonisins are similar to sphingolipids and they may alter sphingolipid biosynthesis and play a significant role in carcinogenesis and DNA damage (Domijan, 2007). The study of Bryla et al. (2017) showed that thermal processing up to 250°C practically does not degrade fumonisins. Other group of mycotoxins found in cereal flours called trichothecenes are non-volatile, low molecular-weight sesquiterpene epoxides, and stable to heating and at neutral and acidic pH (Rocha, 2005). They are well-known inhibitors of protein synthesis (including DNA and RNA synthesis), inhibitors of mitochondrial function and show cell division and membrane effects (Eriksen, 2004).

Another group of toxins of historical importance are the ergot alkaloids. They are synthetized by the fungi from the genus *Claviceps*, with *Claviceps purpurea* being the most frequently found species. Other species capable to synthetize ergot alkaloids and found with less incidence include *Claviceps fusiformis*, *Claviceps paspali* and *Sphacelia sorghi*. Historically, the fungus was known since the Middle Ages when it caused serious poisoning and ergotism of people at that time known as "St. Anthony's fire". Most countries established very strict grain standards, not permitting grains which contain ergot to reach commercial food channels. Yet, studies from different regions worldwide indicate the presence of ergot alkaloids in flour products and with great incidence (Agriopoulou, 2021). Maximum permitted level of ergot sclerotia content in unprocessed cereals (except for corn and rice) for human consumption set by the Commission Regulation (EC) No. 2015/2940 is 0.5 g/kg. The fungus develops sclerotia, which may be harvested with the grain and if are processed together into flours may pose serious threats for human health since it contains toxic ergot alkaloids. The most important ergot alkaloids belong to the chemical groups of clavines, lysergic acid, lysergic acid amides and ergopeptides. They exhibit a wide spectrum of structural diversity with high effectiveness in biological activity. Today, various ergot alkaloids are introduced into therapy and are extensively used in the treatment of migraine, high blood pressure, cerebral circulatory disorder, parkinsonism, acromegalia, etc. The pathogen is prevalent in semi-arid regions such as India and Africa. The contamination of cereal flours is greatly avoided by removal of sclerotia before milling. The stability of ergot alkaloids during baking of bread was investigated by Merkel (2012) and Bryla (2019). Both found reduction of ergot alkaloids by 59 and 22%, respectively.

The hygienic properties of flours are significantly improved during milling process—when germ, pericarp and seed coats are removed from the endosperm—because most of the microbial contaminants are concentrated in the outer grain layers of the kernel (Pagani et al., 2013). This is a good example of a physical biopreservation arisen from cereal processing. Therefore, flours are considered as microbiologically safe since its low water activity cannot support the growth of microorganisms. Nevertheless, during the milling process flours will harbor microorganisms from

136 Novel Approaches in Biopreservation for Food and Clinical Purposes

the inside of the milling equipment and air, especially if the sanitary conditions and cleaning of the equipment are poor. The study of Stolz (1999), among others, found out a total microbial viable counts of wheat grains to be around 10^4–10^7 colony-forming units (CFU)/g (or 4–7 log CFU/g). After milling, the microbial viable counts of flours decreased to such values as 10^4–10^6 CFU/g. Flours will also harbor microorganisms from the immediate surrounding during the storage period. Molds, yeasts, and aerobic psychrophilic, mesophilic and thermophilic bacteria are the most common microorganisms found in cereal flours (Rocha and Malcata, 1999, 2012, 2016a, 2016b).

The most important indicators for microbiological contamination of flours and bread are the total aerobic viable counts (AACC Method 42-11), yeast and mold viable counts (AOAC Official method 997.02), coliform/*Escherichia coli* (AACC Method 42-15), *Salmonella* spp. (AACC Method 42-25B) and *Staphylococcus aureus* (AACC Method 42-30B). Studies from different geographical regions indicated that the viable counts of molds in flours ranged between 2.0–5.3 log CFU/g, yeasts ranged between 1.3–3.7 log CFU/g and aerobic mesophilic bacteria ranged between 1–7 log CFU/g. The pathogenic bacteria *Salmonella* spp. and shiga toxin-producing *E. coli*, are reported in very low viable counts but still may pose serious health risks and cause illnesses (Sabilon et al., 2015; Laca et al., 2006; Aydin et al., 2009; Neil et al., 2011; Berghofer et al., 2003). Several cases of foodborne outbreak with *Salmonella* spp. and shiga toxin-producing *E. coli* indicated the flours as a source of contamination with these pathogenic bacteria (McCallum et al., 2013; Dack, 1961; Eglezos, 2010) and resulted in increased awareness and demand for microbiological specifications of flours (Neil et al., 2011).

Care should be taken in case of consumption of raw cereal flour products, and the use of bioprotective agents as safer alternative of synthetic preservatives in ready to bake products should be considered. Some gram-negative rods such as *Acinetobacter*, *Comamonas*, *Enterobacter*, *Erwinia*, *Pantoea*, *Pseudomonas* and *Sphingomonas* species were also found in flours in lower incidence (Ercolini et al., 2013; Rocha and Malcata, 1999, 2012). Most viable microbial contamination of flours is effectively eliminated during the baking process, yet some thermophilic spores from the genus *Bacillus* may survive at high temperatures and low water activities in dormant state and cause ropiness of bread (Novotni et al., 2020; Rocha, 2011). Rope spores can be easily carried out along with the agriculture commodities and food supply chain, leading to cross-contaminations of flours and other ingredients. Nonetheless, flours (as major constituent of baked goods) have a potential to introduce the greatest number of rope spores in baking dough than any other bread ingredient. *Bacillus subtilis* and *Bacillus mesentericus* are commonly isolated from rope spoiled bread and contaminated flours. Other strains less often found are *Bacillus amyloliquefaciens*, *Bacillus velezensis*, *B. licheniformis*, *B. megaterium* and the pathogenic *Bacillus cereus* (Cauvain, 2015; Berghofer et al., 2003; Aydin et al., 2009). Rope spores are sensitive to low pH, consequently acidification that occurs during the fermentation of bread doughs to give rise to sourdough/leavened dough prevents the growth and activity of rope spores and other acidophobes (Rosenquist and Hansen, 1998; Rocha, 2011; Rocha and Malcata, 1999, 2012, 2016a, 2016b).

Biopreservation in Flours and Bread 137

Since rope spores are thermophilic and optimum temperatures for their growth are from 35 to 45°C, it is obvious that this kind of bread spoilage occurs particularly in summer when the climatic conditions are favorable for the growth of bacteria.

Most common contaminants of flours and bread are molds. Mold spores present in flour may survive for several years, although water activity of dry flours is too low to support growth or toxin production in mycotoxigenic molds. In some cases, changes in moisture content of only 1 or 2% may stimulate molds to grow and produce mycotoxins (Daftary et al., 1970; Eyles et al., 1989). As it was mentioned before, cereal grain microbiota from pre-harvest and postharvest period affects further the microbiota of cereal flours but it does not affect the microbiota of bread since the majority of microorganisms (with the exception of some thermophilic microorganisms) will be inactivated during dough preparation and baking. Consequently, bread loafs after several hours after baking are free of microorganisms and their low water activity ($a_w < 0.6$) are not suitable towards microbial growth (Cauvian, 2015). Nonetheless, the exposure to relative humidity from the surrounding environment (and/or water condensation) during storage, may result in increasing water activity, thus creating suitable conditions for spoilage microorganisms to thrive. Bread becomes contaminated with molds and other spoilage microorganisms during the post-baking handling, such as cooling, slicing, packaging and storage. Mold spores are mostly present in the surrounding atmosphere and on the equipment. *Aspergillus* spp. (*A. niger*, *A. flavus*, *A. candidus*, *A. glaucus*), *Penicillium* spp. (*P. commune*, *P. crustosum*, *P. brevicompactum*, *P. chrisogenum*, *P. paneum* and *P. caneum*). *Eurotium* spp. (*E. amstelodami*, *E. chevalieri*, *E. herbariorum*, *E. rubrum* and *E. Repens*), *Mucor* spp., *Rhizopus* spp. (*R. stolonifer*), *Cladosporium* spp. and *Neurospora sitophila* are found as prevalent mold spoilage contaminants of bread (Císarová et al., 2018; Saranraj and Sivasahthivelan, 2015; Saranraj and Geetha, 2012). The xerophilic ascomycetes fungus *Eurotium* spp. is probably the first mold that will grow on bread since it is capable to develop at low water activity ($a_w = 0.70$) (Beuchat, 1983; Dao and Dantigny, 2011). Other spoilage fungi will thrive when a_w increases above 0.80 (Beuchat, 1983). Temperature has also a great influence on the types of microorganisms that will developed during bread storage and the rate of their occurrence. Spicher (1980) in his research have found that at ambient temperatures (20–25°C), *Aspergillus* spp. and *Erotium* spp. thrive more rapidly while, at lower temperatures, *Penicillium* spp. are more dominant. Also, sourdough breads are more likely to be contaminated with *Penicillium* spp. since the pH (4.5–5.1) of such product is favorable for *Penicillium* spp. development. *Penicillium roqueforti* followed by *Penicillium commune* are prevalent species in sourdough bread post-baking contamination partly because of their tolerance of low pH (Spicher, 1980).

Regarding yeasts, they become inactive at temperatures above 55°C, so bread after baking is free of viable yeasts. Yeast spoilage of bread usually results from post-baking contamination, slicing machines, bread coolers, conveyor belts and racks (Saranraj and Geetha, 2012). Mainly two types of yeasts such as fermentative and filamentous yeasts contribute in bread spoiling. The most troublesome is the filamentous yeast *Hyphopichia burtonii* referred as "chalk mold", since it grows in form of a white powdery patches on the bread surface. *Hyphopichia burtonii* grows rapidly on bread usually before the molds start to thrive (Saranraj and Sivasagthivelan, 2015). Other yeast species commonly isolated from contaminated bread are *Saccharomycopsis fibuligera* and *Wickerhamomyces anomalus* (Barnett et al., 2000; Deschuyffeleer et al., 2011).

Most of the yeasts found in cereal flours are beneficial and together with LAB are of fundamental importance for fermentation and sourdough fermentation, in particular (Ağagündüz et al., 2021, 2022; Bartkiene et al., 2020, 2022; Küley et al., 2020; Matukas et al., 2022; Mockus et al., 2022; Petkova et al., 2022; Păcularu-Burada et al., 2020, 2021; Rathod et al., 2022; Sharma et al., 2021; Trakselyte-Rupsiene et al., 2022; Tolpeznikaite et al., 2022; Yilmaz et al., 2022a, 2022b; Zokaityte et al., 2020). As facultative anaerobes, yeasts are capable of both fermentation and respiratory metabolisms. The process of fermentation is important in bakery industry and is carried out mostly under anaerobic conditions; the oxygenation of dough mainly happens during dough kneading. Yeasts use carbohydrates as their main source of carbon and energy but there are also yeasts that may use some complementary carbon sources (Gobetti and Ganzle, 2013). The preferred carbohydrate for yeasts is glucose. Carbohydrate metabolism in yeasts is regulated by different mechanisms, depending on the type of the carbohydrate, genus, species and strains, as well as according to the different environmental conditions. Well documented studies found in the literature refers to the mechanisms which regulate sugar metabolism in yeasts, *viz.* Pasteur, Kluyver, Custers and Crabtree effect, glucose or catabolite repression and glucose or catabolite inactivation (Barnett and Entian, 2005).

Carbon dioxide (CO_2) which is released by yeasts (also by heterofermentative LAB, in minor amounts) as a product of the fermentation process significantly contributes to dough expansion. In dough, the gluten and similar proteins are stretched into a viscoelastic film that retain gas, which is essential to produce a leavened dough (Rocha, 2011). In sourdough-type fermentations, the synergism of yeasts and LAB contribute to the final flavor and aroma of bread (Corsetti et al., 2007; Rocha, 2011; Rocha and Malcata, 2016a, 2016b; Russo et al., 2017). Yeasts are common microbiota in soil, plant surface, air and, especially, sugary media. In cereal flours they exist in dormant state and become metabolically active when the water activity reaches values of 0.90–0.94, which is optimal for their growth. Yet, some osmiophilic yeasts may grow at water activities as low as 0.60 (Marriott et al., 2018). Yeasts that belong to genus *Kazachstania, Saccharomyces, Cryptococcus, Pichia, Rhodotorula, Torulaspora, Trichosporon* and *Sporobolomyces* are the most frequent microbiota on cereal flours. The most dominating strains are *Saccharomyces cerevisiae, Kazachtania exigua, Kazachtania humilis* and *Candida holmii* (de Vuyst and Neysens, 2005; Corsetti et al., 2001; Galli et al., 1987; Rocha and Malcata, 1999). Metabolic activities of yeasts depend on the available nutrients/sugars of flours and its quality traits, such as ash content, enzymatic activity, starch damage, and coupled with environmental conditions. The type of cereal and distinct conditions will result in predominance of different yeast strains in spontaneous sourdoughs (Sun et al., 2019). Richard-Molard (1994) in his research found that up to 95% of the fermentable carbohydrates present in cereal flours may be transformed to ethanol and carbon dioxide by the baker's yeast *S. cerevisiae*. The remaining fermentable carbohydrates are associated in secondary fermentation reactions and leads to the formation of carbonyl compounds, short chain fatty acids and higher alcohols. Different yeast strains contribute to the formation of distinct compounds including volatile terpenes, aldehydes, ketones, alcohols, organic acids, esters, furans, pyrroles, pyrazines and lactones (Canesin and Cazarin, 2021). Some of them are beneficial and besides contributing to the overall taste and flavor, also contribute to the improvement of shelf-life of final bakery products, due to their antifungal and antioxidant properties. Unlike these favorable metabolites, Mildner-Szkudlarz et al. (2019) identified the formation of acrylamide, a well-known neurotoxin with potential carcinogenic effect on human's health as derivative from the Maillard reaction between amino acids and reducing sugars. Furthermore, authors found that addition of 0,1% of polyphenols such (+)-catechin, quercetin, gallic, ferulic and caffeic acids significantly inhibit the formation of acrylamide.

The effect of fermentation time, yeast amount and steaming time on the volatile profile of Chinese steamed bread was examined by Xi et al. (2021). According to them, the higher amount of yeast strain used and longer fermentation time resulted in the formation of yeast-metabolism-derived compounds, such as 2-methyl-1-propanol, 3-methyl-1-butanol, phenethyl acetate, hexyl acetate and 2-octanone. Prolonged fermentation time decreased the rate of production of lipid oxidation compounds such octanal, nonanal, decanal, (E)-2-nonenal and (E,E)-2,4-decadienal (Xi et al., 2021). Apart from the well-known *Saccharomyces cerevisiae* strains, favorable yeast starters in sourdough which improve the flavor and the taste include *Kazachstania exigua, Pichia kudriavzevii, Pichia norvegensis* and *Wickerhamomyces anomalus* (Salim-ur-Rehman et al., 2006). A recent study on *Saccharomyces cerevisiae* strains confirmed that they became more popular due to the fact that their employment improves the quality, safety and flavor of the final bread (Lee at al., 2022).

As mentioned earlier, the presence of yeasts in cereal flours and sourdoughs is closely related to the existing LAB. Most frequent LAB in flours are obligate homofermentative, facultative heterofermentative and obligate heterofermentative from the genera *Enterococcus, Lactobacillus, Leuconostoc, Pediococcus* and *Weissella*. Strains *Lactobacillus graminis, Lp. plantarum, Pediococcus pentosaceus, Weissella cibaria, Leuconostoc pseudomesenteroides* and *Leuconostoc citreum, Leuconostoc sakei, Leuconostoc mesenterioides* and *Fructolactobacillus sanfranciscensis* are regarded as the most frequent populations of wheat flour (Minervini et al., 2018; Celano et al., 2016; Alfonzo et al., 2013). Another study conducted by Corsetti et al. (2007), showed the prevalence of *Enterococcus faecium* in nonconventional flours (amaranth, chickpea, corn, rice, quinoa and potato flour). Grain and milling equipment provide most of the microbial inoculum of LAB in flours (Rizzelo et al., 2015; Minervini et al., 2015). Also, several studies showed that *Fr. sanfranciscensis* may be carried by fruit flies and weevils (Minervini et al., 2015; Weckx et al., 2019).

The type of available sugars and different metabolic pathways of yeasts and LAB are key-factors to determine the microorganisms that will be associated in spontaneous sourdough fermentations and to establish stable microbial consortia. *Saccharomyces cerevisiae, Kazachtania exigua, Kazachstania humilis, Pichia kudriavzevii, Pichia norvegensis, Wickerhamomyces anomalus* and *Issatchenkia orientalis* are most common yeast species associated with LAB in sourdough

140 *Novel Approaches in Biopreservation for Food and Clinical Purposes*

environment (Yazar and Tavman, 2012; de Vuyst and Neysens, 2005). Because of their highly adapted carbohydrate metabolism, heterofermentative LAB are dominant strains in sourdoughs and the ratio between yeast and LAB is generally 1:100 (Ottogalli et al., 1996; Rocha, 2011). Maltose-negative yeast *Kazachstania exigua* (*Saccharomyces exiguus*) is associated with maltose-positive *Fructolactobacillus sanfranciscensis* in San Francisco sourdough (Sugihara et al., 1971). In their research, Paramithiotis et al. (2005) found that the association between the yeasts *S. cerevisiae* and the LAB species *Fr. sanfranciscensis*, *Levilactobacillus brevis* and/ or *Lactiplantibacillus plantarum* is the most common association between yeasts and LAB species in sourdough.

During dough preparation, the technological processes applied, microbial contamination of surrounding environment and bakery equipment will also influence the microbial composition of the baking dough. Likewise yeasts, LAB in flours exist in dormant state due to the low water activity (4.0–5.0) and become metabolically active after the water is added and mixed and kneaded to result into the unfermented dough. More than 50 LAB isolated from the sourdough have been reported in the literature and generally belong to *Lactobacillus* spp., *Leuconostoc* spp., *Pediococcus* spp. and *Weissella* spp. (Corsetti and Settanni, 2007; Rocha and Malcata, 1999). According to a study of Yazar and Tavman (2012), *Fr. sanfranciscensis*, *Lactiplantibacillus plantarum*, *Limosilactobacillus pontis*, *Furfurilactobacillus rossiae*, *Limosilactobacillus reuteri*, *Lactiplantibacillus pentosus*, *Levilactobacillus brevis*, *Companilactobacillus alimentarius*, *Furfurilactobacillus siliginis* and *Companilactobacillus nantensis* are the most frequently isolated LAB.

It is the type of cereal and the quality of flours that will have most influence on the metabolic activity of lactobacilli and other LAB. The ability of different LAB to metabolize different polysaccharides and proteins and, concomitantly, produce a variety of beneficial substances like acids, viscous exopolysaccharides (EPS), volatile aromatic and bioactive compounds, may also determine the biotechnological interest for their application in different processes in the agri-food industry (Wang et al., 2021). The key sourdough obligate heterofermentative LAB *Fr. sanfranciscensis* possesses the ability to ferment hexoses (glucose, fructose, galactose, etc.) to lactic and acetic acids and CO_2, and pentoses (arabinose, xylose, etc.) to lactic and acetic acids. *Fr. sanfranciscensis* preferentially ferments maltose rather than glucose and produce large amounts of lactic and acetic acids (de Vuyst and Neysens, 2005). Sourdough strains of *Fr. sanfranciscensis* are characterized by a rather long latency phase (lag phase), tolerance to acidity and high maximum-acidification rate, such as high production of lactic and acetic acids. They also possess the highest variability among the heterofermentative species present in sourdoughs (Ercolini et al., 2013). Wu et al. (2022) in their research on sourdough bread aroma found that *Lactococcus lactis* leads to the formation of very potent aroma active compounds, such as (E,E)-2,4-decadienal, 2-pentylfuran, 1-octen-3-ol, 3-methylthio-1-propanol and (E)-2-nonenal. The comparison of the effects when using a single strain of yeast, combination of yeast and acetic acid bacteria, yeast and lactic acid bacteria and the mixture of yeasts, acetic and lactic acid bacteria indicated improved aroma of the bread and gave rise to steamed bread with a better quality through synergism of metabolic pathways in the sourdough fermentation process (Li et al., 2022).

According to the work of de Luca (2021), the fermentation quotient (FQ, molar ratio between lactic and acetic acids) affects the aroma profile and it is also relevant for the structure of final products. The significant impact of the microbial diversity on volatile compounds in steamed bread prepared with traditional Chinese microbial starters was studied by Suo et al. (2020). Their results indicated that predominant bacteria in studied samples were the genera *Pediococcus* and *Lactobacillus*, while the dominant fungi belong to the order Saccharomycetales with the genus *Hyphopichia* being the most prevalent. The presence of higher number of volatile compounds were found in the samples prepared with traditional Chinese starters compared to the sample prepared with a commercial yeast strain (Suo et al., 2020).

Research data from the literature about flour and bread microbiota indicate that some microorganisms, especially those from the orders Saccharomycetales and Lacobacillales, are of crucial importance in breadmaking. They are capable to synthetize different compounds that contribute to rheological and sensory properties of sourdough, making influence on the final quality and sensory characteristics of bread. Indeed, they are responsible for the most diverse types of bread with the vast varieties of rheological characteristics and tastes that occur worldwide. Lactic acid bacteria, as bipopreservatives, are of primary importance even more because of their ability to synthetize bioactive compounds and contribute to the shelf-life extension of breadmaking products.

Biopreservation strategies for enhanced bread safety and extended shelf-life

Protective microbial cultures

Microbial spoilage of bread is a major concern during storage. Apart from the economic losses, bread spoilage may pose a safety hazard for human's health. Traditionally, some chemical preservatives and physical methods have been used to reduce undesired microorganism to develop but strong societal demand for clean label foodstuffs (including less processed and preservative-free food), impose the use of natural alternatives, such as antagonistic microorganisms and/or their inherent antimicrobial compounds (Leyva Salas et al., 2017). They may act indirectly by changing pH or osmotic pressure, or directly by producing numerous antimicrobial compounds by different and complex metabolic pathways. In fact, pH changes are a consequence of the microbial metabolism.

Over the past years, strains from various microbial species and isolated from various food sources have been identified to possess antimicrobial properties (Ağagündüz et al., 2021; Bartkiene et al., 2020, 2022; Küley et al., 2020; Păcularu-Burada et al., 2020, 2021; Rathod et al., 2022; Sharma et al., 2021, Trakselyte-Rupsiene et al., 2022, Yilmaz et al., 2022; Zokaityte et al., 2020). Recently, the isolation of new protective cultures has been extended to marine environments (Wang et al., 2015; Vero et al., 2012). To be used as a bioprotective culture, microorganisms and their metabolites must be non-toxic to humans, have a wide antimicrobial spectral range, be suitable in relatively low doses and do not impair product quality and sensory characteristics at the proposed level of use (Rahman et al., 2022).

142 *Novel Approaches in Biopreservation for Food and Clinical Purposes*

Fermentation is one of the most common and effective forms of food biopreservation and represents one of the oldest processes used by humankind. The antimicrobial capacity of baking sourdough is mainly attributed to the low molecular-weight LAB metabolites such as organic acids (acetic, γ-aminobutyric, butyric, benzoic, formic, fumaric, lactic, propionic, phenyllactic, hydroxyl-phenyllactic and sorbic acids), cyclic dipeptides, propionate, phenyl-lactate, hydroxyphenyllactate, phenyllactic acid, 3-hydroxy fatty acids, exopolysaccharides (glucans, fructans and fructooligosaccharides), bioactive peptides such as bacteriocins (nisin, pediocins, lacticins, enterocins and many others) and some bacteriocin-like inhibitory substances (BLIS), as well as diverse antagonistic compounds (acetoin, carbon dioxide, diacetyls such as butanediole or butane-2,3-dione, aldehydes, esters, carbonyls, ethanol and other alcohols, hydrogen peroxide, fatty acids, reuterin and reutericyclin) (Muhialdin et al., 2011; Rocha, 2011; Reis et al., 2012; Oliveira et al., 2014; Crowley et al., 2013; le Lay, 2016a; Novotni, 2020).

Lactic acid bacteria as protective cultures can be naturally present in sourdough or introduced as pure cultures (so called, microbial starter cultures or simply starters). Biopreservation represent a link between the (biological process of) fermentation and preservation of bread and contribute to the extension of shelf-life of bread and improved food safety through microorganisms and/or their metabolites (Paul Ross et al., 2002). Numerous studies have reported the antimicrobial properties of these Gram-positive bacteria—generally recognized as safe microorganisms—which do not contain lipopolysaccharides attached to the cell membrane and thus avoid the generation of an anaphylactic shock when administered in humans. As a matter of fact, LAB have a long history of safe use in food production and preservation (Rosell et al., 2011).

The most studied and known LAB are the antifungal *Lactobacillus* spp., *Pediococcus* spp. and *Leuconostoc* spp. (Schnürer and Magnusson, 2005). Several LAB species such as *Lactobacillus amylovorous, Limosilactobacillus fermentum, Lactobacillus acidophilus, Lactobacillus paralimentarius, Levilactobacillus brevis, Furfurilactobacillus rossiae, Levilactobacillus hammesii, Lacticaseibacillus paracasei, Pediococcus pentosaceus, Lactiplantibacillus pentosus, Lactiplantibacillus plantarum, Lactobacillus reuteri, Lacticaseibacillus rhamnosus, Pediococcus acidilactici*, as well as *Leuconostoc citreum* have been proposed in the literature as starter protective cultures to enhance the shelf-life of bread and other baking goods (Cizeikiene et al., 2013; Axcel et al., 2016).

Strains of *Lactiplantibacillus plantarum* are widely investigated and have been found to be effective against the molds (filamentous fungi) *Penicillium* spp., *Aspergillus* spp., *Rhizopus* spp., *Alternaria* spp., *Fusarium* spp. and *Erothium* spp. This LAB strain is effective against one of the more resistant fungi to chemical preservatives, the *Penicillium roqueforti* as well as against *Penicillium expansum, Aspergillus niger, Fusarium culmorum*, etc. The extension of the shelf-life of bread using *Lp. plantarum* as a protective culture is doubled or extended to at least 7 days after baking (Coda et al., 2011; Ryan et al., 2008; Sadeghi et al., 2019; Axel et al., 2015; Cizeikiene et al., 2013; Coda et al., 2011). Subsequent studies have showed that *Lp. plantarum* together with some other lactobacilli used as protective cultures such as *Li. ruteri, Co. alimentarius, Fu. rossiae, Lo. coryniformis, Lc. rhamnosus*

Biopreservation in Flours and Bread 143

and *Li. fermentum*, convert the amino acid phenylalanine to phenyllactic acid, which has antifungal properties (Lavermicocca et al., 2000; Broberg et al., 2007; Dal Bello et al., 2007; Prema et al., 2008; Rizzello et al., 2011; Wang et al., 2012; Bronsan et al., 2012). Rizzello et al. (2011) investigated the synergistic antifungal activity of *Fu. rossiae* and *Lp. plantarum* against *P. roqueforti* and observed mycelial development in the wheat germ bread sample after 21 days of inoculation with only a 10% contamination score.

Limosilactobacillus reuteri and *Loigolactobacillus coryniformis* produce reuterin under anaerobic conditions as an intermediate compound during the metabolism of glycerol to 1,3-propanediol (Talarico et al., 1988). *Limosilactobacillus reuteri* also produce negatively charged, highly hydrophobic antagonist reutericyclin which acts as a proton ionophore. Reuterin have bactericidal activities against *Staphylococcus aureus*, *Escherichia coli*, *Salmonella choleraesuis*, *Yersinia enterocolitica*, *Campylobacter jejuni* and *Aeromonas hydrophila* subsp. *hydrophila*, and bacteriostatic activity against *Listeria monocytogenes* (Arqués et al., 2004). The aldehyde group of reuterin is highly reactive and may interact with small molecules and proteins, causing oxidative stress to the cell and growth inhibition (Schaefer et al., 2010). Conversely, reutericyclin is active against Gram-negative bacteria such as *E. coli* and *Salmonella* spp. but also against Gram-positive bacteria including *Lactobacillus* spp., *Bacillus subtilis*, *Bacillus cereus*, *Enterococcus faecalis*, *Staphylococcus aureus* and *Listeria innocua* (Ganzle et al., 2004).

Bioprotective properties of *Li. ruteri* and *Lev. brevis* used as starter cultures in quinoa and rice bread was investigated by Axcel et al. (2016) and revealed the extension of the shelf-life for 2 days in quinoa and rice bread when sourdough was inoculated with *Li. ruteri*, while in the case of *Lev. brevis* the shelf-life was extended for 4 days. Moreover, the same authors observed higher concentrations of acetic, lactic and carboxylic acids in *Li. ruteri* fermented sourdoughs and concluded that these substances contribute to the antifungal affect. Axcel et al. (2015) also investigated the antimicrobial properties of *L. amylovorus* in quinoa flour and observed an extension of the bread shelf-live in 4 days. In this case, the antimicrobial properties are given to 4-hydroxyphenyllactic acid, phloretic acid, 3-phenyllactic acid and hydroferulic acid. In another study of *L. amylovorus*, nine carboxylic acids including three cinnamic acid derivatives, and D-glucuronic and salicylic acids were isolated as antifungal compounds (Ryan et al., 2011). *Pediococcus pentosaceus* was found to be effective against rope spores and extended shelf-life of bread in 6 days (Cizeikiene et al., 2013). In the same study, *Pediococcus acidilactici* and *Pediococcus pentosaceus* sprayed on the surface of the bread inhibited growing of fungi until 8 days of storage in polythene bags. Black et al. (2013) evaluated the antimicrobial properties of fatty acids in *Lev. hammesii* and found that this LAB converts linoleic acid in sourdough and the resulting monohydroxy octadecenoic acid exerts antifungal activity in bread. *Fu. rossiae* and *L. paralimentarius* used in bread and panettone production were found to prevent the growth of *Aspergillus japonicus* from 11 to 32 days (Garofalo et al., 2012).

Other bacterial groups, such as those from the genus *Bacillus* and propionic acid bacteria (PAB) are also attracting attention from the scientific community. They produce a diversity of antimicrobial peptides some of which could also be exploited

144 *Novel Approaches in Biopreservation for Food and Clinical Purposes*

as biopreservatives. Propionic acid bacteria convert lactate to propionate, acetate and CO_2. Propionic acid and its salts are accepted as preservatives in breadmaking. *Propionibacterium freudenreichii* spp. *shermanii* was found to be effective against rope spores in bread. The antimicrobial effect depends on the temperature, moisture and duration of the fermentation (Odame-Darkwah and Marshall, 1993; Suomalainen et al., 1999). Zhang et al. (2010) investigated the co-fermentation of two active propionate providers, *Lentilactobacillus diolivorans* and *Lentilactobacillus buchneri* against *Cladosporium* spp., *Aspergillus clavatus*, *Penicillium roquefortii* and *Mortierella* spp. and observed mold inhibition for more than 12 days. In their study, le Lay (2016b) evaluated 50 propionibacteria for their antifungal properties and concluded that only *Propionibacterium freudenreichii* and *Acidipropionibacterium acidipropionici* species slightly inhibit the growth of *Aspergillus niger* and *Penicillium corylophilum* in milk bread rolls sprayed with antifungal cultures. Appearance and preservation of breads when commercial yeast extracts fermented with immobilized cells of *Propionibacterium freudenreichii* subsp. *shermani* are added in broths and bread formulations were examined by Gardner et al. (2002). The investigation showed that bread from fermented yeast extracts in combination with *Propionibacterium freudenreichii* ssp. *shermani* contained less ethanol and had a better shelf-life against mold growth than bread formulated with non-fermented yeast extract only.

Yeasts also have been involved in food preservation for millennia, through fermentation process. Their antimicrobial activity is substantially enhanced by their synergistic interactions. Yeasts constitute a large and heterogeneous group of microorganisms with antagonistic activities against undesirable bacteria and fungi due to their biological activities such as competitiveness for nutrients, production and tolerance to high concentrations of ethanol and release of antimicrobial compounds (Hatoum et al., 2012). Production of ethanol and organic acids results in the acidification of the growth medium, which is responsible for the effectiveness of yeasts as biopreservatives (Muccilli and Restuccia, 2015). In some cases, the competition for nutrients is considered as a primary mode of action against postharvest fungal pathogens (Mercier and Wilson, 1994; do Carmo-Sousa et al., 1969). The competition mechanisms are extensively studied and, among them, killer toxins seem to play a primary role (Muccilli and Restuccia, 2015). Bortol et al. (1986) in their study investigated 238 wild strains of *Saccharomyces cerevisiae* isolated from different food sources and found only four as killer toxin-producing strains. Furthermore, using hybridization with an industrial strain of *Saccharomyces cerevisiae* by protoplast fusion, they obtained strains that possess good dough-raising activities and simultaneously retain their killing effect.

Coda et al. (2011) investigated the antifungal activity of the yeast *Wickerhamomyces anomalus* and sourdough lactic acid bacteria *Lactiplantibacillus plantarum*—in co-fermentation as well as in single-culture fermentation—towards the extension of shelf-life of wheat flour bread. The outcomes showed delay of fungal contamination of bread for 28 days when both cultures were used together (in co-culture). In another study, *Wickerhamomyces anomalus* was used as a microbial starter culture in bread preparation and was investigated for baking qualities and shelf-life extension compared to the baker's yeast leavened bread. During the storage time of 10 days the number of *Penicillium paneum* colonies were found significantly lower in *W. anomalus* leavened bread. Moreover, authors identified 16 flavor compounds present in *W. anomalus* leavened bread from which phenylethyl alcohol and 2-phenylethyl acetate possessed antifungal properties. Based on these results, authors proposed *W. anomalus* as a feasible leavening agent for production of pan bread with extended shelf-life (Mo and Sung, 2014). *W. anomalus* also possessed biocontrol ability against *Aspergillus flavus* producing 2-phenylethanol, which affects spore germination, growth, toxin production and gene expression in *A. flavus* (Hua et al., 2014).

The minimal inhibition concentration of antimicrobial compounds produced by beneficial LAB and yeasts during the process of fermentation is relatively high and ranges between 0.1 and 10,000 mg/kg (Axel et al., 2017). Since the microorganisms produce antimicrobial compounds in low amount in the dough/sourdough, the opinion of the scientists is that the antimicrobial inhibitory mechanism in microorganisms used as protective cultures is due to the complex mechanisms of synergy between their metabolites (Axel et al., 2017; Ryan et al., 2009; Melini and Melini, 2018). From above it can be concluded that protective cultures especially when compatible species are used together are a promising biopreservation tool in producing natural, high-quality bread and baking goods with substantial extended shelf-life.

Microbial metabolites

Various microorganisms are associated with bread and other bakery products, which through their metabolites may contribute to the spoilage and decreasing of its quality or, conversely, may improve rheological characteristics, aroma and taste of the final product and/or increasing shelf-life. Among microorganisms closely related to bakery products, LAB are of particular importance, especially selected strains that belong to genera of *Lactobacillus*, *Lactococcus*, *Leuconostoc*, *Pediococcus*, *Streptococcus*, *Enterococcus*, *Weissella* and some fermentative yeasts which are known to synthetize primary and secondary metabolites with antimicrobial properties. These microorganisms may act as cells factories for the production of valuable chemical substances and offer natural solutions for biopreservation of bread and enhancing, at the same time, bread structure, aroma and taste. LAB and yeasts produce various organic acids from the fermentation of carbohydrates but also various bacteriocins, vitamins, ethanol, fatty acids, exopolysaccharides, aroma compounds and some other low molecular weight compounds (De Vuyst and Leroy, 2007; Florou-Paneri et al., 2013; Wang et al., 2021). The type and production amount quantity are strain specific, especially of secondary metabolites. Although much progress has been made in understanding the mode of action of these metabolites, there are still many things that remain not fully understood, even more because of the synergistic effect that they exhibit. Scientific opinion is that these synergistic actions contribute to more efficient final antimicrobial activity (Nasrollahzadeh et al., 2022).

Organic acids

Organic acids are mainly produced by LAB and yeasts during carbohydrate fermentation. Lactic (2-hydroxy propionic acid) and acetic (ethanoic acid) acids are

146 *Novel Approaches in Biopreservation for Food and Clinical Purposes*

primary metabolites and recognized as the main products of carbohydrate fermentation by LAB and acetic acid by the yeasts. Lactic acid is widely distributed in nature and exists in L- and D-form. The L-form is recognized as safe to be used as preservative (FDA; Martinez, 2013). In sourdough ecosystems, homofermentative LAB mainly produce lactic acid while heterofermentative LAB (e.g., *Fructolactobacillus sanfranciscensis*) also produce acetic acid and CO_2. Fermentative yeasts produce ethanol or acetic acid and carbon dioxide (Florou-Paneri et al., 2013; Hutkins, 2019).

During the initial stage of dough/sourdough fermentations, lactic and acetic acids are present in small concentrations. Afterwards, the concentration of lactic and acetic acids increases and rich their maximal concentration in mature dough/sourdough because much of the carbohydrates are fermented. Consequently, the pH of the medium drops and inhibit the growth of unwanted acidophilic microorganisms that pose spoilage of bakery products. At pH below 5.0, lactic acid inhibits the growth of spore forming bacteria and molds but does not affect the yeasts (Stoyanova et al., 2012).

Lactic acid bacteria and yeast may produce a vast variety of other organic acids during dough/sourdough fermentation such as oleic acid, linoleic acid, formic acid, propionic acid, sorbic acid, palmitic acid, caproic acid, phenyllactic acid, hydroxyphenyllactic acid, indole lactic acid, stearic acid, pyroglutamic acid, 5-oxo-2-pyrrolidine-carboxylic acid, benzoic acid, azelaic acid, and some microbiologically derived phenolic acids (Sangmanee and Hongpattarakere, 2014; Axel et al., 2016; Brosnan et al., 2012; Guo et al., 2012; Nasrollahzadeh et al., 2022). They are usually produced in very low concentrations that are below the inhibition level of microorganisms but synergistic effects between them (most probably between acetic and lactic acids but also other side fermentation products) enhances the antimicrobial effect. Despite their various chemical structures, most of them act in a similar fashion. According to the currently accepted theory, the antimicrobial effect of organic acid depends in respect to the substrate pH since undissociated organic aids are lipophilic and able to diffuse through the cell plasma membrane that contain lipopolysaccharide bilayer (Stratford and Anslow, 1998). In Gram positive bacteria such as *Bacillus* spp. (rope spoiling bacteria) the entrance is more easily since they do not possess an outer membrane which has a key role in the accessibility of the organic acids and other small molecules (Brul and Coote, 1999). Inside of the cell, the pH of the cytoplasm is neutral which promote dissociation of the organic acid molecules since their dissociation constant (pKa) is at least two units bellow the pH of 7. For example, 3-phenyl-L-lactic acid has a pKa = 3.2; formic acid a pKa = 3.7; lactic acid a pKa = 3.8; benzoic acid a pKa = 4.2; acetic acid a pKa = 4.75; caproic acid a pKa = 4.9, etc. Now, charged anions and protons remain locked inside the cell because are not able to cross the membrane, accumulating and acidifying the cytoplasm. Consequently, the metabolism of the cell particularly the glycolysis pathway is inhibited, decreasing the phosphofructokinase activity (a key enzyme of glycolysis) and, hence, reducing the ATP yield and growth of the cell (Krebs et al., 1983).

Sorbic, propionic, citric and acetic acids are widely used in bread and bakery product preservation (Gioia, 2017; Pattison et al., 2004). Several investigations showed that sorbic acid has more similar effects to alcohols and aldehydes, acting readily as membrane-active substance through the inhibition of the plasma membrane HC-ATPase proton pump and it is effective against bacteria, molds and yeasts (Stratford and Anslow, 1998; Axel et al., 2016).

Another well-known organic acid with natural antibacterial properties, that possesses a similar metabolic pathway as lactic acid, is 2-hydroxy-3-phenyl propionic acid known as phenylactic acid (PLA). Lactic acid bacteria metabolize it during fermentation by the glycolytic enzyme and lactate dehydrogenase (Jung et al., 2019). Phenylactic acid was found to have inhibitory against *Eurotium* spp., *Fusarium* spp., *Penicillium* spp., *Aspergillus* and *Endomyces* spp. (Lavermicocca et al., 2000). In a research from Yan et al. (2016), acetic and phenyllactic acids from *Lactiplantibacillus plantarum* CCFM259 (CCFM259) were found as the most effective organic acids in the inhibition of *Penicillium* roqueforti in Chinese steamed bread. The same authors also noticed that the mixture of acids also exhibited a synergistic effect. Sangmanee and Hongpattarakere (2014) identified several organic acids, such as lactic acid, stearic acid, palmitic acid, oleic acid, linoleic acid, 3-PLA, pyroglutamic acid and 5-oxo-2-pyrrolidine-carboxylic acid, produced by the strain *Lp. plantarum* K35 and demonstrated to be effective against the growth and aflatoxin production of *Aspergillus flavus* and *Aspergillus parasiticus.* Another study conducted by Brosnan et al. (2012) identify 15 organic acids from *Lactobacillus amylovorus FST2.1*, *Lactiplantibacillus plantarum* FST1.7, *Limosilactobacillus reuteri* R2 and *Weisella cibaria* PS2 using LTQ Orbitrap hybrid FT mass spectrometer: vanillic acid, azelaic acid, 4-hydroxybenzoicacid, decanoic acid, benzoic acid, hydrocinnamic acid, ρ-coumaricacid, DL-β-hydroxylauric acid, salicylic acid, DL-ρ-hydroxyphenyllactic acid, 2-hydroxydodecanoic acid, 3-hydroxydecanoic acid and the most abundant (S)-(−)-2-hydroxyisocapric acid and phenyllactic acid. Lactic acid, PLA, 4-hydroxy-phenyllactic acids, and indole lactic acid were also reported in the study of Lavermicocca (2000) as antimicrobial compounds produced by the strain *Lp. plantarum* 21B. Authors found PLA in the highest concentration and with highest activity against *Aspergillus pseudoglaucus*, *Aspergillus rubrum*, *Penicillium expansum*, *Saccharomycopsis fibuligera*, *Aspergillus niger*, *Aspergillus flavus, Monilia sitophila* and *Fusarium graminearum* in a concentration of 50 mg/ml. PLA did not show strong fungicidal activity against *Penicillium corylophilum* IBT6978 and *Penicillium roqueforti* IBT18687.

Other group of organic acids produced by yeasts during assimilation and catabolism of other organic compounds are phenolic acids (PA). The primary source of naturally derived phenolic acids are plant but they can also be produced as secondary metabolites by some LAB and yeasts during dough/sourdough fermentation with multiply benefits for human health which also possesses preservative characteristics (Valenciene et al., 2020). They belong to the large group of aromatic compounds containing one carboxyl group and one or more hydroxyl groups bonded to the aromatic ring. The number and the type of the functional groups attached to the aromatic ring give the bioactive properties of PA (Pereira et al., 2009). According to their chemical structure, they can be divided into hydroxybenzoic and hydroxycinnamic acids. *p*-Hydroxybenzoic acid, gallic acid, vanillic acid, *p*-coumaric acid, caffeic acid, sinapic acid, ferulic acid and hypogallic acid were found to be metabolically derived from yeast and LAB in dough/sourdough fermentation (Klepacka and Fornal, 2006;

148 *Novel Approaches in Biopreservation for Food and Clinical Purposes*

Antognoni et al., 2019; Brosnan et al., 2012). Klepacka and Fornal (2006), found in their study that when wheat bread doughs are enriched with phenolic acids, the quality and properties of the doughs depend on the phenolic acid composition and amount. In another study conducted by Antognoni et al. (2019), it was found that some *Lp. plantarum* strains can enrich the doughs in ferulic acid as well as in other phenolic acids such as *p*-hydroxybenzoic, gallic acid, caffeic acid and sinapic acid but in much lower concentrations.

Release of free phenolic acids by addition of degrading enzymes (mainly xylanase, cellulase, β-glucanase and feruloyl esterase) for 20 h at 20°C led to a fourfold increase in free ferulic acid level. The other phenolic acid *p*-coumaric and sinapic acids, benzoic acids (*p*-OH benzoic, vanillic and syringic) usually are present in smaller amounts. Depending on the type of fermentation, it usually decreases the accumulation of free phenolic acids in breads made of common wheat. Apart of the fermentation type, free benzoic acids are preferentially accumulated in breads, especially *p*-OH benzoic and vanillic acids (Skrajda-Brdak et al., 2019). The level of the phenolic acids in bread depends of the activity of phenolic acid esterases during the fermentation stage. *Lactobacillus acidophilus* LA5, *Lactobacillus johnsonii* LA1 and *Limosilactobacillus reuteri* SD2112 strains can increase the free phenolic acid contents in oat and barley grains, while *Levilactobacillus hammesii* DSM 16381 with two strains of *Lactiplantibacillus plantarum* (LM01 and PM4) effectively reduced the content of bound ferulic acid in wheat sourdough. It has been shown that the bioavailability of ferulic acid, as a structural phenolic compound in cell walls, is increased with the occurrence of sourdough fermentation. Some yeasts, fungi and bacteria have the ability to sequentially degrade ferulic acid to vanillin, vanillic acid and protocatechuic acid. Furthermore, vanillic acid can be transformed to guaiacol and catechol (Huang et al., 1993; Ghosh et al., 2006). Ferulic acid may also be transformed to dihydroferulic acid and then decarboxylated to 4-vinyl-quaiacol or, alternatively, decarboxylated and reduced to 4-ethyl-quaiacol. In these transformations, a crucial role is the ability of the microorganisms to produce phenolic acid decarboxylase (Max et al., 2011).

Three microbial metabolites of phenolic acids (dihydroferulic acid, dihydrocafeic acid and dihydrosinapic acid), and two phenolic acid derivatives (feruloylagmatine and p-coumaroylputrescine) were found to be the most abundant in rye sourdough. Other identified potentially bioactive compounds with significantly increased levels were 2-benzoxazolinone (benzoxazinoid), 2-hydroxyvaleric acid, isorhamnetin (favonoid), N-acetylspermidine and phenylethane (Zannini et al., 2009).

Overall, acidification with selected starters improves the extraction of total phenols which resulted in better antioxidant potential of the final bread. On the other hand, the anti-nutritional factor phytic acid is mainly located in the outer layers, pericarp and germ of wheat kernel. Phytase [myo-inositol hexakis (dihydrogen-phosphate) phosphohydrolase, EC 3.1.3.8] catalyzes hydrolysis of phytic acid into myo-inositol and phosphoric acid and eliminate its harmful effect and makes available phosphate and leads to non-metal chelating compound (Vermeulen et al., 2006). The activity of the endogenous phytases present in flours is usually considered insufficient to significantly decrease the phytic acid content. Low pH (as found in sourdoughs) and acceptable temperatures of fermentation helps the degradation of phytic acid by endogenous microbial phytases, which improve the nutritional quality of the bread. Phytase activity is expressed by several strains and species of wild yeasts and molds, *Bacillus, Escherichia, Klebsiella, Lactobacillus, Leuconostoc* and *Streptococcus* (de Angelis et al., 2003).

Bacteriocins

Bacteriocins are cationic peptides or proteins ribosomally synthesized by Grampositive and Gram-negative bacteria which can be post-translationally modified or not, possess hydrophobic or amphilitic properties, and are effective against various food pathogens (Jack et al., 1995). Depending on the dose and environmental conditions such as temperature, pH, etc., bacteriocins may have bacteriostatic or bactericidal action (Rai et al., 2016; Kraszewska et al., 2016). Four classes of bacteriocins are recognized according to their chemical structure. Small peptides called lantibiotics belong in Class I. They contain from 21–38 modified amino acids (lanthionine or β-methyl lanthionine) and can be linear (Ia), globular (Ib) and multicomponent (Ic). Linear flexible peptides from the group Ia are cationic and pore forming on bacterial membrane. Small and heat stable proteins that do not contain modified amino acids belong to the Class II, which are future divided in pediocin like bacteriocins (subclass IIa), two peptide bacteriocins (subclass IIb) and circular bacteriocins (subclass IIc). Bacteriocins from subclass IIa contain from 37 to 48 amino acids and are usually thermostable compounds. Subsequent IIa bacteriocins are not important in bakery industry since their activity is against the pathogenic strains from the genus *Listeria* which are often found in raw food. Bacteriocins with molecular mass larger than 30 kDa belong to the Class III, while complex bacteriocins containing both proteins and lipids (or carbohydrate component) belong to the Class IV (Zendo and Sonomoto, 2011; Stoyanova et al., 2012; Cintas et al., 2001).

A proteolytic system of LAB enables the hydrolyzes of proteins from the medium to peptides which then, by the combined action of numerous internal peptidases, are degraded into amino acids to be used as a required source of nitrogen for LAB growth and multiplication (Raveschot, 2018). It is the action of enzymes present in LAB called cell envelope proteinases (CEPs) that initiate the protein hydrolyzes and cleave the proteins into peptides ranging from 4 to 30 amino acids (Courtin et al., 2002). Strains from the genus *Lactobacillus* possess higher proteolytic activity than other types of LAB since they possess more than one gene for CEP, and a greater number of genes for di- and tripeptidases as well as for endopeptidase PepO and aminopeptidases (Courtin et al., 2002). Many research studies showed that low-molecular-weight peptides (< 10 KDa) possess higher antimicrobial properties (Nasrollahzadeh et al., 2022). The use of these protein-like compounds as biopreservative agents is favored over the use of acids because their activity is present over a wide range of pH and they are heat stable compounds which make them ideal for use in bread and other bakery products (Muhialdin et al., 2011).

Antimicrobial mechanism of action that peptides display consists in the disruption of cell membrane interacting with the sulfhydryl groups from the spore membrane and preventing its germination or through binding to lipid bilayers in carpet-like and puncturing channels in it, which impairs the membrane function and result in phase separation as well as solubilizing of the membrane.

150   *Novel Approaches in Biopreservation for Food and Clinical Purposes*

Bactericins such as nisin, reuterin, reutericyclin are the most often used as a natural preservative in bread and represent a well alternative to harmful and potentially carcinogenic synthetic additives. Nisin is a polypeptide composed by 34 amino acids and the most famous and first bacteriocin approved as food preservative by the Joint Food and Agriculture Organization/World Health Organization Expert Committee on Food Additives. Its antagonism is sporostatic against rope-forming bacilli (*Bacillus* spp.) in bread and since it is thermostable at acidic conditions, its use in bread manufacture is highly appreciated (Verma et al., 2014). In a study of Martínez-Viedma (2011), the cyclic peptide enterocin AS-48 was found to be sporostatic against vegetative cells of *Bacillus subtilis*, *Bacillus licheniformis*, *Bacillus cereus* and *Bacillus pumilus* at low concentration (14 AU/gm), as well as sporicidal at higher concentration (23 AU/gm) under the used experimental conditions. In order to identify and characterize bacteriocion-producing LAB in sourdoughs and to compare *in vitro* and *in situ* bacteriocin activity of sourdough and non-sourdough LAB, Corsetti et al. (2004) studied antimicrobial compounds produced by 437 *Lactobacillus* strains isolated from 70 sourdoughs. According to their findings, five strains (*Lactiplantibacillus pentosus* 2MF8 and 8CF, *Lactiplantibacillus plantarum* 4DE and 3DM, and *Lactobacillus* spp. CS1) were found to produce distinct bacteriocin-like inhibitory substances. BLIS-producing *Lactococcus lactis* isolated from raw barley showed a wider inhibitory spectrum than LAB isolated from sourdough, and, importantly, they did not inhibit all strains of the key sourdough bacteria *Fructolactobacillus sanfranciscensis*. Antimicrobial production of BLIS by *Lactiplantibacillus pentosus* 2MF8 and *Lactococcus lactis* M30 was also demonstrated *in situ* by Corsetti et al. (2004).

Reuterins are antimicrobial compounds produced by certain strains of *Limosilactobacillus reuteri*. *Limosilactobacillus reuteri* is a heterofermentative LAB found in a variety of fermented food such as sourdoughs for bread production. Reuterins can be found in hydrated, non-hydrated and dimeric forms of 3-hydroxypropionaldehyde (3-HPA) (Stevens et al., 2011). Although reuterins show heat stability when tested *in vitro*, the dough matrix is very complex and provides conditions to reuterin to react in many ways. Chiefly, free thiol groups of dough usually interfere with the antimicrobial potential of reuterins. Additionally, other metabolites present in sourdoughs, such antifungal organic acids produced by *Limosilactobacillus reuteri*, usually decrease the antimicrobial activity of reuterin. A promising solution for such decrease of activity is the supplementation of sourdough with phenylpyruvic acid, a precursor of phenyllactic acid, so that the phenyllactic acid production is increased. This approach revealed to be very effective, both *in vitro* and *in situ*, in maintaining the constant activity of reuterin in sourdoughs (Schmidt et al., 2018). Gänzle et al. (2000) purified the active compound from *Limosilactobacillus reuteri* LTH2584 named reutericyclin and examined its antimicrobial activity against *Lactobacillus* spp., *Bacillus subtilis*, *Bacillus cereus*, *Enterococcus faecalis*, *Staphylococcus aureus* and *Listeria innocua*. The investigation showed that reutericyclin had no effect against Gram-negative bacteria. Furthermore, their results indicated that the fatty acid source, such as wheat germ oil from sourdough, affects the inhibitory activity by *Limosilactobacillus reuteri* LTH2584. Since the overall flavor of the bread depends on the proteolytic release of amino acids during dough

Biopreservation in Flours and Bread 151

fermentation, reutericyclin-induced autolysis of cereal starters may positively affect aroma development in bread (Gänzle et al., 2000). On the other hand, antimicrobial compounds from *Limosilactobacillus reuteri* had significant anti-mold activity on bread surface during 12–15 days of storage and significantly prolong its shelf-life (Jonkuvienė et al., 2016). The mechanism of action of reuterin has been reported to cause oxidative stress to fungal cells. Reuterin exposure *E. coli* increased the expression of genes regulated and expressed in response to periods of oxidative stress. It was determined that the aldehyde group of reuterin binds to thiol groups of small peptides and other molecules, leading to oxidative stress, which is hypothesized as the mechanism of inhibition. Another proposed inhibition mechanism of reuterin is through the suppression of ribonuclease activity, which is the main enzyme mediating the biosynthesis of DNA (Schaefer et al., 2010).

Cyclic dipeptides are metabolites widely synthesized by procaryotic and eucaryotic cells by cyclodipeptide synthases or non-ribosomal peptide synthetases. 2,5-Diketopiperazines are cyclic dipeptides which have been shown to have a variety of effects on the growth and metabolism of fungi, such as inhibition of family C18 chitinase or aflatoxin production in *Aspergillus parasiticus* (Houston et al., 2002; Yan et al., 2004). Ryan et al. (2011) found cyclic dipeptides cyclo (Phe-Pro) and cyclo (Phe-OH-Pro), produced by the *Loigolactobacillus coryniformis* subsp. *coryniformis* Si3, inhibitory to *Aspergillus* spp. (Magnusson, 2003; Ström et al., 2002). Dal Bello et al. (2007) found that sourdough fermented by *Lactiplantibacillus plantarum* FST 1.7 inhibited the outgrowth of *Fusarium* spp. in wheat bread due to the production of lactic acid, PLA and the two cyclic dipeptides cyclo (L-Leu–L-Pro) and cyclo (L-Phe–L Pro). Antimicrobial properties of *Lactiplantibacillus plantarum* FST 1.7 in sourdough were investigated by Ryan et al. (2008). The outcome from their research was that when 20% of *L. plantarum* sourdoughs are added to wheat bread formulations, the growth of *A. niger*, *F. culmorum* and *Penicillium expansum* is inhibited and the final product showed an increase in shelf-life. Furthermore, a combination of *Lp. plantarum* from sourdough and calcium propionate showed remarkable antimicrobial effect even against the troublesome *Penicillium roqueforti*. The authors justified this effect to the synergistic effect between calcium propionate, organic acids, cyclic dipeptides and PLA.

Fatty acids

Generally speaking, antifungal activity of fatty acid highly depends on the structure. In the study of Black et al. (2013) unsaturated monohydroxy fatty acids were found to be antifungally active, while saturated hydroxy fatty acids and unsaturated fatty acids of oleic and stearic acids did not exhibit any activity. This implies that for the fatty acid to function as an antifungal agent, at least one double bond as well as one hydroxyl group along a C18 aliphatic chain should be present in the structure (Black et al., 2013).

Cereals and bread is a good source of fatty acids, mainly mono (e.g., oleic acid) and polyunsaturated (e.g., linolenic acid) but also saturated (e.g., palmitic and stearic) fatty acids (Dziki et al., 2014; Rocha et al., 2012a, 2012b, 2011, 2010a, 2010b). LAB demonstrate that the ability for lipid oxidation during bread preparation gives rise to the formation of volatile compounds and impact the rheology of dough due to

152 *Novel Approaches in Biopreservation for Food and Clinical Purposes*

the consumption of oxygen by endogenous lipoxygenase activity. The mechanism of lipoxygenase begins by oxidation of linoleic acid (and other polyunsaturated fatty acids) to hydroxyperoxy acids and wheat lipoxygenase forms 9-hydroperoxy lineolic acid while rye lipoxygenase forms 13-hydroperoxy isomers. Enzymatic or non-enzymatic reactions degrade hydroperoxydes aldehydes, which are responsible for the overall flavor of the bread. Furthermore, the mechanism of lipoxygenase activity is continuing by reduction of hydroperoxy linoleic acid to hydroxy-linoleic acid and in the presence of cysteine, peroxides are converted to the corresponding hydroxy fatty acids. The final products from this mechanism are hydroxy fatty acids such coriolic acid [13-(S)-hydroxy-9Z,11E-octadecadienoic acid] with potential antifungal activity, increasing the mold-free shelf-life of bread. Unsaturated di- and trihydroxy fatty acids impart a bitter taste to the bread (Gänzle, 2013).

Sjögren et al. (2003) found fatty acids such as 3-hydroxydodecanoic acid, 3-hydroxydecanoic acid, 3-3-hydroxy-5-cis-dodecenoic acid, and hydroxytetradecanoic acid produced by *Lactiplantibacillus plantarum* to be much more effective than cyclic dipeptides against some molds and yeasts. DL-hydroxyphenyl, 3, 3-(4-hydroxyphenyl) propionic, 4-dihydroxyhydrocinnamic, and 3-(4-hydroxy-3-methoxyphenyl) propanoic acids are found the main antifungal compounds in sourdough bread fermented by *Lp. plantarum* CH1, *Lacticaseibacillus paracasei* B20 and *Leuconostoc mesenteroides* L1 (Ouiddir et al., 2019). *Fructolactobacillus sanfranciscensis* has the ability to produce a mixture of fatty acids containing caproic fatty acid (hexanoic acid), which was the organic acid with the highest anti-mold activity. Caproic acid produced by *Fructolactobacillus sanfranciscensis* in combination with acetic, formic, propionic, butyric and *n*-valeric acids play a key role in inhibiting *Fusarium* spp., *Penicillium* spp., *Aspergillus* spp. and *Monilia* spp. growth in bread. Oleic acid, linoleic acid, palmitic acid, 3-PLA, stearic acid, pyroglutamic acid, and 5-oxo-2-pyrrolidine-carboxylic acid from *Lp. plantarum* K35 have a strong antifungal activity by inhibiting the growth and aflatoxin production of *A. flavus* and *A. parasiticus* (Arena et al., 2020).

Exopolysaccharides

Exopolysaccharides are extracellular polysaccharides excreted by some microorganisms. Lactic acid bacteria synthesize EPSs including both homo-(HoPS) and heteropolysaccharides (HePS) in order to survive under environmental stress conditions such as dehydration, phagocytosis, phage attacks and antibiotics, forming a protective barrier around the cells (Nguyen et al., 2020; Badel et al., 2011). Homopolysaccharides contain only one monosaccharide such as glucose [α- (dextran, mutan, reuteran) and β-glucans] or fructose [fructans (inulin-type fructans, levan)]. Heteropolysaccharides contain from 3 to 8 saccharide units (fructose, glucose, galactose and rhamnose) (Oleksy and Klewicka, 2016).

Species of *Lactobacillus*, *Lactococcus*, *Leuconostoc*, *Streptococcus*, *Pediococcus* and *Weissella* are of particular interest for scientists, since their strong ability to produce EPSs with many different structures and without any health risks (Patel et al., 2013; Paul et al., 2011). Many researchers have shown that synthesis of EPSs in LAB is influenced by the available nutrients in the culture medium, such as sugars, nitrogen, carbon dioxide, etc., and the type of cell stress. In general, EPS synthesis in

LAB is stress and species dependent. The oversupply or starvation of some nutrient lead to changes in EPS synthesis. Also, when exposed at specific type of stress, some species may stimulate while other species may inhibit the EPS production (Mbye et al., 2020). The stress contributes to the expression of the genes in the *eps* cluster in bacterial cell which encoded the enzymes to catalyze the synthesis of EPSs (Nguyen et al., 2020). The biological activities of EPSs are related to their monosaccharide composition. Despite their prebiotic, anti-oxidant and anti-inflammatory properties, EPSs also show antibacterial, antifungal and antiviral action. The mechanism of action is not fully understood and attempts to explain the antibacterial activity of EPSs are continuing although it is clear that functional groups in the structure of EPS interact with bacterial cell walls in some manner to yield antimicrobial action (Zhou et al., 2019).

EPSs showed antibacterial activity against Gram-positive and Gram-negative bacteria. Nehal et al. (2019) in their research found that negatively charged EPS from *Lactococcus lactis* strain revealed higher inhibitory action against Gram-positive than Gram-negative pathogens, with the greatest inhibition of *B. cereus*. Authors suggested that negatively charged EPSs revealed higher inhibitory action against Gram-positive bacteria due to their high positive charge on the cell wall. Another proposed potential inhibitory mechanism involves the disruption of the peptidoglycan layer in the bacterial cell walls (Sivasankar et al., 2018). According to the hypothesis of Medrano et al. (2009), the ability of some EPSs to act as a masking or decoy agents can block the receptors or channels on the outer membrane of the Gram-negative bacteria.

Isoca et al. (2022) investigate the anti-mold activity of eight LAB isolated from traditional gluten-free sourdough against some common spoilage microorganisms such as *Aspergillus flavus*, *Aspergillus niger. Penicillium paneum* and some bacteria such as *Escherichia coli*, *Campylobacter jejuni*, *Salmonella typhimurium* and *Listeria monocytogenes.* The obtained results showed that strains *Leuconostoc citreum* UMCC 3011, *Lp. plantarum* UMCC 2996 and *Pediococcus pentosaceus* UMCC 3010 showed wide antibacterial properties. Furthermore, the research disclosed that investigated LAB produce EPSs from the family of glucans and fructans, suggesting their possible interaction in the antimicrobial properties of investigated strains. Reuteran is an EPS produced by the probiotic bacteria *L. reuteri* isolated from the wheat sourdough (Krajl et al., 2005; Meng et al., 2015).

Exopolysaccharides are considered thickeners or hydrocolloids, which represent a good alternative to additives. They also have an effect on many useful technological properties, such as dough viscoelasticity. The formation of EPS *in situ* from sucrose has been reported to promote the production of additional metabolites, such as mannitol, glucose and acetate, which contribute to the quality of the finished products. Some benefits of using sourdough fermentation are a result of the production of EPS by LAB, since they can improve the technological, sensory and nutritional properties. Especially interesting is the increasing effect of EPS on the viscosity of the sourdough (Pepe et al., 2013). Two main technologies are used for addition of EPS in the dough: direct addition of EPS or *in situ* formation of EPS by LAB strains during dough fermentation. A positive relationship between EPS and organic acids formation during fermentation can be observed. In addition, sensory

154 *Novel Approaches in Biopreservation for Food and Clinical Purposes*

attributes of sourdough breads manufactured with added EPS-forming strains were found to be superior in comparison to bread fermented with commercial baker's yeast (Gezginc and Kara, 2019).

Nucleosides

Nucleosides are structural subunits of nucleic acids such as DNA and RNA, composed of a nucleobase [pyrimidine (cytosine, thymine or uracil) or a purine (adenine or guanine)] and a five carbon sugar (ribose or deoxyribose) (Liu et al., 2022). Their use in medicine is invaluable since they act as anti-cancer, anti-viral, anti-fungal, immunosuppressive, antiprotozoal and cardiovascular agent. Nucleosides and their derivatives may also act as herbicides, fungicides and insecticides. Although studies about nucleosides produced by LAB are limited and their antifungal mechanism is still unclear, the antimicrobial purposes of nucleosides make them interesting as bioprotective agents in bakery industry. Cytidine and 2'-deoxycytidine were isolated from the cell-free supernatant of *Lactobacillus amylovorus* LA 19280, *Lactiplantibacillus plantarum*, *Propionibacterium freudenreichii*, *Limosilactobacillus reuteri* and *Levilactobacillus brevis*, and showed antifungal activity against *Aspergillus fumigatus* and *Penicillum expansum* (Ryan et al., 2011; le Lay et al., 2016a; Yépez et al., 2017).

Microbial starter cultures and bread fermentation

All bread formulations have associated as essential ingredients flour or flours, leavening agents, eventually salt, and water. In the early history of bread manufacture, spontaneous fermentation of flour(s) mixed with water was carried out without the addition of any microbial starter culture or, later on, a portion of a mother-dough obtained by back-sloping (i.e., a piece of spontaneously fermented dough transferred from batch to batch). Spontaneous sourdough fermentation is the best example of natural biopreservation of baking goods. Therefore, the fermentation is spontaneously initiated by the endogenous microorganisms naturally present in flour(s) and water as well as microorganisms harbored from the surrounding environment by cross-contamination (Stolz, 2003). At the beginning, the mixture of water and flour(s) has approximately a neutral pH. As the fermentation progresses, the pH steadily declines until reaching a mature dough, the sourdough. The pH and temperature during the sourdough formation together with the available carbohydrates are the main parameters influencing the competitiveness of the natural microorganisms in this microecological niche. In fact, the pH is considered as one of the indicators of the level of sourdough fermentation. Furthermore, stabilization of the starter's pH is an indication that the starter reaches its maturity (Novotni et al., 2020; Rocha and Malcata, 2012, 2016a, 2016b).

Usually, foodborne bacteria from the genera *Enterococcus*, such as *Enterobacter* spp. and *Cronobacter* spp. among others, are commonly present in cereal flours at the beginning of spontaneous fermentation, in addition to beneficial microorganisms, such as *Lactococcus*, *Leuconostoc*, *Pediococcus*, *Streptococcus* and *Weissella* (Novotni et al., 2020), and contribute to the early sourdough formation—being characterized by high carbohydrate availability and pH values ranging 5.0–6.2. In such an environment, heterofermentative cocci *Leuconostoc* spp. and *Weissella* spp. thrive. They are generally good fermenters of simple carbohydrates and may metabolize a wide variety of them, including monosaccharides, disaccharides, oligosaccharides, sugar alcohols and gluconate. Sourdough associated *Leuconostoc* spp. are able to metabolize glucose and fructose via the phosphoketolase pathway producing lactic acid, ethanol (or acetic acid) and carbon dioxide as the end metabolites. They can also metabolize pentoses (arabinose, ribose and xylose) producing lactic and acetic acids as end products (Cogan and Jordan, 1995). Most *Leuconostoc* spp. and *Weissella* spp. grow well in a pH range of 5.0–7.0 but their growth is inhibited when pH decreases below 5 (Cogan and Jordan, 1995; Rocha and Malcata, 2016a, 2016b, 2012, 1999; Novotni et al., 2020).

Afterwards, the microbiota of sourdough changes as the pH become more acidic. Now, more acid tolerant LAB, such as homofermentative *Lactobacillus delbrueckii*, *Lacticaseibacillus casei*, *Companilactobacillus farciminis* and *Lactiplantibacillus plantarum*, become dominant. Other acid tolerant microorganisms such as heterofermentative LAB (*Levilactobacillus brevis*, *Lentilactobacillus buchneri* and *Limosilactobacillus fermentum*), homofermentative pediococci (*Pediococcus acidilactici* and *Pediococcus pentosaceus*) and acid tolerant yeasts (*Saccharomyces turbidans*, *Saccharomyces marchalianus*, *Naganishia albida*, *Kazachstania exigua*, *Saccharomyces cerevisiae* and *Saturnispora saitoi*) may also be present but with less incidence (Rocha, 1999; Stolz et al., 1999). As reference, usually viable LAB and yeasts in sourdoughs are associated in a ratio of *ca.* 100:1, respectively (Ottogali et al., 1996; Rocha and Malcata, 1999, 2012, 2016a, 2016b; Zannini et al., 2014). The spontaneous fermented sourdoughs without the addition of a mother-dough are not practiced in bakery industry since they require significant time for preparation. Instead, the sourdough fermentation involves the back-slopping process. The continuous inoculation of the flour(s) mixed with water with a small piece of sourdough (the sponge dough, mother-dough, sour ferment, mother-sponge or seed dough) from the previous batch leads to the establishment of a relatively stable microbial consortia—thus allowing to get a relatively stable quality and reproducibility of the sourdough bread between baking batches. The sourdough fermentation resorting to a mother-dough is also time consuming and laborious. The carbohydrate metabolism of obligate heterofermentative lactobacilli, which is highly adapted to the energy source of cereal dough, their growth requirements with respect to fermentation pH and temperature, stress response mechanisms and the production of antimicrobial compounds makes from obligate heterofermentative LAB highly competitive and dominant microorganisms in back-slopped sourdoughs (Buron-Moles et al., 2013). In such micro-ecological niche, and depending on the type of sourdough fermentation applied, the most frequently isolated beneficial microorganisms are LAB species such as *Fructolactobacillus sanfranciscensis*, *Limosilactobacillus pontis*, *Limosilactobacillus panis*, *Limosilactobacillus frumenti*, *Companilactobacillus paralimentarius*, and *Companilactobacillus mindensis* (Rocha, 1999; Stolz et al., 1999; Corsetti, 2013).

Three primary types of sourdough starter cultures are recognized taking into account the microbial species composition of the sourdough and the employed baking process. The traditional technique of spontaneous sourdough fermentation previously mentioned is referred as Type I sourdough. Microbial consortia in Type I sourdough depend on the protocol of fermentation which is specific to each type of bread, region and country and may also differ between individual bakers (Brandt, 2004). The most famous Type I sourdough is the San Francisco sourdough. The prevalent strain in this type of sourdough (Type Ia) made from wheat and rye flours when back-slopped daily at ambient temperature is *Fr. sanfranciscensis*, which can produce large amounts of lactic and acetic acids from maltose. Carbon dioxide is also released and helps significantly in dough leavening (Gobbetti and Corsetti, 1997). The microorganisms are kept metabolically active by refreshments once or several times a day at ambient temperature (20–30°C) and the pH of such sourdoughs is around 3.5–4.0. In fact, it is noticed that the temperature and the dough yield (DY, i.e., the mass percentage of dough in flour) of fermentation significantly influence the molar ratio between acetic and lactic acids and, consequently, the aroma of the sourdough. The presence of acetic acid in dough is especially important since it gives the bread desired flavour and at the same time act as biopreservative, inhibiting the rope bacteria and, therefore, contributing to the extension of bread's shelf-life. According to Gobetti et al. (2005), the content of acetic acid in a dough when propagated at ambient temperature (about 25°C) also depends on yeast activity. At ambient temperatures, yeasts hydrolyze fructo-oligosaccharides, especially ketose, producing fructose used as electron acceptor in acetate-kinase pathway of *Fr. sanfranciscensis* and other fructose positive LAB to produce acetic acid. When higher temperatures (> 32°C) during the process of fermentation are applied, the activity of yeasts is inhibited so the fructo-oligosaccharides remain un-hydrolyzed and less fructose is available for acetic acid formation by LAB. Therefore, dough fermented at 25–30°C are expected to contain more acetic acid, while dough fermented on higher temperatures (35–37°C) contains more lactic acid (Decok and Cappelle, 2005; Corsetti, 2013; Rocha and Malcata, 2012; Salovaara and Valjakka 1987).

In this acidic environment *Fr. sanfranciscensis* is associated with maltose negative and acidophilic yeasts such as *Kazachstania humilis* and *K. exigua* (Gänzle and Zheng, 2019). Except wheat and rye flours, the presence of *Fr. sanfranciscensis* is not reported in sourdoughs from other cereals and pseudocereals—such as rice, sorghum, maize or millet—due to their low maltose content which is the preferred substrate for *Fr. sanfranciscensis* (Rocha and Malcata, 1999; Vogel et al., 2011). *Fr. sanfranciscensis* hydrolyses maltose by a maltose phosphorylase pathway and produces one molecule of glucose-1-phosphate and one molecule of glucose. Inside the cell cytoplasm, glucose-1-phosphate is converted into glucose-6-phosphate by the phosphoglucomutase and, afterwards, glucose-6-phosphate is metabolized by the 6-phosphogluconate/phosphoketolase pathway to the final products, *viz.* lactic acid, acetic acid or ethanol and CO_2. The remaining molecule of glucose is excreted outside the cell wall and it is usually used by the existing sourdough yeasts (Yazar and Tavman, 2012). During the consumption of glucose by yeasts, fresh yeast extractives such as unsaturated fatty acids (mainly oleic acid), specific amino acids and peptides act favorably and stimulate synergetically the growth rate of *Fr. sanfranciscensis* (Gobbetti and Corsetti, 1997). The lack of competition for maltose and utilization of glucose and sucrose by *Kazachstania humilis* and *Kazachstania exigua* leads to the release of amino acids, contributing to the establishment of mutualistic stable associations between *Fr. sanfranciscensis* and *Kazachstania exigua/K. humilis* in some type of sourdoughs (Yazar and Tavman, 2012). Those acidophilic yeasts possess a high tolerance to the acetic acid (pH range 3.8–4.5). Also, they are resistant to other substances such as antibiotics produced by sourdough bacteria during fermentation. On the other hand, the baker's yeast *Sacharomyces cerevisiae* mostly prefers maltose and, especially, glucose so the bacterial metabolism of *Fr. sanfranciscensis* will be decreased if it is associated with *S. cerevisiae* (Sugihara et al., 1971).

Another Type I sourdough is the traditional three fermentation steps rye sourdough (Type Ib) including fresh sour, basic sour and full sour. The majority of the microbiota in this sub-type of sourdough consists of *Fr. sanfranciscensis* but depending on the fermentation conditions other species may occur in relevant cell viable counts such as obligate heterofermentative *Levilactobacillus brevis* and its related strains, the *Lentilactobacillus buchneri*, *Fructilactobacillus fructivorans*, *Limosilactobacillus fermentum*, *Limosilactobacillus pontis*, *Limosilactobacillus reuteri* and *Weissella cibaria*, the facultative heterofermentative *Companilactobacillus alimentarius*, *Lacticaseibacillus casei*, *Lactiplantibacillus plantarum* and *Companilactobacillus paralimentarius* and the obligate homofermentative *Lactobacillus acidophilus*, *Lactobacillus delbrueckii*, *Companilactobacillus farciminis* and *Companilactobacillus mindensis* (Yazar and Tavman, 2012). In Type Ib sourdough, yeasts are naturally present. The most commonly isolated yeast is *K. humilis* and it is frequently associated with *Fr. sanfranciscensis* and *Limosilactobacillus pontis* (Ganzle et al., 1998). Moreover, sourdoughs prepared at higher fermentation temperatures (above 35ºC) are referred as Type Ic sourdough and usually consists of obligate heterofermentative *Lactobacillus* spp., *Limosilactobacillus fermentum*, *Limosilactobacillus pontis* and *Limosilactobacillus reuteri*, obligate homofermentative *Lactobacillus amylovorus* and *Issatchenkia orientalis* as the most often associated yeast (Yazar and Tavman, 2012; De Vuyst and Neusens, 2005).

Type II and Type III sourdoughs are developed as a result of the imposed need for faster, more efficient, controllable fermentation processes in culinary industry and industrial breadmaking, and serve as a bakery pre-product for dough acidification. Type II sourdoughs are semi-fluid preparations. The fermentation process last for 2–5 days under elevated temperatures (> 30ºC) to speed up the fermentation process. The pH of such sourdough decreases below 3.5 after 24 hours of fermentation. In such microbial environment, *Fr. sanfranciscensis* is not competitive enough so different microbiota are established. The most frequently isolated LAB from Type II sourdoughs are the obligate homofermentative *Lactobacillus acidophilus*, *Lactobacillus amylovorus*, *Lactobacillus delbrueckii*, *Companilactobacillus farciminis* and *Lactobacillus johnsonii*, and the obligate heterofermentative *Levilactobacillus brevis*, *Limosilactobacillus fermentum*, *Limosilactobacillus frumenti*, *Limosilactobacillus panis*, *Limosilactobacillus pontis*, *Limosilactobacillus reuteri* and *Weissella confusa* (Müller et al., 2001; De Vuyst and Neusens, 2005). Also, such extremely acidic environment prevents contamination of the sourdough by other undesired microorganisms (Sugihara, 2003), thus being an effective natural process of biopreservation.

158 *Novel Approaches in Biopreservation for Food and Clinical Purposes*

Type III sourdoughs are dried using different drying protocols such as freeze-drying, spray granulation, fluidized bed drying and, the most common, spray-drying and drum drying protocols (Corsetti, 2013). In this type of sourdoughs drying-resistant LAB—such as heterofermentative *Levilactobacillus brevis*, and facultative heterofermentative *Pediococcus pentosaceus* and *Lactiplantibacillus plantarum*—are found as dominant strains (De Vuyst and Neusens, 2005). Because of its dried form, Type III sourdoughs are more easily stored and transported. Type II and III sourdoughs require the addition of baker's yeast or any other acid tolerant yeast as leavening agent (Novotni et al., 2020; Meroth et al., 2003). The metabolic activity of the sourdough microbiota is responsible for the nutritional properties and technological performance of the dough and, consequently, the sensory profile, shelf-life and the overall quality of the final bread (Calvert et al., 2021).

Obligate homofermentative LAB (*Pediococcus* spp., *Lactococcus* spp., *Streptococcus* spp. and certain *Lactobacillus* spp.) ferment only hexoses and almost only producing lactic acid ($>85\%$) by the Embden–Meyerhof–Parnas (EMP) pathway. They do not possess the enzyme phosphoketolase, therefore are not able to utilize pentoses and gluconic acid (Novotni et al., 2020; Stolz, 2003). They prefer glucose as a carbon source and metabolize it almost exclusively through the glycolytic pathway (EMP) to pyruvate and then under the action of lactate dehydrogenase produce lactic acid as the only end-product (Gänzle, 2015). Facultative heterofermentative LAB (*Leuconostoc* spp. and certain *Lactobacillus* spp.) possess the enzymes fructose 1,6-biphosphate and are also able to degrade hexoses to lactic acid through the EMP pathway. When an external electron acceptor is absent an extra NAD(P)H, is catalysed by the acetyl-CoA, yielding ethanol as an end-product. The ethanol formation is a limiting step in the hexose fermentation but in presence of electron acceptors [oxygen (O_2), fructose, pyruvate, citrate], NAD(P)H is re-oxidized and results in the formation of acetate instead of ethanol (Zaunmüller et al., 2006). Facultative heterofermentative LAB also possess aldolase and phosphoketolase and, therefore, are able to ferment pentoses, hexoses and gluconate to lactic and acetic acids (Felis and Dellaglio, 2007). Obligatory heterofermentative LAB lack on glycolytic enzyme fructose-1,6-bisphosphate aldolase and, consequently, cannot utilize hexoses through the EMP pathway. Instead, they utilize hexoses by the phosphogluconate pathway, producing not only lactic acid as the end product but also significant amounts of ethanol or acetic acid and carbon dioxide (Pessione, 2012). The genus *Lactobacillus* employ the pentose phosphate pathway to process glucose 6-phosphate to xylulose 5-phosphate, which is a substrate for phosphoketolase or the transketolase pathway and produce lactic acid and acetic acid (or ethanol) as end-products (Ehrmann and Vogel, 2005; Teixeira et al., 2012).

Until the latter half of the 19th century, yeasts were used in association with sourdough but the complexity of sourdough breadmaking process imposed the need of using only baker's yeast (*Saccharomyces cerevisiae*) and, since then, become dominant in the breadmaking industry. *S. cerevisiae* is essential in transforming the fermentable carbohydrates present in the dough (flours mixed with water and, eventually, salt) into ethanol and carbon dioxide, which increases the dough volume during the leavening phase. The optimal temperature for the growth of yeasts is 28–32ºC and optimal pH for fermentation is less than 5. The yeasts can be employed as a compressed cake of 28–32% solids, in the form of a cream which contain approximately 18% solids, or in the form of active dry yeasts which is a dehydrated form of 92–96% solids. The shelf-life of baker's yeast in the form of a compressed cake is 6–8 days, when stored at 4°C, is one day when it is produced in form of a cream, and is around a year when it is in dehydrated form (Hutkins, 2019). The yeast growth starts shortly after flour is combined with water and other ingredients, and mixed into a dough. Baking doughs may contain several sugars, mainly glucose, maltose, fructose and maltotriose, from which glucose is the preferred one. *S. cerevisiae* can use glucose by either aerobic or anaerobic pathways (Marques et al., 2016).

Differences between sourdough bread and baker's yeast bread exist mainly in terms of their qualitative and sensory characteristics as well as in their physical appearance and shelf-life. Sourdough breads have more acidic taste, more distinct flavor profile, and superior nutritional properties compared to the fast yeast leavened breads. Regarding physical appearance, higher crumb compactness and an increased shelf-life are observed in sourdough compared to yeast bread (Catzeddu, 2019; ur-Rehman et al., 2007). The use of sourdough with high acidity is especially appreciated in non-gluten bread preparation such as rye flour-based breads (Brandt, 2007). Gluten proteins present in wheat flours are responsible for the water-binding capacity and dough hydration. Since they are not present in rye flour sourdough fermentation is fundamental to transform water insoluble in water soluble arabinoxylans enhancing the baking properties of non-gluten flours (Novotni et al., 2020).

Bacteriophages and endolysins

Bacteriophages (BPs) and retrieved endolysins are nowadays limelight for researchers due to their unique lytic activity against disease causing/spoilage bacteria. The viruses, which are true parasites in bacteria are known as bacteriophages. They reproduce inside bacterial cells and are not able to damage or infect human or animal cells (Singh, 2018). The majority of bacteriophages employ phage-encoded peptidoglycan hydrolases (also termed phage lysins or endolysins) to enzymatically break down the peptidoglycan layer of the host bacteria. Endolysins are mostly employed to degrade the peptidoglycan layer of a bacterial host at the end of the replication cycle (Oliveira et al., 2013). Phage lysins can be used as antimicrobials, breaking down Gram-positive bacteria from the exterior (Zhang et al., 2013). As the food industry frequently encounters the problem of antibiotic resistance bacteria, the endolysin treatment thus represents a favorable novel strategy of biopreservation (Chang, 2020).

Biopreserving foods with bacteriophages as a natural anti-bacterial agent represents a better option compared to chemical preservatives. Additionally, they target specific bacterial strains which suggests their potential use in food biopreservation (Ramos-Vivas et al., 2021). To our knowledge no data exists on the use of bacteriophages in flours, bread and cereal-based products. However, *Staphylococcus aureus*, *Salmonella* spp. and *Bacillus* spp. (common pathogens in fresh pasta, fermented cereal foods, boiled rice and milk) have been effectively controlled in fruits and dairy products (Modi et al., 2001; Leverentz et al., 2001;

160 *Novel Approaches in Biopreservation for Food and Clinical Purposes*

Kong et al., 2015; Li et al., 2022). Baked items like cakes and bread are easily spoiled by various microorganisms including bacteria and filamentous fungi (Garcia et al., 2019; Morassi et al., 2018). *Bacillus* species (*B. amyloliquefaciens, B. subtilis, B. pumilus, B. licheniformis, B. cereus* and *B. megaterium*) stand out as major spoilage agents for rope or ropiness (Fangio et al., 2010; Valerio et al., 2012). As already referred, *Bacillus* spoilage in bakery products is induced due to their thermal survival even at baking temperatures of 180–200°C, when the bread crumb reaches maximum temperatures of 97–101°C for a few minutes (Valerio et al., 2012).

Nexus to the above, it can be assumed that controlling the activities of *Bacillus* species in baking industry is very important to reduce the wastage and food poisoning incidents. Few studies have been reported in past wherein bacteriophages and their endolysins effectively inhibited the activities of the aforementioned spoilage bacterial group. Decades earlier, three bacteriophages, named Bastille, TP21 and 12826 (*Bacillus cereus* phages), effectively inhibited the activity of various *Bacillus* species but deeper activity was found against *B. cereus* and *B. thuringiensis* (Loessner et al., 1997). LysB4 is endolysin from *B. cereus* phage B4 reported to possess strong lytic activity against *B. cereus, Bacillus subtilis* and *Listeria monocytogenes*, making it a good biocontrol agent. These strains were susceptible to 5 µg LysB4 and the lytic process took only 5 min long (Son et al., 2012). In another study, endolysin (LysBPS13) retrieved from bacteriophage (BPS13) showed broader lytic activity against all of the tested *Bacillus* species, including pathogenic *B. cereus* and *Bacillus thuringiensis*. Among the tested Gram-negative bacteria, LysBPS13 was active against *Salmonella, E. coli, Cronobacter sakazakii*, and *Shigella* strains in contrast to BPS13 only showed activity against *Bacillus* species. Moreover, retrieved endolysin (LysBPS13) outlined notable thermostability in glycerol presence and no difference was being noticed in lytic potential even after being incubated at a temperature of 100°C for thirty minutes (Park et al., 2012).

In a separate line of research, both a siphoviridae virulent phage known as PBC1 and its endolysin were isolated as potential weapons against *B. cereus*. PBC1 was only successful in infecting one out of the 22 studied *B. cereus* strains, but its endolysin, LysPBC1, had a broader lytic spectrum than PBC1. This endolysin was found to be more active against *B. subtilis* than *B. cereus*, but its activity was only specific to *Bacillus* species (Kong et al., 2015). Another phage named vB_BceM-HSE3 and its endolysin were retrieved and tested for possible antibacterial activity. Phage vB_BceM-HSE3 was seen to be effective even at higher temperatures and wide pH range against three strains of *Bacillus cereus* group, *viz. B. thuringiensis, B. cereus* and *B. anthracis*. On the other hand, endolysin PlyHSE$_3$ exhibited broader lytic spectrum, infecting all the strains of *B. cereus* group as well as a strain of another disease inducing bacteria, the *Pseudomonas aeruginosa* (Peng and Yuan, 2018).

Bacillus anthracis was used to isolate *Myoviridae* phage (Bcp1) from landfill soil. Bcp1 infected 11 to 66% of each *B. cereus sensu lato* species tested. In contrast, purified endolysin (PlyB) was bacteriolytic against all 79 *B. cereus sensu lato* isolates tested. PlyB was active across broad temperature, pH and salt ranges, resistant to resistance, and bactericidal as a single agent and also in combination (Schuch et al., 2019). Endolysin named LysPBC2 was retrieved *B. cereus* phage, PBC2 outlined antibacterial activity against tested strains of *Bacillus, Listeria* and *Clostridium* species. All mentioned bacterial groups are responsible for food poisoning incidents and threatening human life. In comparison, PBC2 only infected one strain (ATCC 13061) out of 11 *B. cereus* strains tested and was ineffective against other groups including *Clostridium perfringens, Enterococcus faecalis, Lactococcus lactis, Escherichia coli, C. sakazakii, B. subtilis, B. megaterium, B. circulans, B. licheniformis, B. pumilus, L. monocytogenes, S. aureus, S. epidermidis* (Kong et al., 2019). A study focused on the endolysins of Deep-Blue (PlyB221) and Deep-Purple (PlyP32), two predatory phages of *Bacillus cereus*. To evaluate their antimicrobial properties and binding specificity, both endolysins were expressed and purified. PlyB221 and PlyP32 efficiently recognized and lysed all the tested strains of *B. cereus* group (Leprince et al., 2020).

Phages PW2 and PW4 were identified in another study that infected *B. cereus* isolates as well as four cereulide-producing *Bacillus weihenstephanensis* strains and *Bacillus paranthracis* isolates. Retrieved endolysins named LysPW2 and LysPW4, especially the former, showed a much broader host range than the phages (Wan et al., 2021). Another group of researchers conducted a study to isolate BPs from fish wastewater, retail food, and soil samples to assess their activity against *B. cereus, B. subtilis* and *Shewanella putrefaciens*, knowns food deterioration agents. Bacteriophage activity was also checked in food products including pasta and rice. Using *B. cereus, B. subtilis* and *S. putrefaciens* as hosts, a total of four BPs, which have been given the names S1-BC, S2-BC, S1-BS and SW-SP, were successfully isolated. During the host range determination, S1-BC, S2-BC and S1-BS showed activity against *Bacillus* species, such as *B. cereus* and *B. subtilis*, but ineffective against other bacteria. In order to determine the antibacterial activity in food products, pasta and rice were purposefully tainted with the host bacteria. Bacteriophages (S1-BC, S2-BC, S1-BS) significantly decreased the number of bacteria present in samples by more than 90 percent (Roseline and Waturangi, 2021).

A total of four bacteriophages named S1-BC, S2-BC, S1-BS, SW-SP were isolated with *B. cereus, B. subtilis* and *S. putrefaciensas* host bacteria. S1-BC, S2-BC and S1-BS reduced the *Bacillus* species, i.e., *B. cereus, B. subtilis*, in host range determination and found ineffective against others. Pasta and rice were artificially contaminated with host bacteria to calculate bacteria reduction if samples were added with bacteriophages. S1-BC, S2-BC, and S1-BS bacteriophages significantly reduced bacterial concentration in samples by more than 90%. Refers to their activity, the isolated bacteriophages in this study might have a great prospect to be used as food biocontrol and also can be further tested to make a phage cocktail (Roseline and Waturangi, 2021).

Recently, a novel bacteriophage, named DLn1, was isolated and characterized, and its endolysin was expressed. In addition to *B. cereus* 2177, phage DLn1 lysed 16 out of the 122 *B. cereus* strains including emetic strains. When tested in milk, the purified endolysin of phage DLn1 had a much wider lytic range and the inhibitory effect against *B. cereus* in milk was more efficient (Li et al., 2022). Literature suggests bacteriophages can tolerate a wide range of temperatures and pH values, making them an ideal choice for biopreservation. Importantly, it can be seen that phages and their retrieved endolysins outlined better antibacterial potential in products, i.e., in dairy (Li et al., 2022), rice (Kong et al., 2015; Roseline and Waturangi, 2021)

162 *Novel Approaches in Biopreservation for Food and Clinical Purposes*

as compared to *in vitro* analysis. Correspondingly, using bacteriophages and their endolysins, *B. cereus* and *B. subtilis* can also be controlled in bread and flours as both bacteria were effectively lysed in the reported studies (Li et al., 2022; Son et al., 2012; Kong et al., 2015). These results not only contribute to a deeper comprehension of the diversity of phages but also provided a foundation upon which to build future phage therapies. Table 1 summarizes the bacterial groups, phages and their activity, retrieved endolysins and their activity.

Table 1. Bacterial phages and retrieved endolysins activity against various bacterial groups.

Target bacterial group	Phage	Activity	Retrieved endolysin	Activity	References
Bacillus *Listeria*	B4	Not tested	LysB4	Lysed Six *B. cereus* strains, *B. subtilis*, and two *L. monocytogenes* strains	Son et al., 2012
Bacillus	BPS13	Lysed *B. cereus* and *B. thuringiensis*	LysBPS1	*B. cereus*, *B. thuringiensis*, *Salmonella*, *E. coli*, *Cronobacter sakazakii*	Park et al., 2012
Bacillus	PBC1	Lysed 1/22 *B. cereus* strains	LysPBC1	lower against the *B. cereus* group but higher against the *Bacillus subtilis* group	Kong et al., 2015
Bacillus	vB_BceM-HSE$_3$	Lysed *B. cereus*, *B. anthracis*, *B. thuringiensis*	PlyHSE$_3$	Lysed two strains of *B. anthracis*, seven strains of *B. cereus*, and seven strains of *B. thuringiensis*	Peng and Yuan, 2018
Bacillus	Bcp1	Lysed 15/44 *B. cereus*, 6/25 *B. thuringiensis*, 2/7 *Bacillus mycoides*, and 2/3 *B. anthracis* strains	PlyB	Lysed all tested strains	Schuch et al., 2019
Bacillus *Listeria* *Clostridium*	PBC2	Lysed 1/11 *B. cereus* strains	LysPBC2	Against all tested *Bacillus* *Listeria*, *Clostridium* strains	Kong et al., 2019
Bacillus	Deep-Blue Deep-Purple	lysed 3/14 *B. cereus* strains	PlyB221 PlyP32	lysed all *B. cereus* tested strains	Leprince et al., 2020
Bacillus	PW2 PW4	Lysed *B. weihenstephanensis* and *B. paranthracis*	LysPW2 LysPW4	lysed all *B. cereus* tested strains	Wan et al., 2021
Bacillus	S1-BC, S2-BC, S1-BS, SW-SP	Lysed *B. cereus*, and *B. subtilis*	-	-	Roseline and Waturangi, 2021
Bacillus	DLn1	lysed 16/122 *B. cereus* strains	LysDLn1	Lysed 82/102 *B. cereus* strains	Li et al., 2022

Additives, packaging technologies and physical processes in bakery industry

Bread as an intermediate moisture product and pH around 6 is highly perishable food which is a prime issue in bakery industry since it may deteriorate people's health and cause economy losses. Main spoilage concerns of bread are addressed to some physical and chemical processes and particularly to the activity of microbial agents (Rahman et al., 2022; Alpers, 2021). Bread is prone to loss or absorb moisture from the surrounding environment. Moisture in bread may also migrate from crumb to crust. These variations in moisture are the most severe physical changes in bread, which results in a series of other processes that cause staling and spoiling of bread. During staling bread loses its nutritional value (loss of vitamin A and E, protein degradation) and chemical stability such as lipid degradation, starch re-crystallization, polymer reorganization, etc., resulting in changes in texture, off-odors and flavors. These chemical changes in bread are known as rancidity and are favored by the presence of endogenous enzymes such as lipases and lipoxygenase (Pareyt, 2011; Taglieri, 2021; Smith, 2004; Pateras, 2007; Alpers, 2021).

The quality of dough and bread is traditionally improved by additives and benefits from their use in breadmaking industry are multiple. They simplify and accelerate the process of breadmaking, adjusting the flour characteristics to the process of breadmaking, improve bread quality, reduce costs and extend the shelf-life of the product by delaying staling and inhibiting the development of microorganisms. Historically, some natural products such as salt, vinegar, alcohol, etc., have been used as additives. Today, the supply of additives is much greater. They are coded by the international food safety guidelines and should be approved and permitted by Food and Drug Authorities to be used globally in the breadmaking industry. The development of the bakery industry imposed the need of additives that are easier to handle, store, readily available and, at the same time, are very economical.

There are currently 18 different types of additives permitted in EU for breadmaking solely from wheat flour, water, yeast or leaven and salt: acetic acid (E 260), ascorbic acid (E 300), fatty acid esters of ascorbic acid (E 304), calcium acetate (E 263), calcium ascorbate (E 302), calcium lactate (E327), lactic acid (E 270), lecithins (E 322), potassium acetates (E 261) (approved for use after February 6th 2023), potassium lactate (E 26), sodium acetate (E 262), sodium ascorbate (E 301), sodium lactate (E 325), mono- and diglycerides of fatty acids (E 471), acetic acid esters of mono- and diglycerides of fatty acids (E 472a), tartaric acid esters of mono- and diglycerides of fatty acids (E 472d), mono and diacetyl tartaric acid esters of mono- and diglycerides of fatty acids (E 472e), and mixed acetic and tartaric acid esters of mono- and diglycerides of fatty acids (E 472f). In other bakery wares, some other additives such as propionates, sorbates, *etc*, are also permitted (EU Food Additives Database).

According to their purpose of use, additives are classified into four main classes: (i) oxidants/reductants; (ii) emulsifiers; (iii) hydrocolloids; and (iv) preservatives (Gioia et al., 2017). Primarily, oxidants in breadmaking industry are used to assist with gluten network development. The most common substance used as oxidant agent is L-ascorbic acid (E 300). Some oxidant agents used in the past such as azodicarbonamide (E 927), potassium bromate (E 924), potassium iodate (E 917) and acetone peroxide (E 929) are not permitted in EU and many other countries due to their toxicity and harmfulness. Ascorbic acid (AA), commonly known as vitamin C, has strong antioxidant properties and it is the most acceptable additive in bread. Actually, AA is a reducing agent but in the presence of oxygen from the air and the enzyme ascorbic acid-oxidase (an enzyme which is naturally present in wheat flour), it is converted into dehydroascorbic acid which has the potential to take part in oxidative reactions during flour-water mixing and it is the dehydroascorbic acid which is the effective reagent. The dehydroascorbic acid possesses strong gluten strengthening effect. It oxidases the thiol groups of gluten-forming proteins and forms disulphide bonds causing gluten cross-linking and polymerization. This contributes to increasing elasticity, improve volume and shape, and give more uniform texture to the finished breads (Cauvain, 2015; Grosch and Wieser, 1999). Khan et al. (2011) investigated the effect of ascorbic acid (AA) on the staling and spoilage of chapatti and find that it reduced the mold growth during the storage period of 24 h compared to the control. Another study performed by Eacute Sar et al. (2016) showed that the combination of α-amylase, xylanase and ascorbic acid extended the shelf-life of bread for 13 days on average.

Reductants possess inverse effect to oxidants, reducing the number of cross-links between gluten-forming proteins and convert high molecular weight glutenins into smaller molecules (Msagati, 2013). The most famous reductant is l-cysteine (E 930).

Emulsifiers are another group of additives used in breadmaking industry, primary as crumb softeners, dough conditioners or gluten strengtheners. Mono- and diglycerides are mostly used as crumb softeners, while diacetyl tartaric acid (DATA), esters of mono- and diglycerides (DATEM) and polysorbate are used as dough conditioners or gluten strengtheners. Distilled monoglycerides act, simultaneously, as anti-staling agents in breads. During baking, when dough is exposed to high temperatures, monoglycerides bind to the amylose fraction of the starch and slow down retrogradation of the starch during cooling and sub-sequent storage (Cauvain, 2015; Moonen, 2015). Sodium stearoyl-lactylate (SSL) acts similarly to distilled monoglycerides, binding to starch amylose and prolonging the shelf-life of bread (Moonen, 2015).

Hydrocolloids play an important role in breadmaking industry since in very low doses (< 0.1% of flour basis) improve the dough-handling properties, the quality of fresh bread and extend the shelf-life of stored breads. They usually act by reducing the dehydration rate of bread during storage (Guarda et al., 2004) Hydrocolloids reported to inhibit staling include exudate gums [Arabic gum (AG), brea gum (BG), tragacanth gum (TG)] gums from seaweeds [sodium alginate (AL), Kappa carrageenan (k-CAR), Iota carrageenan (i-CAR), Lambda carrageenan (l-CAR)], modified celluloses [carboxymethyl cellulose (CMC), hydroxypropylmethyl cellulose (HPMC), hydroxypropyl cellulose (HPC), microcrystalline cellulose (MCC)], galactomannans from leguminous seeds [locust bean gum (LBG), guar gum (GG), tamarind seed gum (TSG)], exopolysaccharides from microbial fermentation [xanthan gum (XG), Dextran], pectins from plant cell wall, etc. (Ferrero, 2017).

The use of preservatives in manufacturing bakery products have been driven by the need for longer shelf-life of the products. Preservatives are substances added to the mixture of flour and water to prevent food spoilage by microorganisms during storage and transportation. Commonly used preservatives in bakery industry belong to the groups of propionates, sorbates, acetates and fermentates (Gioia, 2017). Aliphatic acids [propionic acid (E 280), sorbic acid (E 202) and benzoic acids (E 210)] are lipophilic in nature and can penetrate the cell membrane in the undissociated form. As soon as the undissociated acid enters in the cell wall, the higher pH environment inside the cell of spoilage microorganism favor dissociation of the molecule, resulting in the release of charged anions and protons which cannot cross the plasma membrane. Since these acids are quite volatile and corrosive, their sodium, potassium and calcium salts, which are more soluble in water, more stable and easier to handle, are used. They retard the rate of mold development and at the same time prevent the rope spores from certain *Bacillus* spp. to thrive. In the past, the invariable practice was to use calcium propionate in 0.2–0.4% of the flour weight to prevent rope spoilage of bread. Also, propionates do not have influence on yeast growth and development, which make them highly acceptable in yeast raised breads (Sauer, 1977). However, the use of calcium propionate proved to be unsatisfactory because of its poor effectiveness against *Penicillium* spp., especially *P. expansum* and *P. roqueforti* (Lavermicocca et al., 2000). High levels of propionic acid in dietary intake are associated with propionic acidemia in children. Some of the symptoms include learning disabilities, arrhythmia, seizures, gastrointestinal symptoms and recurrent infections (Pena and Burton, 2012). Sorbic acid and its salts are found more effective than propionates against mold spoilage of bread (Samapundo, 2017). Also, its half-life in human body is around 40–110 minutes before it is completely oxidized to CO_2 and H_2O (Liewen and Marth, 1985). Acetic acid (E 260) is another weak acid which is found effective against molds and bacteria, especially against the rope spoilage of bread. In USA its use is limited to equivalents of 0.25% of flour bases to avoid the vinegar odor that appears because of its use (US FDA, 2022). It can be obtained naturally trough fermentation of carbohydrates, as already deeply referred in this chapter, or by chemically synthetic methods. However, the use of propionates, sorbates and acetates have detrimental effects on dough and bread characteristics since they lower the pH of the dough under the acceptable level. Some solutions are proposed in the literature as alternatives such as the use of encapsulated formulations of these acids and their salts or spraying the acid on the surface of the bread after baking (as already mentioned above) (Suhr and Nielsen, 2004). Sodium bisulfite (E 222) and sulfur dioxide (E 220) are usually used as preservatives in products that have to pass through a long transport or have to spend a long time exposed to the air such as dry biscuits. Their use is restricted on 50 mg/kg since they can alter the taste of the end-product and might cause allergies at people who have certain ingredient-based sensitivity (EFSA, 2016).

The spoilage of bread is a quite complex process, thus much of the research focus has been devoted on determination of the intrinsic and extrinsic factors that support microorganisms' growth in bread. Additionally, mathematical models are generated which take into account environmental and physical factors like temperature and humidity of storage and are able to predict the remaining shelf-life of the product (Ellis and Goodacre, 2006). Highly nutritional value of bread, as well as the absorption of moisture and increase of water activity, has the most significant impact on the microbial spoilage of bread. When the moisture of bread is moderate (a_w = 0.6 – 0.84), osmophilic yeasts and molds are primary spoilage pathogens but when water activity became higher (a_w = 0.94 – 0.99) mainly all yeasts, molds and bacteria can thrive. This happens especially when hot bread is wrapped, which cause moisture to precipitate on the surface of the inner area of the package (Rahman et al., 2022). The pH of bread is found also as a determining factor for the types of microorganisms which will thrive during bread spoilage. Baker's yeast bread with pH usually from 5.5–6.0 and sourdough bread with pH from 4.5 for rye breads and 5.1 for sourdough wheat bread are luckily to be colonized with contaminating molds during post-baking processing (Magan and Aldred, 2006). Therefore, keeping clean environment and controlling the environmental factors such as product's a_w and pH and also the temperature of storage are common practice in inhibiting microorganisms' growth and activity on the bread itself after baking.

Another strategy to control bread spoilage include surface sterilization with some physical process which include radiation, such as ultraviolet (UV), microwave radiation (MW) and infrared (IR) radiation. This kind of sterilization destroys spores and bacteria present on bread. The UV irradiation can be used immediately after baking and during the post-baking time at the wavelength of 260 nm. It does not use heat which can destroy wrapping material or promote condensation. UV light have germicidal action on spores inactivating the microorganisms by DNA mutations as a result of UV light absorption by DNA (Gayán et al., 2015). It breaks down the DNA and cause inactivation of microorganisms (Campos et al., 2014). Nevertheless, the UV light is harmful for workers and protective measures should be applied which may complicate the processing. Also, it is effective against bacteria and mold spores only on bread surface since its poor penetrative capacity does not affected microorganisms from the bread interior (Seiler, 1989; Cauvain, 2015). These disadvantages are overcome in microwave radiation since microwaves heat rapidly even in bread interior and without significant changes in temperature between the different bread layers. In his research, Seiler (1983) found that when wrapped bread is heated in a microwave for 30–60 s, the surface reach 75°C which is sufficient to render the product free of molds and, together, not interrupting the integrity of the packaging material. Anyway, the problem with condensation remains due to the heating effect. When IR radiation is used the condensation is minimized because only outside surfaces are heated to 75°C. The negative side of this sterilization is the high cost and the need of product rotation in order product to be heated on each side (Seiler, 1989; Cauvain, 2015).

Today's bakery industry faces a tremendous challenge in producing bread that should not only be wholesome and tasteful but much more long-lasting and safer, produced in more natural way, free of chemical additives and preservatives. The uses of chemical preservatives which may cause health hazard problems can be reduced significantly by using proper packaging. The primary purpose of bread packaging is to serve as an obstacle between the spoilage microorganisms from the surrounding environment and the products, to protect the product from damaging and other contaminants and to facilitate their transportation and distribution (Salgado et al., 2021, Teixeira-Costa and Andrade, 2021). Nowadays, packaging is much more than that. The expansion of new scientific achievements, especially in the field of nanotechnology, resulted in the creation of advanced packaging technologies such as modified atmosphere packaging (MAP), and some nanotechnology-derived packages such as active packaging (AP) and intelligent packaging (IP). Encapsulated nanomaterials might be used coupled to antimicrobial compounds such as preservatives and distributed onto the food product surface where its action is needed, and which will resulted in significant reduction of the amount of preservative compared to when it is distributed throughout the whole product (Bradley et al., 2011; Lorite et al., 2016).

MAP use gases, usually CO_2 and/or nitrogen (N_2), to replace air in the bread packaging before it is sealed. The resultant anaerobic environment in the sealed packaging does not support the aerobic metabolism of spoilage microorganisms. Additionally, CO_2 can dissolve in water from the product and form carbonic acid that lowers the pH and cause inhibition of bacterial growth (Upasen and Wattanacha, 2018). Therefore, MAP with CO_2 is especially suitable for packaging partly baked bread since it has higher water content. The research of Seiler (1984) showed that the extension of mold-free shelf-life of bread is proportional to the concentration of CO_2 in the atmosphere inside the packaging and depends on the equilibrium relative humidity (ERH). In bread with ERH > 90% and CO_2 content greater than 65%, the extension of the shelf-life was almost 200% (Seiler 1984, 1989). Another study showed that when the level of O_2 is below 0.4%, the development of spoilage molds is delayed in 5–10 days (Smith et al., 1986). Rodriguez et al. (2000) investigated the efficacy of CO_2 and N_2 and their combinations in MAP of sliced wheat bread and found that the combination of 50% CO_2 and 50% N_2 prove to be most efficient against mold and yeast growth, increasing the shelf-life in 117% at 22–25°C and 158% at 15–20°C. Very similar results were obtained in the research of Ooraikul (1982) when combination of 60% CO_2 and 40% N_2 has been found the most effective. Degirmencioglu et al. (2010) found that MAP of 100% carbon dioxide in combination with 0.15% potassium sorbate as bread preservative, inhibits yeast and mold growth in sliced bread, tested during 21 days of storage at ambient conditions (20±2°C and 60±2% RH). The application of MAP in bread and other bakery products is proved to be not economically viable because of the gas cost and special wrapping equipment that use laminated films. Therefore, MAP is limited only on high-value breads and bakery products intended to have long shelf-life (Cauvian, 2015). Currently, a decisive conclusion for the possibility of negative consequences because of the CO_2 used in MAP of bread is not possible to give due to different outcomes found in the research investigations available in the literature data (Hasan et al., 2014; Hematian et al., 2010; Khoshakhlagh et al., 2014). According to Upasen and Wattanacha (2018) carbon dioxide is not a necessity present gas in the MAP as long as the amount of oxygen was minimized up to 5% (v/v) by balancing with nitrogen concentration and using a three-layer low-density laminated polyethylene films. This polymer film coated with iron-based oxygen scavenger nanoparticles is high oxygen barer, acting as antimicrobial agent that prolongs the shelf-life of the bread up to 5–7 days.

One of the latest classifications of nanotechnology-based packages recognize five different technologies of package, that are active packaging, intelligent packaging, nano-coatings, surface biocides and nanocomposites. Active packaging act as a protective barrier and, at same time, incorporate nanomaterials with antimicrobial, antioxidant, moisture and oxygen absorbing properties, among others. The antimicrobial AP is usually prepared by incorporating the antimicrobial agent into the packaging material by coating the active compound onto the surface of the packaging material or by adding a sachet with antimicrobial compounds into the packaging from which the active compound is released during storage (Devlieghere et al., 2004). The concept of AP was first developed in Japan in late 1970s when sachets, containing active ingredients, such as moisture absorber, oxygen absorber and ethylene scavenger, were used in food packaging. Antimicrobials involved in AP can be synthetic or natural agents (Kuorwel et al., 2011). Synthetic antimicrobial agents commonly used in bread packaging are organic acids and their salts, alcohols, antibiotics, sulphides and nitrites (Jideani and Vogt, 2015). Organic acids can also be delivered from natural sources, such as the organic acids produced during sourdough fermentation. Other natural compounds reported as efficient against bread spoilage microorganisms include essential oils, spice extracts, honey and propolis extracts, etc., and their purified compounds such as carvacrol, eugenol, linalool, thymol, p-cymene, cinnamaldehyde, nisin, lacticin and other bacteriocins (Hulin et al., 1998; Davidson and Naidu, 2000; Dorman and Deans, 2000; Aboukhalaf et al., 2020, 2022; Neves et al., 2021; Özogul et al., 2022).

Oxygen absorbers must meet specific criteria to be effective and gain commercial visibility. They should absorb oxygen at an appropriate rate, should be compact and uniform in size, should not be toxic, nor produce unfavorable side reactions. The choice of oxygen absorbers is influenced by food properties, such as size, shape, weight and a_w of the food, the amount of dissolved oxygen in the food, the desired shelf-life of the product and the permeability of the packaging material to oxygen. Oxygen absorbers prevent bread staling and spoilage since oxygen is essential for mold and aerobic microorganisms to grow. Also, they do not have any effect on the sensory quality of bread over the storage (Galić et al., 2009). Sachet-based packaging and plastic film-based packaging are most commonly used AP (Cooksey, 2005).

Nanomaterials incorporated with antimicrobial substances release the active substance during the post-packaging period in the head-space between the package and the product or interact with the product to protect it from spoiling. Nano-coatings incorporate nanomaterials inside or outside the packaging surface to improve barrier properties. Also, nanomaterials can be incorporated as a layer in laminate. Surface biocides are intended to prevent the microbial growth on the food contact surface without having a preservative effect by incorporating nanomaterials with antimicrobial properties on the packaging surface. The application of surface agents in the single-use disposable packaging is still questionable but it is found useful in food processing equipment, especially in the case of conveyor belts that are difficult to clean, in reusable food containers such as crates and boxes, and in the inner lining of freezers and refrigerators (Bradley et al., 2011).

Nanocomposites consist on nanoparticles having at least one dimension in the nanometer range of 1–100 nm coupled to a biopolymer matrix and incorporated into the packaging to improve its barrier properties, physical performances, biodegradation and durability (Bradley et al., 2011). Nanocomposites in the food packaging sector refer to materials containing modified nanoclays of approximately 1–7 wt.% (Lagaron and Lopez-Rubio, 2011).

Various antimicrobial compounds and technologies have been tested to find the most suitable agent and corresponding technology to create effective antimicrobial packaging of bread. Balaguer et al. (2013) investigated the efficacy of gliadin films incorporated with 1.5, 3 and 5% cinnamaldehyde in *Penicillium expansum* and *Aspergillus niger* suppression on sliced bread under *in vitro* conditions and found 1.5% cinnamaldehyde effective, in the case of *Penicillium expansum*, and 3% cinnamaldehyde, in the case of *Aspergillus niger*. When 5% cinnamaldehyde was incorporated in AP, the mold growth was observed after 27 days of storage at 23°C compared to the control where, in the last, the mold growth appeared after 4 days of storage. In the study of Rodriguez et al. (2008), the incorporation of 6% of cinnamon oil with solid wax paraffin as an active coating in packages of sliced bread products have been shown that completely inhibit the growth of *Rhizopus stolonifer* after three days of storage. Suhr and Nielsen (2005) investigated the effectiveness of the MAP in wheat and rye bread in 0, 50, 75, and 100% CO_2 modified atmosphere, when residual O_2 was 1, 0.03 and < 0.01% in the presence of O_2 absorber. Wheat and rye bread were artificially inoculated with common spoilage fungi *Penicillium commune*, *Penicillium solitum*, *Penicillium polonicum*, *Aspergillus flavus*, *Penicillium roqueforti*, *Penicillium corylophilum*, *Eurotium repens* and *Endomyces fibuliger*. The obtained results showed that the most troublesome mold of rye bread was *Penicillium roqueforti*, being the most resistant to CO_2 and only the use of O_2 absorber could prevent its growth. *Penicillium commune* was able to grow in 99% CO_2 when high residual O_2 was present and *Aspergillus flavus* was able to grow at lowest O_2 concentration in the presence of 75% CO_2. The causer of the chalk mold *Endomyces fibuliger* proved to be the least affected by the changes in O_2 and the tested treatments. Furthermore, authors investigate the effectiveness of the AP when mustard oil in concentrations of 0, 1, 3, 5 and 10 µL per bread was added in atmospheres of 100 and 80% CO_2 with 5% O_2. In this case the most mustard oil-resistant species were *A. flavus* and *Eurotium repens*. The most efficient concentrations were between 2 and 3 µL of mustard oil per bread. Gutiérrez et al. (2011) investigate the efficiency in term of microbial and sensory criteria of active packaging, MAP and a combination of both in the shelf-life extension of gluten-free bread. The AP contained the essential oil of cinnamon (*Cinnamomum zeylanicum*) coated in nitrocellulose base in amount of 0.0215 and 0.0374 g per package and the MAP was a mixture of 60% CO_2 and 40% N_2. The outcome of this research was that the AP was more efficient in microbial inhibition and, at the same time, in maintaining the sensorial properties of the gluten-free bread. Much research on active packages containing mustard essential oil or its component allyl isothiocyanate showed their efficacy in inhibition of the growth and development of microorganism and extension of the shelf-life of bread (Nielsen and Rios, 2000; Han, 2005; Appendini and Hotchkiss, 2002).

While AP is usually designed to inhibit growth of undesirable microorganisms and control the oxidation and moisture, the function of IP or smart packaging is to monitor the food quality or the environment surrounding the food, and alert when any changes occur. Indeed, the technology of IP includes the incorporation of nanosensors which can detect the presence of harmful microorganism and alert the consumer about the ending of the shelf-life by triggering the changes in color, for example. Furthermore, the IP concept has been upgraded based on the concept of the electronic tongue to release preservatives when food begins to spoil (Otlers and Yalcin, 2014). Several types of sensors are proposed in the literature, such as microbial indicators, chemical sensors, ripeness indicators, radio frequency identification and time-temperature indicators but only latest two have got commercial application (Kuswandi et al., 2011). Although still not mostly commercially important so far, the chemical sensors especially their subgroup of biosensors represent a promising tool for monitoring of spoilage microorganisms. They represent a real time simultaneous detection of multiple biohazardous agents providing immediate information about the sample being tested. This quickly enables recognition of impending threats and taking corrective measures on time (Bhadoria and Chaudhary, 2011; Pereira-Barros et al., 2019). The sensitivity of the biosensors developed is in the rage of ng/mL for microbial toxins and < 100 CFU/mL for bacteria. Different biosensors are developed to be used in food packaging technology such as electrochemical, piezoelectric, acoustical and optical sensors, among which the electrochemical biosensors (ECB) are the most developed and used (Lam et al., 2013). Based on the observed parameters, such as current, potential, impedance and conductance electrochemical, biosensors can be further classified as amperometric, potentiometric, impedimetric and conductometric, respectfully (Sadik et al., 2009). In ECB, antibodies, enzyms, nucleic acids, bacterial phages, aptamers, biomimics (Singh et al., 2013) as biorecognition agents are coupled to an electrode array or solid electrode surfaces (Pt-, Au-, Ag-, graphite- or carbon-based conductors). As soon as the biosensor detect the presence of the microorganism it generates signal using a transducer (Caygill et al., 2010). The widespread use of antibodies and nucleic acids as biosensors is limited due to a high assay cost. Aptamers are recently recognized as more acceptable for biosensor applications. They are small synthetic oligonucleotides that contained around 40–100 bases. They are particularly suitable as biorecognition elements in biosensor applications since they can specifically recognize and bind to any kind of target (ions, cells, drugs, toxins, low–molecular weight ligands, peptides, proteins, etc.). Biosensors based on aptamers are highly accurate and reproducible even at very low concentrations. Their use is simple and selective (Zelada-Guillén et al., 2009; Pereira-Barros et al., 2019). Bacteriophage-based bioreceptors provide selective detection of food born bacteria. As referred earlier, bacteriophages are viruses in size of 20–200 nm. They recognize the specific bacterial receptor on the bacterial surface through their tail spike proteins and bind to this receptor injecting its genetic material inside the bacterial cell (Singh et al., 2013).

Several risks are associated with innovative packages such as possible migration of nanomaterials and/or absorbers into the product which might be accidentally ingested by the consumer causing toxicological risk, environmental pollution after disposal of the package and possibility of negative effects during recycling and recovery in the process of making new packaging materials (Otles and Yalcin, 2014). To overcome these issues, polymer films and labels have been developed (Ahvenainen, 2003).

Products in the market

A lot of studies discussed in this review outlined the importance of biopreservation. Fermentation using LAB imparts several features including specific aroma, flavor and texture alongside with extension of shelf-life. Studies given below necessitate the practical application of biopreservation techniques. Recently, around 270 LAB were isolated and checked for possible antifungal potential against *Penicillium expansum* MUCL2919240. Fourteen isolates outlined potential against seven molds but three LAB strains including 4F, 4JC (*Lactiplantibacillus plantarum*) and 3MI3 (*Lacticaseibacillus paracasei*) showed highest activity. When these three strains were tested in bread as biopreservative, imparted antifungal effects suggesting its practical application in baking industry (El oirdi et al., 2021).

The results of another investigation suggested the successful application of starter bacterial cultures in breadmaking. In that study, wholemeal rye bread made with sourdough containing culture of *Lactiplantibacillus plantarum* 2MI8 and exopolysaccharide EPS-producing *Weissella confusa/cibaria* 6PI3 strains showed better results as compared to bread made with commercial culture. Adding yeast to sourdough breads, particularly those made from the aforementioned starting culture, enhanced their organoleptic aspects, volume and hardness. Fermenting wholemeal rye dough, particularly without yeast, reported to reduce the quantity of higher inositol phosphates, that enhanced mineral bioavailability when compared to applied flour (Litwinek et al., 2022). When compared to bread fermented using only baker's yeast, sourdough fermented with different LAB species (*Companilactobacillus farciminis, Lactiplantibacillus plantarum, Limosilactobacillus fermentum, Levilactobacillus brevis, Leuconostoc citreum, Weissella minor*) bread showed a notable decline in fructans, increased acidity, volume, and improved shelf-life. Including specific cultures as starters in sourdough reduced fructans content by > 92%, thereby producing a low fermentable oligo, di-, and monosaccharides and polyols (FODMAP) bread suitable for irritable bowel syndrome (IBS) patients with improved nutritional and technological properties (Menezes et al., 2021).

In another study, 194 LAB isolated from several Algerian raw milk samples and Amoredj (conventional fermented product) were evaluated for their biopreservation potential against mold spoilers in dairy and baking items. During *in vitro* studies, all the strains were tested against *Aspergillus flavus* T5, *Mucor racemosus* UBOCC-A-109155, *Paecilomyces formosus* AT, *Yarrowia lipolytica* UBOCC-A-216006, *Penicillium commune* UBOCC-A-116003 and *Aspergillus tubingensis* AN, and only 3 strains named *Leuconostoc mesenteroides* L1, *Lactiplantibacillus plantarum* CH1, and *Lactiplantibacillus plantarum* CH1 inhibited mold activity in a notable manner. The preservation efficiency of these strains was subsequently confirmed in two actual products, *viz.* sourdough bread and sour cream that has been exposed to fungal spoilers. Results indicated that antifungal LAB were able to inhibit the development of their fungal targets and may have promising industrial uses (Ouiddir et al., 2019).

Decades earlier, 4 different LAB strains (*Levilactobacillus brevis* CRL 772, *Lp. plantarum* CRL 778, *Lev. brevis* CRL 796, *L. reuteri* CRL 1100) were reported to retard the growth of *Penicillium* spp. and doubled bread shelf-life compared to breads made with just *Saccharomyces cerevisiae* having shelf-life of 2 days. The higher biopreservation effect was seen with *Lp. plantarum* CRL 778, wherein shelf-life of 5 days was observed, comparable to the effect of 0.2% calcium propionate. The preservation properties of SL778 were supposed to be associated with the release of acetic and phenyllactic acid, and lactic acid, which decreased the pH of the dough. Moreover, combining SL778 with 0.4% calcium propionate doubled the shelf-life of packaged bread to 24 days that was almost 2.6-folds more as compared to the breads made with only 0.4% calcium propionate (Gerez et al., 2010).

Recently, 18 day's shelf-life of packed bread was noted using phenyllactic acid produced by *Pediococcus acidilactici* CRL 1753 (Bustos et al., 2018). *Lp. plantarum* FST 1.7 was reported to inhibit the activity of bread spoilage agents (*Fusarium culmorum* and *F. graminearum*) wherein lactic acid, phenyllactic acid, and two cyclic dipeptides cyclo (L-Leu-L-Pro) and cyclo (L-Phe-L-Pro) were discovered as the strain's primary antifungal chemicals (Dal Bello et al., 2007). *Lp. plantarum* 21B strain showed antifungal activity against *Aspergillus niger* in sourdough and bread made from it. The activity was reported to be due to the existence of phenyllactic acetic and lactic acids. The same research also showed that LAB may be more effective in preserving food than certain chemical preservatives (Lavermicocca et al., 2000).

Importantly, biopreserved bread can be labeled as "clean label" product as it does not contain any chemical preservatives. Around the world, consumers now look at the label and see the ingredients. In Europe, the number of customers who consider the ingredient list vital rose from 3% in 2011 to 78% in 2013 (Sweetman, 2016). Breads made with sourdough (or bacterial starter cultures) are now available in the markets with longer shelf-life and gaining the attention of consumers, despite sometimes of high prices. Sourdough product descriptions attracting the consumer with various slogans like: extended shelf-life; rich taste; high in fibers, protein, vitamins, minerals; suitable for anyone who are looking for low calories; vegan friendly; keto friendly; no artificial colors, flavors or preservatives; organic. A few examples of market available sourdough breads can be seen in table below. It contains product name, description, ingredients, country of origin and market available picture. A few examples of market available sourdough products (bread, pasta, biscuits, noodles, crackers) are shown in Table 2.

Regulatory framework

The general principles of the EU legislation on procedures for contaminants in food were laid down in 1993 by the Council Regulation (EEC) No 315/93 (The Council of the European Communities, 1993). This legal act authorized the European Commission (EC) to implement measures for public safety and health, containing the implementation of maximum levels. It is crucial to preserve public health, thus it is necessary to maintain contaminants at low levels, i.e., acceptable on a toxicological level. As it is difficult to completely understand and eradicate food contamination, maximum levels of contaminants must be established at a strict standard that is practically reachable using sound farming practices while considering the danger associated with food intake. The EC set

Table 2. Examples of sourdough breads available in the market.

Product	Description	Ingredients/other features	Country
FREYA'S	Freya's German sourdough bread	Inspired by artisan sourdough bread, Freya's German sourdough is made with a sourdough starter from the North Rhine region of Germany. This gives the bread a delicious tangy flavor, with the soft texture of Freya's that you love.	Germany
Rudolph's	The finest rye breads baked in the age-old Bavarian style	Unbleached wheat flour, water, sour, (water, dark rye flour and bacterial culture), coarse whole rye grain, whole rye seed, sesame seeds, flax seed, honey, yeast, salt, whey powder (milk), calcium propionate.	Canada
Sainsbury's Sourdough medium sliced white bread	Sliced fermented white sourdough bread. Made with a slow-fermented starter dough	Fortified British wheat flour, water, sea salt, rapeseed oil, extra virgin olive oil (0.5%), soya flour, rice flour, wheat gluten, fermented wheat flour, flour treatment agent, ascorbic acid.	United Kingdom
San Luis sourdough bread	Truly authentic sourdough taste with bubbling flavorful crust	Unbleached enriched wheat flour, water, sourdough starter (wheat flour, water), salt, cultured wheat flour.	USA
Alma sandwich bread loaf	Alma Baking Sourdough Bread Sliced - Handmade, Artisinal, Small Batch - Sandwich Bread Loaf	Water, wheat flour, whole wheat flour, rye flour, salt/baked products packaging by utilizing natural enzymes created through fermentation.	USA
Laurent sourdough bread	30-hour recipe Stone baked sourdough	It's made in Australia using locally milled flour and a natural starter culture, which is over 25 years old and based on a traditional recipe.	Australia
Inked organics	Inked organics Rosie's San Francisco bay sourdough bread OU kosher	Made naturally with an aged organic sourdough starter.	USA
San Francisco style sourdough bread	Izzio Artisan bakery	Wheat flour, water, sourdough culture (wheat flour, water), whole wheat flour, sea salt, malted barley flour, enzymes.	USA
No.1 Spelt Sourdough Bread	Sliced White and Whole meal spelt sourdough bread made with slow fermented sourdough	Fortified spelt flour (spelt (wheat) flour, calcium carbonate, iron, niacin, thiamin), water, whole meal spelt (wheat) flour, salt.	United Kingdom

Table 2 contd. ...

...Table 2 contd..

Product	Description	Ingredients/other features	Country
Kaslo Sourdough's Pasta Fermentata	A Promise of Quality and Flavor	Our sourdough pasta is not only flavorful but also packed with nutrients including proteins from quinoa, buckwheat, and other types of grain. It also contains dietary fibers, B-vitamins, potassium, and iron.	Canada
Sourdough Pasta - Khorasan Macaroni	A rich taste and higher quantities of fiber, protein, vitamins and minerals compared to modern wheat varieties	Sustainable Durum Wheat Flour, Sustainable Khorasan Flour and Filtered Water.	Australia
Sourdough Pasta - Spelt Shells	The pasta is long-fermented which improves the digestibility and nutrient availability. Sourdough Pasta is nutrient dense and good source of protein and dietary fiber.	Sustainable Durum Wheat Flour, Stoneground Sustainable Spelt Flour and Filtered Water.	Australia
Sourdough pasta emmer twists	The offering changes daily and is influenced by the seasons and the desire to provide people with better choices for their health	Certified sustainable stoneground emmer flour, certified sustainable durum wheat flour, sourdough starter.	Australia
Arnott's Sourdough crispms	Crackers with real sourdough starter	A delicious crisp made with real sourdough starter and oven-roasted to perfection to deliver a flavorsome snack you'll love.	USA
Spinach Sourdough Noodles	Our sourdough noodles are suitable for anyone who are looking for low calories, vegan friendly and keto friendly noodles.	Sourdough Culture, Unbleached Wheat Flour, Water, Salt, Spinach.	Malaysia
Sourdough Cookies Sunflower & Pumpkin Seeds	Our commitment on minimum food processing to optimum the food nutrients and away from artificial elements, yet the food is delicious, healthy and safe to consume.	Sourdough Starter, Butter, Brown Sugar, Unbleached Flour, Salt, Egg, Pumpkin Seeds, Sunflower Seeds.	Malaysia
Smoked sea salt sourdough cracker	Our light and crisp Smoked Sea Salt Sourdough Crackers are lightly seasoned with a hint of smoked sea salt to add depth and complexity to the palate.	Organic Flour, Organic Sourdough Starter (Organic Flour and Water), Organic Olive Oil, Sea Salt.	Australia

Product	Description	Ingredients	Country
No.1 Rye Sourdough Crackers	Crackers made with rye sourdough and honey. With the distinctive flavour of rye, these crisp, crunchy sourdough crackers are from a Dorset bakery founded in 1916. They're delicious with blue cheese.	Wholemeal, rye flour (39%), wholemeal wheat flour, fortified wheat flour (wheat flour, calcium carbonate, iron, niacin, thiamin), sourdough (12%) (rye flour, water, malted wheat flour, starter culture, yeast), honey (6%), black treacle, sunflower oil, fermented rye flour, salt, yeast.	United Kingdom
Peter's Yard Fig & Spelt Sourdough Crackers	Great taste 2020, Baked to Imperfection, Good Things Take Time, No Palm Oil, High in Fibre, No Artificial Colours, Flavours or Preservatives, Suitable for Vegetarians	Spelt Wheat Flour 79%, Milk, Dried Fig Pieces 16%, Sourdough 10% (Rye Flour, Water), Honey, Sea Salt.	United Kingdom
Rosemary & Thyme Sourdough Chips	-	Flours (wheat flour, whole-wheat flour), extra-virgin olive oil, filtered water, rosemary, thyme, salt.	Canada

176 *Novel Approaches in Biopreservation for Food and Clinical Purposes*

maximum permitted concentrations that has direct influence on food/feed traders throughout Europe. The maximum limits for mycotoxins and nitrates in food is present in Commission Regulation (EC) No 194/97 and Commission Regulation (EC) No 466/2001, respectively (European Commission, 1997, 2001). In order to ensure both quality and safety of food all through the food production chain, the guidelines of *Codex Alimentarius* (Codex Alimentarius, 1995a) must be followed. This contains the enactment of systems such as good hygienic practices (GHPs), good agricultural practices (GAPs), good manufacturing practices (GMPs) and Hazard Analysis and Critical Control Point (HACCP) systems. These practices and systems are collectively referred as "good practices".

After a series of updates and repeals, Commission Regulation (EU) 1881/2006 (European Commission, 2006) established maximum limits for nitrates, mycotoxins, metals and dioxins and polychlorinated biphenyls (PCBs) as well as polycyclic aromatic hydrocarbons (PAHs), melamine and its structural analogues in various foodstuffs, as well as a number of other pollutants. As previously mentioned, mycotoxins are among the main contaminant and most significant hazards linked to cereal safety (Codex Alimentarius, 1991). Maximum permitted level of various mycotoxins including B1, sum of B1 and B2, sum of G1 and G2, deoxynivalenol, zearalenone and ochratoxin A is defined in EU regulation (No. 1881/2006) (EEC, 2006) as 2, 4, 4, 500, 50 and 3 µg/kg, respectively, in bread and other bakery items. Moreover, it also describes the maximum permitted levels of these mycotoxins in flours (Table 3).

As previously described, chemical preservatives are used in baking industry because they are useful but they also have negative effects on the consumers. As chemical preservatives, weak organic acids like propionic and sorbic acids are often

Table 3. Maximum permitted level of different types of mycotoxins (European Commission, 2013).

Type of Mycotoxin	Baking items	Permitted level
B1	Cereals and cereal-based products	2.0 µg/kg
Sum B1 & B2 G1 & G2		4 µg/kg
Deoxynivalenol	Bakery products including bread, cookies and pastries, etc.	500 µg/kg
	Cereals offered for direct human consumption, flour, bran, and germ	750 µg/kg
Zearalenone	Bread, pastries, cookies, cereal snacks but omitting maize-based snacks and cereals	50 µg/kg
	Cereals offered for direct human consumption, flour, bran, and germ	75 µg/kg
Ochratoxin A	All items derived from unprocessed cereals, including processed and human-edible cereals	3 µg/kg
Fumonisin B1+ B2	Not specified	
Sum T-2 & HT-2 toxin	Not specified	

added to stop the growth of unwanted microorganisms and make bakery products last longer. Most of the time, when the pH is low these acids are in their non-dissociated form and can easily get through the plasma membrane. In Europe, synthetic preservatives can only be used in certain amounts in baked goods to make sure the food is safe (EEC, 2008). Most of the time, potassium, sodium and calcium salts of propionic and ascorbic acids are used because they are soluble in water and are easy to work with, contrasting to their corrosive acids (Magan et al., 2003). In European regulatory framework the maximum permitted level of sorbate and propionate is mentioned as 0.2 and 0.3%, respectively (EEC, 2008). European Union has also set maximum limits of propionic acid and propionate rye bread and pre-packed sliced bread as 3000 mg/kg, in pre-baked and energy reduced breads as 2000 mg/kg, and in pre-packed bread as 1000 mg/kg (EEC, 2008).

No doubt, higher concentrations of sorbate or propionate are desirable for their antifungal action. However, this might also affect the sensory quality of the product. If we continue to use the same preservatives over a lengthy period of time, the fungus we are trying to protect against may eventually become resistant to those chemical preservatives (Levinskaite, 2012; Stratford et al., 2013b; Suhr and Nielsen, 2004). In addition, concentrations of the preservative that are lower than the maximum level must be selected with caution. It has been shown that concentrations that are less than 0.03% may lead to an increase in the formation of mycotoxins (Arroyo et al., 2005) as well as an in acceleration of fungal development (Marin et al., 2002). The problems associated with the use of chemical preservatives have prompted researchers to seek alternatives to prevent spoilage organisms in food items, e.g., the use of fermentation, plant derived compounds, protective packaging and ethanol.

The use of LAB as starter cultures in sourdough breadmaking has been thoroughly discussed in previous sections. LAB release biopreservation compounds *in situ*. Sourdough, in place of artificial preservatives, guarantees a clean label, enhance flavour, texture, nutritional aspects and consumer acceptability (Rocha, 2011; Pawlowska et al., 2012). Many LAB species including of the genera *Lactobacillus*, *Pediococcus*, *Lactococcus* and *Leuconostoc* have been submitted for safety review to the European Food Safety Authority (EFSA) without raising any safety concerns. As a result, they have been added to the qualified presumption of safety (QPS) list for use in the food and feed chain of the European Union (EFSA, 2012). They have the GRAS certification in USA, which is administered by the Food and Drug Administration (FDA). The use of *W. anomalus* seems to be safe, since this species is also on the QPS list, but only for enzyme production (EFSA, 2012). *Bacillus subtilis* is also QPS, and its usage in food has not been linked to any safety issues (EFSA, 2012). As a matter of fact, endospore-forming Gram-positive rods belonging to the family of Bacillaceae are likely to be find in sourdoughs (Rocha and Malcata, 1999, 2012, 2016a, 2016b).

EFSA requested the Panel on Biological Hazards (BIOHAZ) for a scientific opinion on maintaining the list of QPS biological agents intentionally added to food or feed. In food and feed production, bacteria and fungi are utilized directly or as additives or enzymes. Some have a lengthy history of safe usage, while others may be risky for consumers. The scientific committee assessed microorganisms likely to

178 *Novel Approaches in Biopreservation for Food and Clinical Purposes*

be the subject of an EFSA opinion and prepared a QPS list (Arora et al., 2011; Antas et al., 2012). Generally describing, of the 70 microorganisms notified to EFSA, 64 were not evaluated, wherein 11 were filamentous fungi, 5 *E. coli*, one each *Clostridium butyricum, Enterococcus faecium, Streptomyces* spp. and *Bacillus nakamurai*, and 43 taxonomic units that already had a QPS status. However, 6 taxonomic units were evaluated wherein *Paenibacillus lentus* was reassessed because an update was requested for the current mandate. *Enterococcus lactis, Aurantiochytrium mangrovei, Schizochytrium aggregatum, Chlamydomonas reinhardtii* and *Haematococcus lacustris* were assessed for the first time. After evaluation the taxonomic units *A. mangrovei, C. reinhardtii* and *S. aggregatum* were not recommended for QPS status due to the lack of information in feed and food chain, and *E. lactis* and *P. lentus* because of limited information and safety issues, respectively. *Haematococcus lacustris* was recommended for QPS status with the qualification 'for production purposes only' (EFSA, 2022).

Qualified presumption of safety has recently entered EU law with the publication of a new Commission Implementing Regulation (EU) No 562/2012 amending Commission Regulation (EU) No 234/2011 (Commission Implementing Regulation, 2012) with regard to specific data required for risk assessment of food enzymes. Importantly, no toxicity data for enzyme application will be needed prior production of a food enzyme biological agent has QPS status according to the most recent list of EFSA. On the other hand, if safety concerns including impurities, residues and degradation products connected with production, recovery and purification of total enzyme, then the EFSA, pursuant to Article 6(1) of Regulation (EC) No 1331/2008, may seek further data for assessment of risk. The 1-monoglyceride of medium chain fatty acid belong to food additives in accordance with the principle of "*quantum satis*" addition for good manufacturing practice and commonly acts as emulsifier in breads (EEC, 2008).

Regulation (EU) No. 10/2011 of the Commission (EEC, 2011), also termed as Plastic Implementation Measure (PIM), governs the type of material that can be employed for food packaging and sets limits on migration in order to keep the food safe. In said regulation, three different types of materials mentioned in Table 2 of aforementioned regulation titled "food category specific assignment of food simulants" with serial number 08.06, 02.06 and 02.05 can be used for bakery items, mainly in bread and flours. In the USA, ethanol is known as GRAS owing to its low toxicity. The law on food additives in the European Union does not cover ethanol (EEC, 2008). So, there are no rules about how ethanol can be used as a preservative. If it is added to a product, the label must say "ethanol" or "ethyl alcohol". Natamycin produced by *Streptomyces natalensis* was reported to possess antimicrobial properties also listed as preservative (E235) in the EU regulation on food additives (EEC, 2008).

This section covers the maximum level for aflatoxins described by European Commission and the use of chemical preservatives in bakery items mainly bread and flours owing to their negative impacts towards health. Alternatively, various bacterial strains having long history of safe use acting as biopreservation agent mentioned in QPS list need to undergo no legal requirements.

Conclusions

A vast variety of microorganisms are present in the ecological niche of bread production. Some of them are beneficial and some may pose negative impact for human's health. Therefore, it is crucial to have a good insight of their diversity which may be helpful in finding new and evolving biopreservation processes to prolong shelf-life of bakery products and gain economical and public health benefits. Latest findings in microbial diversity during all stages in flour and bread production are described in this chapter, with special emphasis on the most troublesome ones as well as the dangers that arise due to their presence in food goods. Although a great amount of viable microorganisms will be decreased or even vanished during the milling process, cereal grain microbiota still affects further the microbiota and quality of flours which also harbor microorganisms from the inside of the milling equipment and other surrounding environment. The most unwanted microorganisms in these stages are mycotoxin producing molds of *Aspergillus* spp., *Penicillium* spp., *Fusarium* spp. and *Caviceps* spp. These secondary metabolites are thermally stable and persistent and, therefore, it is important to prevent molds to synthesize them. During the dough/sourdough formation other microbial communities become predominant and is influenced mostly by available carbohydrates, temperature and water activity of the medium. The microbiota of dough/sourdough is not constant and change in time with respect to the temperature and decreasing of pH. Homo-e and heterofermentative LAB as well as fermentative yeasts are the main microorganisms present in developed sourdoughs. More than 50 LAB species have been reported in the literature mostly belonging to *Lactobacillus* spp., *Leuconostoc* spp., *Pediococcus* spp. and *Weissella* spp. Most of them are beneficial and contribute to the rheological and sensory traits of bread and also some of them possess antimicrobial and antioxidant properties, playing an important role in safety of the final product. Available sugars and metabolic pathways of microorganisms are key-factors during establishing stable microbial consortia in mature dough/sourdough.

Bread after baking is free of microorganisms since high temperatures during baking inactivate most of them. Also, the acidic environment during dough/sourdough preparation will inactivate a great number of microorganisms. The exception are some thermophilic microorganisms such as *Bacillus* spp. that cause rope spoilage of bread and the mycotoxins. Microbiological changes of bread occur mostly in post baking period during staining due to microbial contamination that arise from surrounding environment and bakery equipment and contribute in bread spoilage. Usually, molds from the genus *Eurotium* spp., *Penicillium* spp., *Alternaria* spp., *Rhizopus* spp., rope spores (*Bacillus* spp.) and some yeasts such as *Saccharomycopsis fibuligera*, *Wickerhamomyces anomalus* and *Hyphopichia burtonii* causing "chalk mold" are the major organisms responsible for post-processing contamination of bread showing easily visible growth on bread surface.

The shelf-life of bread is a dynamic system which undergo physical, chemical and microbiological changes during standing. Finding the existing models for occurrence of the main spoilage group of microorganisms and developing new models of biopreservation for the same or other spoilage organisms is crucial in development of future perspective methods for shelf-life prolongation, food security and better food safety bakery product.

Various microbial species from different kind of sources are identified and cited in the literature as possessing antagonistic and antimicrobial properties but only few of them meet the conditions to be used as protective cultures in bread biopreservation. In fact, to be used as bioprotective culture, microorganisms and their metabolites, must be non-toxic for humans and possess wide antimicrobial spectral range in relatively low doses that do not impair final product quality and sensory characteristics. The chapter gives an insight of microbial species with antagonistic and antimicrobial action well known in the literature with potential to be used as protective cultures in bread biopreservation. Fermentation is one of the oldest and at the same time completely natural and safe process for biopreservation of bread and other kind of food. Fermentation of baking dough/sourdough is conditioned by the presence of LAB and yeasts, which may be naturally present or intentionally introduced. Most of the LAB used as protective cultures belong to the genera of *Lactobacillus*, *Pediococcus* and *Leuconostoc*. Among yeasts the most studied and involved in food preservation are the baker's yeast *S. cerevisiae* and *W. anomalus*. Both groups of microorganisms can act in a vast array of ways and inactivate unwanted microorganisms. They compete in the utilization of nutrients and synthetize compounds necessary for the rheological and sensory characteristics of bread and, at the same time, produce metabolites with antimicrobial effects against common undesirable microorganisms that contribute to bread spoilage. In fact, the main compounds produced in the process of fermentation such as lactic and acetic acids act as the most important biopreservation agents. Other bacterial groups, such as some *Bacillus* spp. and propionic acid bacteria produce a diversity of antimicrobial peptides and propionic acid, which are accepted as preservatives in breadmaking industry. The variety of synthetized metabolites with antimicrobial effect is most diverse although most of them are not synthesized in sufficient quantity. It is assumed that complex synergistic action of these compounds in the medium play crucial role in preservation properties of protective cultures. The research on the metabolic pathways and metabolic characteristics particularly in LAB as well as understanding the synergism and mechanism of proto-cooperation between LAB and yeasts will help further to improve the existing strategies on biopreservation. Even more, to create new biosynthetic metabolic pathways to guide the development of gene editing systems to carry out targeted transformation in selected strains and obtain constructed LAB engineering strains able to synthetize valuable antimicrobial compounds in higher yield.

Long-term consumption of the chemical preservatives used to extend the shelf-life of bakery products may increase the risk of chronic diseases in humans and animals. A wide variety of primary and secondary metabolites synthetized by microorganisms especially LAB are a promising tool to find alternative and more natural solution to this problem.

The benefits of LAB when used in breadmaking are numerous and their application is extensive. These bacteria act as a cell factory to produce different substances with wide antimicrobial action, which are harmless for humans such as various organic acids, bacteriocins, fatty acids, exopolysaccharides and some other low weigh molecular substances that at the same time contribute to the rheological, sensory and nutritional characteristics of dough and consequently bread as a final product.

The use of individual substance isolated from LAB to be used for its bioactive properties is cost effective due to the isolation and purification procedure and often not so effective especially when it is used for its antimicrobial properties. Therefore, the use of sourdough enriched with particular LAB strains enables control over the characteristics of the bread as well as more effective protection against spoilage microorganisms as a result of the synergistic mode of action of the LAB primary and secondary metabolites, especially the synergistic effect which they show with other preservatives.

The bakery industry is facing a big challenge since the consumer's demand of products free of artificial additives and harmful chemicals and at the same time products with superior flavor that keep its quality and shelf-life during longer period of time. Fermentation is a completely natural solution which satisfies simultaneously all these conditions. This chapter also focused the characteristics of the microorganisms used as starter cultures in bread dough fermentation, different types of fermentation processes giving special attention to the biopreservation processes that arise from LAB and yeast carbohydrate fermentation. The process of fermentation is species and substrate specific and, at the same time, dependent on the temperature, humidity and pH of the substrate, among other factors. Metabolic activity of dough/sourdough microbiota during fermentation is responsible for rheological and sensory performance of dough and, consequently, of the rheological, sensory characteristics and the shelf-life of the final product. During the dough/sourdough fermentation the pH decreases mainly due to lactic and acetic acid production as the end products of the metabolic pathways of LAB and yeasts. This reduce the pH value of the substrate bellow the metabolic inhibition as well as the growth rate of undesired microorganism. The pH ranges around 3.5–4.0 in mature sourdough and around 4–5 in dough. Moreover, some of the metabolites obtained during the processes of fermentation, especially low molecular compounds possess important antimicrobial activity.

This chapter also provides potential application data of bacteriophages and retrieved endolysins as biopreservation agent in bread and related baking items. It can be seen that *Bacillus* group is mainly responsible for the deterioration of bakery items and a lot of studies showed their control using bacteriophages and their endolysins replacing chemical preservatives. Nexus to the mentioned antibacterial activities, the isolated bacteriophages have a great prospect to be used as food biocontrol agent.

Bread can be naturally protected from spoiling during preparation by the processes of fermentation triggered by selected and beneficial LAB and yeasts. During the storage time several hours after baking, bread is prone to contamination by spoiling organisms present in the surrounding environment. This spoiling and staling can be postponed if bread is wrapped as soon as it is cooled after baking. The packaging nowadays is not just a simple barrier between microorganisms and the product. Trends in the technology of food packaging are designed to meet three concepts: (i) rational application and reduction of the amount of the packing materials, (ii) implementation of environmentally friendly packing materials, and (iii) improvement of the packaging performances. The last concept refers to some innovative technologies developed that modify the atmosphere in the package and/or utilize antimicrobial compounds, antioxidant compounds, moisture and oxygen absorbers formulated in nanoparticle delivery systems to control changes in bread quality and/or alert when these changes appear. Such innovative approaches of packaging are known as modified atmosphere packaging, active packaging and intelligent or smart packaging. Integrating several methods of these innovative packaging technologies shows better results in prolongation of the shelf-life of bread. Polymers are still the most common packaging materials, greatly justified by their desired features such as lightness, softness and transparency. But due to their non-biodegradability, they pose a serious ecological problem. Therefore, a new class of materials with much improved properties such as bio-nanocomposites are being developed. In the recent years, intelligent and smart food packaging technologies gained popularity because they extend shelf-life and safety, improve quality, enhance absorption of nutrients and control external and internal conditions of packed food products. Other benefits of intelligent packaging are their ability to switch on and off in response to the changes in internal and external conditions, and to communicate with the customers offering information about the status of the product. New packaging materials created to incorporate antifungal compounds formulated in nanoparticle delivery systems to control the spoilage microorganisms in bread represents a significant advance in the use of nanobiotechnology in bread packaging towards the biopreservation of food. Antibodies formulated in nanoparticles and incorporated in bread packages is currently the main approach of pathogen detection in innovative/smart technology. The isolation of aptamers for number of pathogens showed to have numerous advantages over the antibodies such as improved stability, low cost, and ease of production. Bio-barcode assays using target-specific antibody or aptamer-coated nanoparticles offered highly sensitive capabilities in the detection of microorganisms with potential of multiplexing. If health concerns addressed to the use of nanomaterials are solved, then the bio-barcode assays could be a promising approach for the future of food packaging. This advances in nanobiotechnology offer bread contaminants to be monitored during the all stages of the food chain until consuming, which will result at the end of the day in significant decrease of the incidence of foodborne diseases and increasing consumer's confidence in the quality and safety of food.

This chapter also described the importance of starter cultures mainly LAB. Bacterial strains showed host range during *in vitro* analysis also outlined lytic activity against hose spoilage agents in food products where tested. Practical applications and scope of biopreservation were also tabulated, mentioning the market available products. Sourdough breads, pasta, crackers, biscuits and noodles are available in the market with the caption "more shelf-life and no preservative" which is very attractive for the consumers.

Acknowledgements

This work is based upon the work from COST Action 18101 SOURDOMICS—Sourdough biotechnology network towards novel, healthier and sustainable food and bioprocesses (https://sourdomics.com/; https://www.cost.eu/actions/CA18101/, accessed in 2022-11-05), where the authors B.K., S.K., T.E., A.C., M.Q., J.M.R. are members of different working groups of SOURDOMICS (see https://sourdomics. com/en/working-groups), and the author J.M.R. is the Chair and Grant Holder Scientific Representative, and is supported by COST (European Cooperation in Science and Technology) (https://www.cost.eu/, accessed in 2022-11-05). COST is a funding agency for research and innovation networks. Regarding to the author J.M.R., this work was also financially supported by LA/P/0045/2020 (ALiCE) and UIDB/00511/2020 - UIDP/00511/2020 (LEPABE) funded by National funds through FCT/MCTES (PIDDAC).

Abbreviations

3-hydroxypropionaldehyde (3-HPA), Active packaging (AP), Aflatoxins (AF), Arabic gum (AG), Ascorbic acid (AA), Bacteriocin-like inhibitory substances (BLIS), Bacteriophages (BPs), Biological Hazards (BIOHAZ), Bread gum (BG), Carbon dioxide (CO_2), Carboxymethyl cellulose (CMC), Colony-forming units (CFU), Diacetyl tartaric acid (DATA), Dough yield (DY), Electrochemical biosensors (ECB), Embden–Meyerhof–Parnas pathway (EMP), Equilibrium relative humidity (ERH), Esters of mono- and diglycerides (DATEM), European Commission (EC), European Committee Regulations (ECR), European Food Safety Authority (EFSA), European Union (EU), Exopolysaccharides (EPS), Fermentable oligo, di-, and monosaccharides and polyols (FODMAP), Fermentation quotient (FQ), Food and Drug Administration (FDA), Fumonisins (F), Generally recognized as safe (GRAS), Good agricultural practices (GAPs), Good hygienic practices (GHPs), Good manufacturing practices (GMPs), Guar gum (GG), Hazard Analysis and Critical Control Point (HACCP), Heteropolysaccharides (HePS), Homopolysaccharides (HoPS), Hydroxypropyl cellulose (HPC), Hydroxypropylmethyl cellulose (HPMC), Infrared (IR), Intelligent packaging (IP), Iota carrageenan (i-CAR), Irritable bowel syndrome (IBS), Kappa carrageenan (k-CAR), Lactic acid bacteria (LAB), Lambda carrageenan (l-CAR), Locust bean gum (LBG), Median lethal dose (LD50), Microcrystalline cellulose (MCC), Microwave radiation (MW), Modified atmosphere packaging (MAP), Molecular oxygen (O_2), Nitrogen (N_2), Ochratoxins (OT), Phenolic acid (PA), Phenylactic acid (PLA), Plastic Implementation Measure (PIM), Polychlorinated biphenyls (PCBs), Polycyclic aromatic hydrocarbons (PAHs), Propionic acid bacteria (PAB), Qualified presumption of safety (QPS), Rapid Alert System for Food and Feed (RASFF), Sodium alginate (AL), Sodium stearoyl-lactylate (SSL), Tamarind seed gum (TSG), Tragacanth gum (TG), Trichothecenes (T), Ultraviolet (UV), United Nations (UN), United States of America (USA), Water activity (a_w), World Health Organization (WHO), Xanthan gum (XG)

Bibliography

Aboukhalaf, A., El Amraoui, B., Tabatou, M., Rocha, J.M. and Belahsen, R. 2020. Screening of the antimicrobial activity of some extracts of edible wild plants in Morocco. *Journal of Functional Food in Health Disease*, 6(10): 265–273. doi: 10.31989/ffhd.v10i6.718. ISSN 2574-0334, Publication date: 25/06/2020.

184   *Novel Approaches in Biopreservation for Food and Clinical Purposes*

Aboukhalaf, A., Tbatou, M., Kalili, A., Naciri, K., Moujabbir, S., Sahel, K., Rocha, J.M., Belahsen, R. 2022. Traditional knowledge and use of wild edible plants in Sidi Bennour region (Central Morocco. *Ethnobotany Research and Applications* (University of Hawaii Press), 23(11): 1–18. doi: https://dx.doi.org/10.32859/era.23.11.1-18.

Ağagündüz, D., Yılmaz, B., Şahin, T.O., Güneşliol, B.E., Ayten, S., Russo, P., Spano, G., Rocha, J.M., Bartkiene, E., Özogul, F. 2021. Dairy Lactic Acid Bacteria and Their Potential Function in Dietetics: The Food–Gut-Health Axis. *Foods* (MDPI), 10(12): 3099. doi: https://doi.org/10.3390/foods10123099. Publication date: 14/12/2021.

Agriopoulou, S. 2021. Ergot alkaloids mycotoxins in cereals and cereal-derived food products: characteristics, toxicity, prevalence, and control strategies. *Agronomy*, 11(5): 931. https://doi.org/10.3390/agronomy11050931.

Ahvenainen, R. 2003. Novel Food Packaging Techniques; Woodhead Publishing: Boca Raton, FL, USA.

Alfonzo, A., Ventimiglia, G., Corona, O., di Gerlando, R., Gaglio, R., Francesca, N., Moschetti, G. and Settanni, L. 2013. Diversity and technological potential of lactic acid bacteria of wheat flours. *Food Microbiology*, 36(2): 343–354. https://doi.org/10.1016/j.fm.2013.07.003.

Alpers, T., Kerpes, R., Frioli, M., Nobis, A., Hoi, K.I., Bach, A., ... and Becker, T. 2021. Impact of storing condition on staling and microbial spoilage behavior of bread and their contribution to prevent food waste. *Foods*, 10(1): 76.

Antas, P., Brito, M., Peixoto, E., Ponte, C. and Borba, C. 2012. Neglected and emerging fungal infections: review of hyalohyphomycosis by *Paecilomyces lilacinus* focusing in disease burden, *in vitro* antifungal susceptibility and management. *Microbes and Infection*, 14: 1–8.

Antognoni, F., Mandrioli, R., Potente, G., Taneyo Saa, D.L. and Gianotti, A. 2019. Changes in carotenoids, phenolic acids and antioxidant capacity in bread wheat doughs fermented with different lactic acid bacteria strains. *Food Chem.*, 292: 211–216.

Appendini, P. and Hotchkiss, J.H. 2002. Review of antimicrobial food packaging. *Innovative Food Science and Emerging Technologies*, 3: 113–126.

Arena, M.P., Russo, P., Giuseppe Spano, G. and Capozzi, V. 2020. From microbial ecology to innovative applications in food quality improvements: the case of sourdough as a model matrix. *J. Multidiscip. Res.*, 3: 9–19;

Arora, S., Saini, H.S. and Singh, K. 2011. Biological decolorization of industrial dyes by *Candida tropicalis* and *Bacillus firmus*. *Water Science and Technology*, 63(4): 761–768.

Arqués, J.L., Fernández, J., Gaya, P., Nuñez, M., RodríGuez, E. and Medina, M. 2004. Antimicrobial activity of reuterin in combination with nisin against food-borne pathogens. *International Journal of Food Microbiology*, 95(2): 225–229. https://doi.org/10.1016/j.ijfoodmicro.2004.03.009.

Arroyo, M., Aldred, D. and Magan, N. 2005. Environmental factors and weak organic acid interactions have differential effects on control of growth and ochratoxin A production by *Penicillium verrucosum* isolates in bread. *International Journal of Food Microbiology*, 98: 223–31.

Ashiq, S. 2015. Natural occurrence of mycotoxins in food and feed: Pakistan perspective. *Comprehensive Reviews in Food Science and Food Safety*, 14(2): 159–175.

Axel, C., Röcker, B., Brosnan, B., Zannini, E., Furey, A., Coffey, A. and Arendt, E.K. 2015. Application of *Lactobacillus amylovorus* DSM19280 in gluten-free sourdough bread to improve the microbial shelf life. *Food Microbiology*, 47: 36–44. https://doi.org/10.1016/j.fm.2014.10.005.

Axel, C., Brosnan, B., Zannini, E., Furey, A., Coffey, A. and Arendt, E.K. 2016. Antifungal sourdough lactic acid bacteria as biopreservation tool in quinoa and rice bread. *International Journal of Food Microbiology*, 239: 86–94. https://doi.org/10.1016/j.ijfoodmicro.2016.05.006.

Axel, C., Zannini, E., Elke, K. and Arendt, E.K. 2017a. A review of current strategies for bread shelf life extension. *Crit. Rev. Food Sci. Nutr.*, 57: 16, 3528–3542;

Axel, C., Zannini, E., Arendt and E.K. 2017b. Mold spoilage of bread and its biopreservation: a review of current strategies for bread shelf-life extension. *Crit. Rev. Food Sci. Nutr.*, 57: 3528–3542.

Aydin, A., Paulsen, P. and Smulders, F.J.M. 2009. The physico-chemical and microbiological properties of wheat flour in Thrace. *Turkish Journal of Agriculture and Forestry*. https://doi.org/10.3906/tar-0901-20.

Badel, S., Bernardi, T. and Michaud, P. 2011. New perspectives for *Lactobacilli exopolysaccharides*. *Biotechnology Advances*, 29(1): 54–66. https://doi.org/10.1016/j.biotechadv.2010.08.011.

Balaguer, M.P., Lopez-Carballo, G., Catala, R., Gavara, R. and Hernandez-Munoz, P. 2013. Antifungal properties of gliadin films incorporating cinnamaldehyde and application in active food packaging of bread and cheese spread foodstuffs. *International journal of food microbiology*, 166(3): 369–377. https://doi.org/10.1016/j.ijfoodmicro.2013.08.012.

Bankefa, O.E., Oladeji, S.J., Ayilara-Akande, S.O. and Lasisi, M.M. 2021. Microbial redemption of "evil" days: a global appraisal to food security. *Journal of Food Science and Technology*, 58(6): 2041–2053.

Barnett, J.A., Payne, R.W., Yarrow, D. and Barnett, L. 2000. Yeasts: characteristics and Identification (3rd ed. Cambridge University Press.

Barnett, J.A. and Entian, K.D. 2005. A history of research on yeasts 9: regulation of sugar metabolism. Yeast, 22(11): 835–894. https://doi.org/10.1002/yea.1249.

Bartkiene, E., Lele, V., Ruzauskas, M., Mayrhofer, S., Domig, K., Starkute1, V., Zavistanaviciute, P., Bartkevics, V-, Pugajeva, I., Klupsaite, D., Juodeikiene, G., Mickiene, R. and Rocha, J.M. 2020. Lactic acid bacteria isolation from spontaneous sourdough and their characterization including antimicrobial and antifungal properties evaluation. *Microorganisms (MDPI)*, 8(1): 64; Special Issue: Microbial Safety of Fermented Products; https://doi.org/10.3390/microorganisms8010064. ISSN 2076-2607,

Bartkiene, E., Özogul, F. and Rocha, J.M. 2022. Bread sourdough lactic acid bacteria—technological, antimicrobial, toxin-degrading, immune system- and faecal microbiota-modelling biological agents for the preparation of food, nutraceuticals and feed. *Foods (MDPI)*, 11: 452. doi: https://doi.org/10.3390/foods11030452. Special issue: Regulation of Food Fermentations by Bacteria, Yeasts and Filamentous Fungi): https://www.mdpi.com/journal/foods/special_issues/regulation_fermentation. Publication date: 2022-02-03.

Belz, M.C., Mairinger, R., Zannini, E., Ryan, L.A., Cashman, K.D. and Arendt, E.K. 2012. The effect of sourdough and calcium propionate on the microbial shelf-life of salt reduced bread. *Applied Microbiology and Biotechnology*, 96(2): 493–501.

Benkerroum, N. 2020. Aflatoxins: producing-molds, structure, health issues and incidence in Southeast Asian and Sub-Saharan African Countries. *International Journal of Environmental Research and Public Health*, 17(4): 1215. https://doi.org/10.3390/ijerph17041215.

Bennett, J.W. and Klich, M. 2003. Mycotoxins. *Clin. Microbiol.*, 16: 497–516. https://doi: 10.1128/CMR.16.3.497-516.2003.

Bensassi, F., Chennaoui, M., Hassen, B. and Mohamed, R. 2011. Survey of the mycobiota of freshly harvested wheat grains in the main production areas of Tunisia. *Afr. J. Food Sci.*, 5: 292–298.

Berghofer, L.K., Hocking, A.D., Miskelly, D. and Jansson, E. 2003. Microbiology of wheat and flour milling in Australia. *International Journal of Food Microbiology*, 85(1-2): 137–149. https://doi.org/10.1016/s0168-1605(02)00507-x.

Beuchat, L.R. 1983. Influence of water activity on growth, metabolic activities and survival of yeasts and molds. *Journal of Food Protection*, 46(2): 135–141. https://doi.org/10.4315/0362-028x-46.2.135.

Bhadoria, R. and H.S. Chaudhary. 2011. Recent advances of biosensors in biomedical sciences. *International Journal of Drug Delivery*, 3(4): 571–585.

Black, B.A., Zannini, E., Curtis, J.M. and Gänzle, M.G. 2013. Antifungal hydroxy fatty acids produced during sourdough fermentation: microbial and enzymatic pathways, and antifungal activity in bread. *Applied and Environmental Microbiology*, 79(6): 1866–1873. https://doi.org/10.1128/aem.03784-12.

Bortol, A., Nudel, C., Fraile, E., de Torres, R., Giulietti, A., Spencer, J.F.T. and Spencer, D. 1986. Isolation of yeast with killer activity and its breeding with an industrial baking strain by protoplast fusion. *Applied Microbiology and Biotechnology*, 24(5): 414–416. https://doi.org/10.1007/bf00294599.

Boudra, H., le Bars, P. and le Bars, J. 1995. Thermostability of Ochratoxin A in wheat under two moisture conditions. *Applied and Environmental Microbiology*, 61(3): 1156–1158. https://doi.org/10.1128/aem.61.3.1156-1158.1995.

Bradley, E.L., Castle, L. and Chaudhry, Q. 2011. Applications of nanomaterials in food packaging with a consideration of opportunities for developing countries. *Trends in Food Science & Technology*, 22: 604–610.

Brandt, M.J. 2007. Sourdough products for convenient use in baking. *Food Microbiol.*, 24: 161–164.

186 *Novel Approaches in Biopreservation for Food and Clinical Purposes*

Brandt, M.J., Hammes, W.P. and Gänzle, M.G. 2004. Effects of process parameters on growth and metabolism of *Lactobacillus sanfranciscensis* and *Candida humilis* during rye sourdough fermentation. *European Food Research and Technology*, 218: 333–338 DOI 10.1007/s00217-003-0867-0.

Broberg, A., Jacobsson, K., StröM, K. and SchnüRer, J. 2007. Metabolite profiles of lactic acid bacteria in grass silage. *Applied and Environmental Microbiology*, 73(17): 5547–5552. https://doi.org/10.1128/aem.02939-06.

Brosnan, B., Coffey, A., Arendt, E.K. and Furey, A. 2012. Rapid identification, by use of the LTQ Orbitrap hybrid FT mass spectrometer, of antifungal compounds produced by lactic acid bacteria. *Analytical and Bioanalytical Chemistry*, 403(10): 2983–2995. https://doi.org/10.1007/s00216-012-5955-1.

Brul, S. and Coote, P. 1999. Preservative agents in foods Mode of action and microbial resistance mechanisms. *International Journal of Food Microbiology*, 50(1-2): 1–17. https://doi.org/10.1016/s0168-1605(99)00072-0.

Bryła, M., Ksieniewicz-Woźniak, E., Waśkiewicz, A., Podolska, G. and Szymczyk, K. 2019. Stability of ergot alkaloids during the process of baking rye bread. LWT, 110: 269–274. https://doi.org/10.1016/j.lwt.2019.04.065.

Bryła, M., Waśkiewicz, A., Szymczyk, K. and Jędrzejczak, R. 2017. Effects of pH and temperature on the stability of fumonisins in maize products. *Toxins*, 9(3): 88. https://doi.org/10.3390/toxins9030088.

Buron-Moles, G., Chailyan, A., Dolejs, I., Forster, J. and Mikš, M.H. 2019. Uncovering carbohydrate metabolism through a genotype-phenotype association study of 56 lactic acid bacteria genomes. *Applied Microbiology and Biotechnology*, 103(7): 3135–3152. https://doi.org/10.1007/s00253-019-09701-6.

Bustos, A.Y., de Valdez, G.F. and Gerez, C.L. 2018. Optimization of phenyllactic acid production by *Pediococcus acidilactici* CRL 1753. Application of the formulated bio-preserver culture in bread. *Biological Control*, 123: 137–143.

Cabañas, R., Bragulat, M., Abarca, M., Castellá, G. and Cabañes, F. 2008. Occurrence of *Penicillium verrucosum* in retail wheat flours from the Spanish market. *Food Microbiology*, 25(5): 642–647. https://doi.org/10.1016/j.fm.2008.04.003.

Calvert, M.D., Madden, A.A., Nichols, L.M., Haddad, N.M., Lahne, J., Dunn, R.R. and McKenney, E.A. 2021. A review of sour- dough starters: ecology, practices, and sensory quality with applications for baking and recommendations for future research. *PeerJ*, 9:e11389 http://doi.org/10.7717/peerj.11389.

Campos, C.A., Gliemmo, M.F. and Castro, P.M. 2014. Strategies for controlling the growth of spoilage yeasts in foods. pp. 497–511. *In*: Rai, R.V. and Bai, A.J. (Eds.). *Microbial Food Safety and Preservation Techniques.* Tailor and Francis, CRC Press. ISBN 9780429168291.

Canesin, M.R. and Cazarin, C.B.B. 2021. Nutritional quality and nutrient bioaccessibility in sourdough bread. *Current Opinion in Food Science*, 40: 81–86. https://doi.org/10.1016/j.cofs.2021.02.007.

Catzeddu, P. 2019. Sourdough breads. pp. 177–187. *In:* Preedy, R.V. and Watson, R.R. (Eds.). *Flour and Breads and their Fortification in Health and Disease Prevention.* Second ed. Elsevier Academic press.

Cauvain, S.P. 2015. Technology of Breadmaking. 3rd ed. Switzerland: Springer; 408 p. DOI: 10.1007/9783319146874.

Caygill, R.L., G.E. Blair and P.A. Millner. 2010. A review on viral biosensors to detect human pathogens. *Analytica Chimica Acta*, 681(1-2): 8–15.

Celano, G., de Angelis, M., Minervini, F. and Gobbetti, M. 2016. Different flour microbial communities drive to sourdoughs characterized by diverse bacterial strains and free amino acid profiles. *Frontiers in Microbiology*, 7. https://doi.org/10.3389/fmicb.2016.01770.

Chang, Y. 2020. Bacteriophage-derived endolysins applied as potent biocontrol agents to enhance food safety. *Microorganisms*, 8(5): 724.

Cho, I.H. and Peterson, D.G. 2010. Chemistry of bread aroma: a review. *Food Science and Biotechnology*, 19(3): 575–582.

Cintas, L.M., Casaus, M.P., Herranz, C., Nes, I.F. and Hernandez, P.E. 2001. Review: bacteriocins of lactic acid bacteria. *Food Sci. Tech. Int.*, 7: 281–305.

Císarová, M., Hleba, L., Tančinová, D., Florková, M., Foltinová, D., Charousová, I., Vrbová, K., Božik, M. and Klouček, P. 2018. Inhibitory effect of essential oils from some *Lamiaceae* species on growth of *Eurotium* spp. isolated from bread. *Journal of Microbiology, Biotechnology and Food Sciences*, 8(2): 857–862. https://doi.org/10.15414/jmbfs.2018.8.2.857-862.

Cizeikiene, D., Juodeikiene, G., Paskevicius, A. and Bartkiene, E. 2013. Antimicrobial activity of lactic acid bacteria against pathogenic and spoilage microorganism isolated from food and their control in wheat bread. *Food Control*, 31(2): 539–545. https://doi.org/10.1016/j.foodcont.2012.12.004.

Coda, R., Cassone, A., Rizzello, C.G., Nionelli, L., Cardinali, G. and Gobbetti, M. 2011. Antifungal Activity of Wickerhamomyces anomalus and *Lactobacillus plantarum* during sourdough fermentation: identification of novel compounds and long-term effect during storage of wheat bread. *Applied and Environmental Microbiology*, 77(10): 3484–3492. https://doi.org/10.1128/aem.02669-10.

Codex Alimentarius. 1991. Codex standard for durum wheat semolina and durum wheat flour 178–1991 (Rev. 1-1995). Rome: FAO/WHO.

Codex Alimentarius. 1995a. General Standard for Contaminants and Toxins in Food and Feed (Codex Stan 193–1995).

Cogan, T.M. and Jordan, K.N. 1995. Metabolism of Leuconostoc bacteria. *Journal of Dairy Science*, 77: 2704–2717.

Collar, C., Mascaros, A.F., Prieto, J.A. and De Barber, C.B. 1991. Changes in free amino acids during fermentation of wheat doughs started with pure culture of lactic acid bacteria. *Cereal Chem*, 68(1): 66–72.

Commission Implementing Regulation, 2012. Commission Implementing Regulation (EU) No 562/2012 of 27 June 2012 amending Commission Regulation (EU) No 234/2011 with regard to specific data required for risk assessment of food enzymes. *Official Journal of the European Union*, L 168/1.

Cooksey, K. 2005. Effectiveness of antimicrobial food packaging materials. *Food Additive Contaminants*. 22(10): 980–98.

Corsetti, A., Lavermicocca, P., Morea, M., Baruzzi, F., Tosti, N. and Gobbetti, M. 2001. Phenotypic and molecular identification and clustering of lactic acid bacteria and yeasts from wheat (species *Triticum durum* and *Triticum aestivum*) sourdoughs of Southern Italy. *International Journal of Food Microbiology*, 64(1-2): 95–104. https://doi.org/10.1016/s0168-1605(00)00447-5.

Corsetti, A., Settanni, L. and Van Sinderen, D. 2004. Characterization of bacteriocin-like inhibitory substances (BLIS) from sourdough lactic acid bacteria and evaluation of their *in vitro* and *in situ* activity. J. Appl. Microbiol. 96(3): 521–34.

Corsetti, A. and Settanni, L. 2007. Lactobacilli in sourdough fermentation. *Food Research International*, 40(5): 539–558. https://doi.org/10.1016/j.foodres.2006.11.001.

Corsetti, A., Settanni, L., Chaves López, C., Felis, G.E., Mastrangelo, M. and Suzzi, G. 2007. A taxonomic survey of lactic acid bacteria isolated from wheat (*Triticum durum*) kernels and non-conventional flours. *Systematic and Applied Microbiology*, 30(7): 561–571. https://doi.org/10.1016/j.syapm.2007.07.001.

Corsetti, A. 2013. Technology of sourdough fermentation and sourdough applications. pp. 85–103. *In*: Gobbetti, M. and Gänzle, M. (Eds.). *Handbook on Sourdough Biotechnology*. New York: Springer Science and Business Media.

Creppy, E.E. 2002. Update of survey, regulation and toxic effects of mycotoxins in Europe. *Toxicology Letters*, 127(1–3): 19–28. https://doi.org/10.1016/s0378-4274(01)00479-9.

Crowley, S., Mahony, J. and van Sinderen, D. 2013. Current perspectives on antifungal lactic acid bacteria as natural bio-preservatives. *Trends in Food Science & Technology*, 33(2): 93–109. https://doi.org/10.1016/j.tifs.2013.07.004.

da Cruz Cabral, L., Pinto, V.F. and Patriarca, A. 2013. Application of plant derived compounds to control fungal spoilage and mycotoxin production in foods. *International Journal of Food Microbiology*, 166(1): 1–14.

Dack, G.M. 1961. Public health significance of flour bacteriology. *Cereal Sci. Today*, 6: 9–12.

Daftary, R.D., Pomeranz, Y. and Sauer, D.B. 1970. Changes in wheat flour damaged by mold during storage. Effects on lipid, lipoprotein, and protein. *Journal of Agricultural and Food Chemistry*, 18(4): 613–616. https://doi.org/10.1021/jf60170a022.

Dal Bello, F., Clarke, C.I., Ryan, L.A.M., Ulmer, H., Schober, T.J., Ström, K., ... and Arendt, E.K. 2007. Improvement of the quality and shelf life of wheat bread by fermentation with the antifungal strain *Lactobacillus plantarum* FST 1.7. *Journal of Cereal Science*, 45(3): 309–318. https://doi.org/10.1016/j.jcs.2006.09.004.

188 *Novel Approaches in Biopreservation for Food and Clinical Purposes*

Dao, T. and Dantigny, P. 2011. Control of food spoilage fungi by ethanol. *Food Control*, 22(3-4): 360–368. https://doi.org/10.1016/j.foodcont.2010.09.019.

Davidson, P.M. and Naidu, A.S. 2000. Phyto-phenol. pp. 265–294. *In*: *Natural Food Antimicrobial Systems*. Naidu, A.S. (Ed.): CRC Press, Boca Raton, Fl.

de Angelis, M., Gallo, G., Corbo, M.R., McSweeney, P.L.H., Faccia, M., Giovine, M. and Gobbetti, M. 2003. Phytase activity in sourdough lactic acid bacteria: purification and characterization of a phytase from *Lactobacillus sanfranciscensis* CB1. *Int. J. Food Microbiol.*, 87: 259–270.

de Luca, L., Aiello, A., Pizzolongo, F., Blaiotta, G., Aponte, M. and Romano, R. 2021. Volatile organic compounds in breads prepared with different sourdoughs. *Applied Sciences*, 11(3): 1330. https://doi.org/10.3390/app11031330.

de Vuyst, L. and Neysens, P. 2005. The sourdough microflora: biodiversity and metabolic interactions. *Trends in Food Science & Technology*, 16(1–3): 43–56. https://doi.org/10.1016/j.tifs.2004.02.012.

De Vuyst, L. and Leroy, F. 2007. Bacteriocins from lactic acid bacteria: production, purification, and food applications. *J. mol. Microbiol. Biotechnol.*, 13: 194–199.

Decock, P. and Cappelle, S. 2005. Bread technology and sourdough technology. *Trends Food Sci. Tech.*, 16: 113–120.

Degirmencioglu, N., Göcmen, D., Inkaya, A.N., Aydin, E., Guldas, M. and Gonenc, S. 2010. Influence of modified atmosphere packaging and potassium sorbate on microbiological characteristics of sliced bread. *Journal of Food Science and Technology*, 48(2): 236–241. https://doi.org/10.1007/s13197-010-0156-4

Deschuyffeleer, N., Audenaert, K., Samapundo, S., Ameye, S., Eeckhout, M. and Devlieghere, F. 2011. Identification and characterization of yeasts causing chalk mould defects on par-baked bread. *Food Microbiology*, 28(5): 1019–1027. https://doi.org/10.1016/j.fm.2011.02.002.

Devlieghere, F., Vermeiren, L. and Debevere, J. 2004. New preservation technologies: possibilities and limitations. *International Dairy Journal*, 14(4): 273–285. https://doi.org/10.1016/j.idairyj.2003.07.002.

di Stefano, V. 2019. Occurrence & Risk of OTA in Food and Feed. pp. 420–423. *In:* Melton, L. and Varelis, P. (Eds.): *Encyclopedia of Food Chemistry*, Elsevier Inc., Volume I. ISBN 978-0-12-814045-1.

do Carmo-Sousa, L. 1969. Distribution of yeasts in nature. pp. 79–105. *In*: Rose, A.H. and Harrison, J.S. (Eds.). *The Yeasts* (Vol. 1). Academic Press: London, UK.

Domijan, A.M., ŽElježić, D., Milić, M. and Peraica, M. 2007. Fumonisin B1: oxidative status and DNA damage in rats. *Toxicology*, 232(3): 163–169. https://doi.org/10.1016/j.tox.2007.01.007.

Dorman, H.J.D. and Deans, S.G. 2000. Antimicrobial agents from plants: antibacterial activity of plant volatile oils. *Journal of Applied Microbiology*, 88: 308–316.

Duarte, S., Pena, A. and Lino, C. 2010. A review on ochratoxin A occurrence and effects of processing of cereal and cereal derived food products. *Food Microbiology*, 27(2): 187–198. https://doi.org/10.1016/j.fm.2009.11.016.

Dziki, D., Różyło, R., Gawlik-Dziki, U. and Świeca, M. 2014. Current trends in the enhancement of antioxidant activity of wheat bread by the addition of plant materials rich in phenolic compounds. *Trends Food Sci. Technol.*, 40(1): 48–61.

Eacute Sar, M.B., Natalia, B.B., Jane, M.L.N.G. and Sydnei, M.S. 2016. Influence of enzymes and ascorbic acid on dough rheology and wheat bread quality. *African Journal of Biotechnology*, 15(3): 55–61. https://doi.org/10.5897/ajb2015.14931.

EC Commission. 2006. Setting of maximum levels for certain contaminants in foodstuffs. *Regulation*, 1881, 5–24.

EEC. 2006. Regulation (EC) No 1881/2006 of 19 December 2006 setting maximum levels for certain contaminants in foodstuffs. *Official Journal of the European Union* L 364, 5.

EEC. 2008. Regulation (EC) No 1333/2008 of the European Parliament and of the Council of 16 December 2008 on food additives. *Official Journal of the European Union* L 354, 16.

EFSA. 2012. Scientific Opinion on the maintenance of the list of QPS biological agents intentionally added to food and feed (2012 update). *EFSA Journal*, 10: 3020.

EFSA. 2016. Scientific opinion on the re-evaluation of sulfur dioxide (E 220): sodium sulfite (E 221) sodium bisulfite (E 222): sodium metabisulfite (E 223): potassium metabisulfite (E 224): calcium sulfite (E 226): calcium bisulfite (E 227) and potassium bisulfite (E 228) as food additives. European Food Safety Authority Journal, 14, 4: 4438.

EFSA Panel on Biological Hazards (BIOHAZ): Koutsoumanis, K., Allende, A., Alvarez-Ordóñez, A., Bolton, D., Bover-Cid, S., ... and Herman, L. 2022. Update of the list of QPS-recommended biological agents intentionally added to food or feed as notified to EFSA 15: suitability of taxonomic units notified to EFSA until September 2021. *EFSA Journal*, 20(1): e07045.

Eglezos, S. 2010. Microbiological quality of wheat grain and flour from two mills in queensland, Australia. *Journal of Food Protection*, 73(8): 1533–1536. https://doi.org/10.4315/0362-028x-73.8.1533.

Ehrmann, M.A. and Vogel, R. 2005. Molecular taxonomy and genetics of sourdough lactic acid bacteria. *Trends Food Sci. Technol.*, 16: 31–42.

el Khoury, A. and Atoui, A. 2010. Ochratoxin A: general overview and actual molecular status. *Toxins*, 2(4): 461–493. https://doi.org/10.3390/toxins2040461.

El oirdi, S., Lakhlifi, T., Bahar, A.A., Yatim, M., Rachid, Z. and Belhaj, A. 2021. Isolation and identification of *Lactobacillus plantarum* 4F, a strain with high antifungal activity, fungicidal effect, and biopreservation properties of food. *Journal of Food Processing and Preservation*, 45(6): e15517.

El Sheikha, A.F. 2015. Bread and its Fortification: Nutrition and Health Benefits.

Ellis, D.I. and Goodacre, R. 2006. Quantitative detection and identification methods for microbial spoilage. *In*: Clive de W. Blackburn (Ed.). *Food Spoilage Microorganisms*. Woodhead Publishing Series in Food Science, Technology and Nutrition, Woodhead Publishing, 2006, Pages 194-212, ISBN 9781855739666, https://doi.org/10.1533/9781845691417.2.194. DOI:10.1533/9781845691417.2.194.

Ercolini, D., Pontonio, E., de Filippis, F., Minervini, F., La Storia, A., Gobbetti, M. and di Cagno, R. 2013. Microbial ecology dynamics during rye and wheat sourdough preparation. *Applied and Environmental Microbiology*, 79(24): 7827–7836. https://doi.org/10.1128/aem.02955-13.

Eriksen, G.S. and Pettersson, H. 2004. Toxicological evaluation of trichothecenes in animal feed. *Animal Feed Science and Technology*, 114(1–4): 205–239. https://doi.org/10.1016/j.anifeedsci.2003.08.008.

EU Food Additives Database. Available at https://webgate.ec.europa.eu/foods_system/main/?event=category.view&identifier=88##. Accessed 01.09.2022.

European Commission, Directorate-General for Health and Food Safety. 2020. RASFF annual report 2019, Publications Office. https://data.europa.eu/doi/10.2875/993888.

European Commission. 1997. Commission Regulation (EC) No 194/97 of 31 January 1997 setting maximum levels for certain contaminants in foodstuffs. *Official Journal of the European Communities*, 31: 48–50.

European Commission. 2001. Commission Regulation (EC) No 466/2001 of 8 March 2001 setting maximum levels for certain contaminants in foodstuffs. *Official Journal of the European Communities*, L 77, 1e13.

European Commission. 2013. Commission Recommendation No 2013/165/EU of 27 March 2013 on the presence of T-2 and HT-2 toxin in cereals and cereal products. *Official Journal of the European Union*, L 91, 12e15.

Eyles, M.J., Moss, R. and Hocking, A.D. 1989. The microbiological status of Australian flour and the effects of milling procedures on the microflora of wheat and flour. Food Australia - Official Journal of CAFTA and AIFST, 41(4): 704–708.

Fangio, M.F., Roura, S.I. and Fritz, R. 2010. Isolation and identification of *Bacillus* spp. and related genera from different starchy foods. *Journal of Food Science*, 75(4): M218–M221.

Felis, G.E. and Dellaglio, F. 2007. Taxonomy of Lactobacilli and Bifidobacteria. *Current Issues in Intestinal Microbiology*, 8(2): 44–61.

Ferrero, C. 2017. Hydrocolloids in wheat breadmaking: a concise review. *Food Hydrocolloids*, 68: 15–22. https://doi.org/10.1016/j.foodhyd.2016.11.044.

Florou-Paneri, P., Christaki, E. and Bonos, E. 2013. Lactic acid bacteria as source of functional ingredients. pp. 589–614. *In*: Kongo, M. (Ed.): *Lactic Acid Bacteria - R & D for Food, Health and Livestock Purposes*. IntechOpen. https://doi.org/10.5772/47766.

Freire, F.C.O.A. 2011. A deterioração fúngica de produtos de panificação no Brasil. *Embrapa Agroindústria Tropical-Comunicado Técnico (INFOTECA-E)*.

Galić, K., Curić, D. and Gabrić, D. 2009. Shelf life of packaged bakery goods—a review. *Crit. Rev. Food Sci. Nutr.*, 49: 405–426.

190 *Novel Approaches in Biopreservation for Food and Clinical Purposes*

Galli, A., Franzetti, L. and Fortina, M.G. 1987. Isolation and identification of yeasts and lactic bacteria in wheat flour. *Microbiologie Aliments Nutrition*, 5: 3–9.

Gänzle, M.G., Ehmann, M. and Hammes, W.P. 1998. Modeling of growth of *Lactobacillus sanfranciscensis* and *Candida milleri* in response to process parameters of sourdough fermentation. *Appl. Environ. Microbiol.* 64: 2616–2623.

Gänzle, M.G., Höltzel, A., Walter, J., Jung, G. and Hammes, W.P. 2000. Characterization of reutericyclin produced by *Lactobacillus reuteri* LTH2584. *Appl. Environ. Microbiol.* 66(10): 4325–4333.

Ganzle, M.G. 2004. Reutericyclin: biological activity, mode of action, and potential applications. *Applied Microbiology and Biotechnology*, 64(3): 326–332. https://doi.org/10.1007/s00253-003-1536-8.

Gänzle, M.G. 2013. Enzymatic and bacterial conversions during sourdough fermentation. *Food Microbiol.*, 37: 2–10.

Gänzle, M.G. 2015. Lactic metabolism revisited: metabolism of lactic acid bacteria in food fermentations and food spoilage, *COFS* (2015): http://dx.doi.org/10.1016/j.cofs.2015.03.001.

Gänzle, M.G. and Zheng, J. 2019. Lifestyles of sourdough lactobacilli—do they matter for microbial ecology and bread quality? *International Journal of Food Microbiology*, 302: 15–23 DOI 10.1016/j.ijfoodmicro.2018.08.019.

Garcia, M.V., Bernardi, A.O. and Copetti, M.V. 2019a. The fungal problem in bread production: insights of causes, consequences, and control methods. *Current Opinion in Food Science*, 29: 1–6.

Garcia, M.V., da Pia, A.K.R., Freire, L., Copetti, M.V. and Sant'Ana, A.S. 2019b. Effect of temperature on inactivation kinetics of three strains of *Penicillium paneum* and *P. roqueforti* during bread baking. *Food Control*, 96: 456–462.

Gardner, N., Champagne, C. and Gelinas, P. 2002. Effect of yeast extracts containing propionic acid on bread dough fermentation and bread properties. *Journal of Food Science*, 67(5): 1855–1858. https://doi.org/10.1111/j.1365-2621.2002.tb08735.x.

Garofalo, C., Zannini, E., Aquilanti, L., Silvestri, G., Fierro, O., Picariello, G. and Clementi, F. 2012. Selection of sourdough Lactobacilli with antifungal activity for use as biopreservatives in bakery products. *Journal of Agricultural and Food Chemistry*, 60(31): 7719–7728. https://doi.org/10.1021/jf301173u.

Gashgari, R.M., Shebany, Y.M. and Gherbawy, Y.A. 2010. Molecular characterization of mycobiota and aflatoxin contamination of retail wheat flours from Jeddah markets. *Foodborne Pathogens and Disease*, 7(9): 1047–1054. https://doi.org/10.1089/fpd.2009.0506.

Gayán, E., Serrano, M.J., PagánI, R., Álvarez, I. and Condón, S. 2015. Environmental and biological factors influencing the UV-C resistance of *Listeria monocytogenes. Food Microbiology*, 46: 246–253. https://doi.org/10.1016/j.fm.2014.08.011.

Gerez, C.L., Torino, M.I., Obregozo, M.D. and De Valdez, G.F. 2010. A ready-to-use antifungal starter culture improves the shelf life of packaged bread. *Journal of Food Protection*, 73(4): 758–762.

Gezginc, Y. and Kara, Ü. 2019. The effect of exopolysaccharide producing *Lactobacillus plantarum* strain addition on sourdough and wheat bread quality. *Qual. Assur. Saf. Crops Foods*, 2019; 11(1): 95–106.

Ghosh, S., Sachan, A., Sen, S.K. and Mitra, A. 2006. Microbial transformation of ferulic acid to vanillic acid by *Streptomyces sannanensis* MTCC 6637. *Journal of Industrial Microbiology & Biotechnology*, 34(2): 131–138. https://doi.org/10.1007/s10295-006-0177-1.

Gioia, L.C., Ganancio, J.R. and Steel, C.J. 2017. Food additives and processing aids used in breadmaking. *Food Additives*. https://doi.org/10.5772/intechopen.70087.

Giray, B., Girgin, G., Engin, A.B., Aydın, S. and Sahin, G. 2007. Aflatoxin levels in wheat samples consumed in some regions of Turkey. *Food Control*, 18(1): 23–29. https://doi.org/10.1016/j. foodcont.2005.08.002.

Gobbetti, M. and Corsetti, A. 1997. *Lactobacillus sanfrancisco* a key sourdough lactic acid bacterium: a review. *Food Microbiol*, 14: 175–187.

Gobbetti, M., De Angelis, M., Corsetti, A. and Di Cagno, R. 2005. Biochemistry and physiology of sourdough lactic acid bacteria. *Trends Food Sci. Tech.* 16: 57–69.

Gobetti, M. and Ganzle, M. 2013. Handbook of sourdough biotechnology. Springer, Boston, MA. https://doi.org/10.1007/978-1-4614-5425-0.

Grosch, W. and Wieser, H. 1999. Redox reactions in wheat dough as affected by ascorbic acid. *Journal of Cereal Science*, 29.1: 1–16.

Guarda, A., Rosell, C., Benedito, C. and Galotto, M. 2004. Different hydrocolloids as bread improvers and antistaling agents. *Food Hydrocolloids*, 18(2): 241–247. https://doi.org/10.1016/s0268-005x(03)00080-8.

Gutiérrez, L., Batlle, R., Andújar, S., Sánchez, C. and Nerín, C. 2011. Evaluation of antimicrobial active packaging to increase shelf life of gluten-free sliced bread. *Packaging Technology Science*, 24: 485–494.

Han, J.H. 2005. Antimicrobial packaging systems. pp. 80–107. *In:* Han, J.H. (Ed.). *Innovations in Food Packaging*. Academic Press: San Diego.

Hartikainen, K. and Katina, K. 2012. Improving the quality of high-fibre breads. Breadmaking, 736–753.

Hasan, S.M., Naje, S.A. and Abosalloum, S. 2014. Shelf life extrusion of pita bread by modified atmosphere packaging. *J. Food Dairy Sci. Mansoura Univ.*, 5(2): 55–62.

Hatoum, R., Labrie, S. and Fliss, I. 2012. Antimicrobial and probiotic properties of yeasts: from fundamental to novel applications. *Frontiers in Microbiology*, 3. https://doi.org/10.3389/fmicb.2012.00421.

Hematian Sourki, A., Yazdi, F.T., Ghiafeh Davoodi, M., Mortazavi, S.A., Karimi, M., Jahromi, S.H.R. and Pourfarzad, A. 2010. Staling and quality of Iranian flat bread stored at modified atmosphere in different packaging. *Int. J. Soc. Behav. Educ. Econ. Bus. Ind. Eng.*, 4(5): 567–572.

Houston, D.R., Eggleston, I., Synstad, B., Eijsink, V.G.H. and Van Aalten, D.M.F. 2002. The cyclic dipeptide CI-4 [cyclo-(L-Arg-D-Pro)] inhibits family 18 chitinases by structural mimicry of a reaction intermediate. *Biochem. J.*, 368: 23–27.

Hua, S.S.T., Beck, J.J., Sarreal, S.B.L. and Gee, W. 2014. The major volatile compound 2-phenylethanol from the biocontrol yeast, Pichia anomala, inhibits growth and expression of aflatoxin biosynthetic genes of *Aspergillus flavus*. *Mycotoxin Research*, 30(2): 71–78. https://doi.org/10.1007/s12550-014-0189-z.

Hulin, V., Mathot, A., Mafart, P. and Dufossй, L. 1998. Les proprйtйs anti-microbiennes des huiles essentielles et composйs d'aromes. *Sci. Aliments*, 18: 563–582.

Hussein, H. 2001. Toxicity, metabolism, and impact of mycotoxins on humans and animals. *Toxicology*, 167(2): 101–134. https://doi.org/10.1016/s0300-483x(01)00471-1.

Hutkins, W.R. 2019. Microbiology and technology of fermented food. Wiley Blackwell, IFT Press.

Hwang, J. and Lee, K. 2006. Reduction of aflatoxin B1 contamination in wheat by various cooking treatments. *Food Chemistry*, 98(1): 71–75. https://doi.org/10.1016/j.foodchem.2005.04.038.

IARC. 1993. Toxins derived from *Fusarium moniliforme*: fumonisins B1 and B2 and fusarin C. 1993. IARC Monographs on the Evaluation of Carcinogenic Risks to Humans, 56: 445–466.

Jack, R.W., Tagg, J.R. and Ray, B. 1995. Bacteriocins of Gram-positive bacteria. *Microbiol. Mol. Biol. Rev.*, 59: 171–200.

Jideani, V.A. and Vogt, K. 2015. Antimicrobial packaging for extending the shelf life of bread – a review. *Critical Reviews in Food Science and Nutrition*, DOI: 10.1080/10408398.2013.768198.

Jonkuvienė, D., Vaičiulytė-Funk, L., Šalomskienė, J., Alenčikienė, G. and Aldona Mieželienė, A. 2016. Potential of *Lactobacillus reuteri* from spontaneous sourdough as a starter additive for improving quality parameters of bread. *Food Technol Biotechnol.* 54(3): 342–350.

Ju, J., Xu, X., Xie, Y., Guo, Y., Cheng, Y., Qian, H. and Yao, W. 2018. Inhibitory effects of cinnamon and clove essential oils on mold growth on baked foods. *Food Chemistry*, 240: 850–855.

Jung, S., Hwang, H. and Lee, J.H. 2019. Effect of lactic acid bacteria on phenyllactic acid production in kimchi. *Food Control*, 106.

Khan, A.N., Ahmed, A., Bhatti, M.S., Randhawa, M.A., Ahmad, A. and Yousaf, A.A. 2011. Effect of Additives on the Shelf Life Extension of Chapatti. *Food Science and Technology Research*, 17(3): 203–208. https://doi.org/10.3136/fstr.17.203.

Khanom, A., Shammi, T. and Kabir, M.S. 2016. Determination of microbiological quality of packed and unpacked bread. *Stamford Journal of Microbiology*, 6(1): 24–29.

Khoshakhlagh, K., Hamdami, N., Shahedi, M. and Le-Bail, A. 2014. Quality and microbial characteristics of part-baked Sangak bread packaged in modified atmosphere during storage. *J. Cereal Sci.*, 60(1): 42–47.

Klepacka, J. and Fornal, L. 2006. Ferulic acid and its position among the phenolic compounds of wheat. *Crit. Rev. Food Sci. Nutr.*, 46: 639–647. doi: 10.1080/10408390500511821.

Koistinen, V.M., Mattila, O., Katina, K., Poutanen, K., Anna-Marja Aura, A.M. and Hanhineva, K. 2018. Metabolic profiling of sourdough fermented wheat and rye bread. *Sci. Rep.*, 8: 5684.

192 *Novel Approaches in Biopreservation for Food and Clinical Purposes*

Kong, M. and Ryu, S. 2015. Bacteriophage PBC1 and its endolysin as an antimicrobial agent against *Bacillus cereus. Applied and Environmental Microbiology*, 81(7): 2274–2283.

Kong, M., Na, H., Ha, N.C. and Ryu, S. 2019. LysPBC2, a novel endolysin harboring a *Bacillus cereus* spore binding domain. *Applied and Environmental Microbiology*, 85(5): e02462-18.

Kraszewska, J., Beckett, M.C., James, T.C. and Bond, U. 2016. Comparative analysis of the antimicrobial activities of plant defensin-like and ultrashort peptides against food-spoiling bacteria. *Appl. Environ. Microbiol.* doi:10.1128/AEM.00558-16.

Krebs, H.A., Wiggins, D., Stubs, M., Sols, A. and Bedoya, F. 1983. Studies on the mechanism of the antifungal action of benzoate. *Biochemical Journal*, 214: 657–663.

Küley, E., Özyurt, G., Özogul, I., Boga, M., Akyol, I., Rocha, J.M. and Özogul, F. 2020. The role of selected lactic acid bacteria on organic acid accumulation during wet and spray-dried fish-based silages. Contributions to the Winning Combination of Microbial Food Safety and Environmental Sustainability. *Microorganisms (MDPI)*, 8(2): 172; Special Issue: Microbial Safety of Fermented Products; https://doi.org/10.3390/microorganisms8020172; ISSN 2076-2607, ISSN 2076-2607.

Kuorwel, K.K., Cran, M.J., Sonneveld, K., Miltz, J. and Bigger, S.W. 2011. Essential oils and their principal constituents as antimicrobial agents for synthetic packaging films. *Journal of Food Science*, 76(9): R164–R177.

Kuswandi, B., Wicaksono, Y., Jayus, Abdullah, A., Heng, L.Y. and Ahmad, M. 2011. Smart packaging: sensors for monitoring of food quality and safety. *Sensing and Instrumentation for Food Quality and Safety*, 5(3-4): 137–146. https://doi.org/10.1007/s11694-011-9120-x.

Laca, A., Mousia, Z., DíAz, M., Webb, C. and Pandiella, S.S. 2006. Distribution of microbial contamination within cereal grains. *Journal of Food Engineering*, 72(4): 332–338. https://doi.org/10.1016/j.jfoodeng.2004.12.012.

Lagaron, J.M. and Lopez-Rubio, A. 2011. Nanotechnology for bioplastics: Opportunities, challenges and strategies. *Trends in Food Science & Technology*, 22: 611–617.

Lam, B., J. Das, R.D. Holmes, L. Live, A. Sage, E.H. Sargent and S.O. Kelley. 2013. Solution-based circuits enable rapid and multiplexed pathogen detection. *Nature Communications*, 4: 2001.

Lavermicocca, P., Valerio, F., De Bellis, P., Sisto, A. and Leguérinel, I. 2016. Sporeforming bacteria associated with bread production: spoilage and toxigenic potential. pp. 275–293. *In*: *Food hygiene and toxicology in ready-to-eat foods*. Academic Press.

Lavermicocca, P., Valerio, F., Evidente, A., Lazzaroni, S., Corsetti, A. and Gobbetti, M. 2000. Purification and characterization of novel antifungal compounds from the sourdough *Lactobacillus plantarum* Strain 21B. *Applied and Environmental Microbiology*, 66(9): 4084–4090. https://doi.org/10.1128/aem.66.9.4084-4090.2000.

le Lay, C., Coton, E., le Blay, G., Chobert, J. M., Haertlé, T., Choiset, Y., van Long, N.N., Meslet-Cladière, L. and Mounier, J. 2016a. Identification and quantification of antifungal compounds produced by lactic acid bacteria and propionibacteria. *International Journal of Food Microbiology*, 239: 79–85. https://doi.org/10.1016/j.ijfoodmicro.2016.06.020.

le Lay, C., Mounier, J., Vasseur, V., Weill, A., le Blay, G., Barbier, G. and Coton, E. 2016b. *In vitro* and *in situ* screening of lactic acid bacteria and propionibacteria antifungal activities against bakery product spoilage molds. *Food Control*, 60: 247–255. https://doi.org/10.1016/j.foodcont.2015.07.034.

Lee, Y.G., Kim, B.Y., Bae, J.M., Wang, Y. and Jin, Y.S. 2022. Genome-edited *Saccharomyces cerevisiae* strains for improving quality, safety, and flavor of fermented foods. *Food Microbiology*, 104: 103971. https://doi.org/10.1016/j.fm.2021.103971.

Leprince, A., Nuytten, M., Gillis, A. and Mahillon, J. 2020. Characterization of PlyB221 and PlyP32, Two novel endolysins encoded by phages preying on the *Bacillus cereus* Group. Viruses, 12(9): 1052.

Leverentz, B., Conway, W.S., Alavidze, Z., Janisiewicz, W.J., Fuchs, Y., Camp, M.J., ... and Sulakvelidze, A. 2001. Examination of bacteriophage as a biocontrol method for Salmonella on fresh-cut fruit: a model study. *Journal of Food Protection*, 64(8): 1116–1121.

Levinskaite, L. 2012. Susceptibility of food-contaminating *Penicillium genus* fungi to some preservatives and disinfectants. *Annals of Agricultural and Environmental Medicine*: *AAEM*, 19: 85–9.

Leyva Salas, M., Mounier, J., Valence, F., Coton, M., Thierry, A. and Coton, E. 2017. Antifungal Microbial Agents for Food Biopreservation—A Review. *Microorganisms*, 5(3): 37. https://doi.org/10.3390/microorganisms5030037.

Lhomme, E., Onno, B., Chuat, V., Durand, K., Orain, S., Valence, F., Dousset, X. and Jacques, M.A. 2016. Genotypic diversity of *Lactobacillus sanfranciscensis* strains isolated from French organic sourdoughs. *International Journal of Food Microbiology*, 226: 13–19. https://doi.org/10.1016/j.ijfoodmicro.2016.03.008.

Li, H., Hu, S. and Fu, J. 2022. Effects of acetic acid bacteria in starter culture on the properties of sourdough and steamed bread. *Grain & Oil Science and Technology*, 5(1): 13–21. https://doi.org/10.1016/j.gaost.2021.11.003.

Li, N., Yuan, X., Li, C., Chen, N., Wang, J., Chen, B., ... and Ding, Y. 2022. A novel *Bacillus cereus* bacteriophage DLn1 and its endolysin as biocontrol agents against *Bacillus cereus* in milk. *International Journal of Food Microbiology*, 369: 109615.

Liewen, M.B. and Marth, E.H. 1985. Growth and inhibition of microorganisms in the presence of sorbic acid. *J. Food Protection*, 48: 364–75.

Litwinek, D., Boreczek, J., Gambuś, H., Buksa, K., Berski, W. and Kowalczyk, M. 2022. Developing lactic acid bacteria starter cultures for wholemeal rye flour bread with improved functionality, nutritional value, taste, appearance and safety. *Plos one*, 17(1): e0261677.

Liu, A., Xu, R., Zhang, S., Wang, Y., Hu, B., Ao, X., Li, Q., Li, J., Hu, K., Yang, Y. and Liu, S. 2022. Antifungal mechanisms and application of lactic acid bacteria in bakery products: a review. *Frontiers in Microbiology*, 13. https://doi.org/10.3389/fmicb.2022.924398.

Loessner, M.J., Maier, S.K., Daubek-Puza, H., Wendlinger, G. and Scherer, S. 1997. Three *Bacillus cereus* bacteriophage endolysins are unrelated but reveal high homology to cell wall hydrolases from different bacilli. *Journal of Bacteriology*, 179(9): 2845–2851.

Lorite, G.S., Rocha, J.M., Miilumäki, N., Saavalainen, P., Selkälä, T., Morales-Cid, G., Gonçalves, M.P., Pongrácz, E., Rocha, C.M.R. and Toth, G. 2016. Evaluation of physiochemical/microbial properties and life cycle assessment (LCA) of PLA-based nanocomposite active packaging. *LWT – Food Science and Technology*, 75: 305–315. doi: 10.1016/j.lwt.2016.09.004.

Magan, N., Arroyo, M. and Aldred, D. 2003. 24 - Mould prevention in bread. pp. 482–494. *In*: Cauvain, S.P. (Ed.). *Bread Making, Woodhead Publishing Series in Food Science, Technology and Nutrition*. Woodhead Publishing.

Magan, N. and Aldred, D. 2006. 8 – Managing microbial spoilage in cereal and baking products. pp. 194–212 *In*: Clive de W. Blackburn (Ed.). *Food Spoilage Microorganisms*. Woodhead Publishing Series in Food Science, Technology and Nutrition, Woodhead Publishing, 2006, ISBN 9781855739666, https://doi.org/10.1533/9781845691417.2.194. DOI:10.1533/9781845691417.2.194.

Mannaa, M. and Kim, K.D. 2017. Influence of temperature and water activity on deleterious fungi and mycotoxin production during grain storage. *Mycobiology*, 45(4): 240–254. https://doi.org/10.5941/myco.2017.45.4.240.

Marin, S., Guynot, M.E., Neira, P., Bernado, M., Sanchis, V. and Ramos, A.J. 2002. Risk assessment of the use of sub-optimal levels of weak-acid preservatives in the control of mould growth on bakery products. *International Journal of Food Microbiology* 79: 203–211.

Marin, S., Sanchis, V., Vinas, I., Canela, R. and Magan, N. 1995. Effect of water activity and temperature on growth and fumonisin B1and B2production by *Fusarium proliferatum* and *F. moniliforme* on maize grain. *Letters in Applied Microbiology*, 21(5): 298–301. https://doi.org/10.1111/j.1472-765x.1995.tb01064.x.

Marques, W.L., Raghavendran, V., Stambuk, B.U. and Gombert, A.K. 2016. Sucrose and *Saccharomyces cerevisiae*: a relationship most sweet. *FEMS Yeast Res*. 16.

Marriott, N.G., Schilling, W.M. and Gravani, R.B. 2018. Principles of Food Sanitation (Food Science Text Series) (6th ed. 2018 ed.). Springer.

Martinez, F.A.C., Balciunas, E.M., Salgado, J.M., González, J.M.D., Converti, A. and Oliveira, R.P.S. 2013. Lactic acid properties, applications and production: a review. *Trends Food Sci. Technol.*, 30: 70–83.

Martınez-Viedma, P., Abriouel, H., Ben Omar, N., Lucas-Lopez, R. and Ga´lvez, A. 2011. Inhibition of spoilage and toxinogenic *Bacillus* species in dough from wheat flour by the cyclic peptide enterocin AS-48. *Food Control*, 22(5): 756–761.

Matukas, M., Starkute, V., Zokaityte, E., Zokaityte, G., Klupsaite, D., Mockus, E., Rocha, J.M. and Bartkiene, E. 2022. Effect of different yeast strains on beer characteristics including biogenic amines and volatile compounds formation. *Foods*, 11(15): 2317. doi: https://doi.org/10.3390/foods11152317. Special issue: Regulation of Food Fermentations by Bacteria, Yeasts and Filamentous Fungi: https://www.mdpi.com/journal/foods/special_issues/regulation_fermentation.

Max, B., Carballo, J., Cortés, S. and Domínguez, J.M. 2011. Decarboxylation of Ferulic Acid to 4-Vinyl Guaiacol by *Streptomyces setonii*. *Applied Biochemistry and Biotechnology*, 166(2): 289–299. https://doi.org/10.1007/s12010-011-9424-7.

McCallum, L., Paine, S., Sexton, K., Dufour, M., Dyet, K., Wilson, M., Campbell, D., Bandaranayake, D. and Hope, V. 2013. An outbreak of *Salmonella typhimurium* phage Type 42 associated with the consumption of raw flour. *Foodborne Pathogens and Disease*, 10(2): 159–164. https://doi.org/10.1089/fpd.2012.1282.

Melini, F., Melini, V., Luziatelli, F. and Ruzzi, M. 2017. Current and forward-looking approaches to technological and nutritional improvements of gluten-free bread with legume flours: a critical review. *Comprehensive Reviews in Food Science and Food Safety*, 16(5): 1101–1122.

Melini, V. and Melini, F. 2018. Strategies to extend bread and GF bread shelf-life: from sourdough to antimicrobial active packaging and nanotechnology. *Fermentation*, 4(1): 9. https://doi.org/10.3390/fermentation4010009.

Méndez-Albores, A., Veles-Medina, J., Urbina-Álvarez, E., Martínez-Bustos, F. and Moreno-Martínez, E. 2009. Effect of citric acid on aflatoxin degradation and on functional and textural properties of extruded sorghum. *Animal Feed Science and Technology*, 150(3-4): 316–329. https://doi.org/10.1016/j.anifeedsci.2008.10.007.

Menezes, L.A.A., De Marco, I., Neves Oliveira dos Santos, N., Costa Nunes, C., Leite Cartabiano, C.E., Molognoni, L., ... and De Dea Lindner, J. 2021. Reducing FODMAPs and improving bread quality using type II sourdough with selected starter cultures. *International Journal of Food Sciences and Nutrition*, 72(7): 912–922.

Mercier, J. and Wilson, C. 1994. Colonization of apple wounds by naturally occurring microflora and introduced *Candida oleophila* and their effect on infection by *Botrytis cinerea* during storage. *Biological Control*, 4(2): 138–144. https://doi.org/10.1006/bcon.1994.1022.

Merkel, S., Dib, B., Maul, R., Köppen, R., Koch, M. and Nehls, I. 2012. Degradation and epimerization of ergot alkaloids after baking and in vitro digestion. *Analytical and Bioanalytical Chemistry*, 404(8): 2489–2497. https://doi.org/10.1007/s00216-012-6386-8.

Meroth, C.B., Walter, J., Hertel, C., Brandt, M.J. and Hammes, W.P. 2003. Monitoring the bacterial population dynamics in sourdough fermentation processes by using PCR-denaturing gradient gel electrophoresis. *Appl. Environ. Microbiol.*, 69: 475–482.

Milani, J., Nazari, S.S.S, Bamyar, E. and Maleki, G. 2014. Effect of bread making process on aflatoxin level changes. *Journal of Chemical Health Risk*, 4(4): 1–7.

Milani, J. and Heidari, S. 2016. Stability of ochratoxin a during bread making process. *Journal of Food Safety*, 37(1): e12283. https://doi.org/10.1111/jfs.12283.

Mildner-Szkudlarz, S., Różańska, M., Piechowska, P., Waśkiewicz, A. and Zawirska-Wojtasiak, R. 2019. Effects of polyphenols on volatile profile and acrylamide formation in a model wheat bread system. *Food Chemistry*, 297: 125008. https://doi.org/10.1016/j.foodchem.2019.125008.

Minervini, F., Dinardo, F. R., Celano, G., de Angelis, M. and Gobbetti, M. 2018. Lactic acid bacterium population dynamics in artisan sourdoughs over one year of daily propagations is mainly driven by flour microbiota and nutrients. *Frontiers in Microbiology*, 9. https://doi.org/10.3389/fmicb.2018.01984.

Minervini, F., Lattanzi, A., de Angelis, M., Celano, G. and Gobbetti, M. 2015. House microbiotas as sources of lactic acid bacteria and yeasts in traditional Italian sourdoughs. *Food Microbiology*, 52: 66–76. https://doi.org/10.1016/j.fm.2015.06.009.

Mo, E.K. and Sung, C.K. 2014. Production of white pan bread leavened by *Pichia anomala* SKM-T. *Food Science and Biotechnology*, 23(2): 431–437. https://doi.org/10.1007/s10068-014-0059-7.

Mockus, E., Starkute, V., Zokaityte, E., Klupsaite, D., Bartkevics, V., Borisova, A., Rocha, J.M., Ruibys, R., Liatukas, Z., Ruzgas, V. and Bartkiene, E. 2022. The potential of traditional 'Gaja' and new breed lines of waxy, blue and purple wheat in wholemeal flour fermentation. *Fermentation*, 8(10): 563. doi: https://doi.org/10.3390/fermentation8100563. Special issue: The Role of Antioxidant

Compounds in Fermented Foods, https://www.mdpi.com/journal/fermentation/special_issues/antioxidant_compounds1.

Modi, R., Hirvi, Y., Hill, A. and Griffiths, M.W. 2001. Effect of phage on survival of *Salmonella enteritidis* during manufacture and storage of cheddar cheese made from raw and pasteurized milk. *Journal of Food Protection*, 64(7): 927–933.

Moonen, H. and Bas, H. 2015. Mono- and diglycerides. pp. 73–91. *In*: Norn, V. (Ed.). *Emulsifiers in Food Technology*. 2nd ed. Chichester: Wiley-Blackwell; 2015.

Morassi, L.L., Bernardi, A.O., Amaral, A.L., Chaves, R.D., Santos, J.L., Copetti, M.V. and Sant'Ana, A.S. 2018. Fungi in cake production chain: occurrence and evaluation of growth potential in different cake formulations during storage. *Food Research International*, 106: 141–148.

Msagati, T.A.M. 2013. The Chemistry of Food Additives and Preservatives. 1st ed. Chichester: Wiley-Blackwell, 322 p.

Muccilli, S. and Restuccia, C. 2015. Bioprotective role of yeasts. *Microorganisms*, 3(4): 588–611. https://doi.org/10.3390/microorganisms3040588.

Muhialdin, B.J., Hassan, Z. and Sadon, S.Kh. 2011. Antifungal activity of *Lactobacillus fermentum* Te007, *Pediococcus pentosaceus* Te010, *Lactobacillus pentosus* G004, and *L. paracasi* D5 on selected Foods. *J. Food. Sci.*, 76.7: 493–499.

Müller, M.R.A., Wolfrum, G., Stolz, P., Ehrmann, M.A. and Vogel R.F. 2001. Monitoring the growth of *Lactobacillus* species during a rye flour fermentation. *Food Microbiology*, 18: 217– 227.

Nasrollahzadeh, A., Mokhtari, S., Khomeiri, M. and Saris, P.E.J. 2022. Antifungal preservation of food by lactic acid bacteria. *Foods*, 11(3): 395. https://doi.org/10.3390/foods11030395.

Nazhand, A., Durazzo, A., Lucarini, M., Souto, E.B. and Santini, A. 2020. Characteristics, occurrence, detection and detoxification of aflatoxins in foods and feeds. Foods, 9(5): 644. https://doi.org/10.3390/foods9050644.

Nehal, F., Sahnoun, M., Smaoui, S., Jaouadi, B., Bejar, S. and Mohammed, S. 2019. Characterization, high production and antimicrobial activity of exopolysaccharides from *Lactococcus lactis* F-mou. *Microb. Pathogen.*, 132: 10–19. doi: 10.1016/j.micpath.2019.04.018.

Neil, K.P., Biggerstaff, G., MacDonald, J.K., Trees, E., Medus, C., Musser, K.A., Stroika, S.G., Zink, D. and Sotir, M.J. 2011. A novel vehicle for transmission of *Escherichia coli* O157:H7 to humans: multistate outbreak of *E. coli* O157:H7 infections associated with consumption of ready-to-bake commercial prepackaged cookie dough--United States, 2009. *Clinical Infectious Diseases*, 54(4): 511–518. https://doi.org/10.1093/cid/cir831.

Neves, M., Antunes, M., Fernandes, W., Campos, M.J., Azevedo, Z.M., Freitas, V., Rocha, J.M., |Tecelão, C. 2021. Physicochemical and nutritional profile of leaves, flowers, and fruits of the edible halophyte chorão-da-praia (*Carpobrotus edulis*) on Portuguese west shores. *Food Bioscience* (Elsevier): 43: 101288. Doi: https://doi.org/10.1016/j.fbio.2021.101288. ISSN 2212-4292.

Nguyen, P.T., Nguyen, T.T., Bui, D.C., Hong, P.T., Hoang, Q.K. and Nguyen, H.T. 2020. Exopolysaccharide production by lactic acid bacteria: the manipulation of environmental stresses for industrial applications. *AIMS Microbiol.*, 2020 Nov 17;6(4): 451–469. doi: 10.3934/microbiol.2020027. PMID: 33364538; PMCID: PMC7755584.

Nielsen, P.V. and Rios, R. 2000. Inhibition of fungal growth on bread by volatile components from spices and herbs, and the possible application in active packaging, with special emphasis on mustard essential oil. *International Journal of Food Microbiology*, 60: 219–229

Novotni, D., Gänzle, M., Rocha, J.M. 2020. Chapter 5. Composition and activity of microbiota in sourdough and their effect on bread quality and safety. Charis M. Galanakis (Ed.). In Trends in Wheat and Bread Making, Elsevier-Academic Press, 469 pp. Cambridge, MA, USA, 2020. Charis M. Galanaksis (Editor): https://doi.org/10.1016/B978-0-12-821048-2.00005-2, Galanaksis-TWBM-1632435, ISBN 978-0-12-821048-2.

Odame-Darkwah, J.K. and Marshall, D.L. 1993. Interactive behavior of *Saccharomyces cerevisiae*, *Bacillus pumilus* and *Propionibacterium freudenreichii* subsp. shermanii. *International Journal of Food Microbiology*, 19(4): 259–269. https://doi.org/10.1016/0168-1605(93)90018-c.

Oleksy, M. and Klewicka, E. 2016. Exopolysaccharides produced by *Lactobacillus*sp.: Biosynthesis and applications. *Critical Reviews in Food Science and Nutrition*, 1–13. https://doi.org/10.1080/10408398.2016.1187112.

Oliveira, H., Melo, L.D., Santos, S.B., Nóbrega, F.L., Ferreira, E.C., Cerca, N., ... and Kluskens, L.D. 2013. Molecular aspects and comparative genomics of bacteriophage endolysins. *Journal of Virology*, 87(8): 4558–4570.

Oliveira, P.M., Zannini, E. and Arendt, E.K. 2014. Cereal fungal infection, mycotoxins, and lactic acid bacteria mediated bioprotection: from crop farming to cereal products. *Food Microbiology*, 37: 78–95. https://doi.org/10.1016/j.fm.2013.06.003.

Ooraikul, B. 1982. Gas packaging for a bakery product. *Canadian Institute of Food Science and Technology*, 15(4): 313–315.

Osborne, B. 1980. The occurrence of Ochratoxin A in mouldy bread and flour. *Food and Cosmetics Toxicology*, 18(6): 615–617. https://doi.org/10.1016/s0015-6264(80)80009-5.

Otles, S. and Yalcin, B. 2014. Smart/intelligent nanopackaging technologies for the food sector. *In*: Rai, V.R. and Bai, J.A. (Eds.). *Microbial Food Safety and Preservation Techniques.* CRC Press. ISBN 9780429168291. https://doi.org/10.1201/b17465.

Ottogalli, G., Galli, A. and Foschino R. 1996. Italian bakery products obtained with sour dough: characterization of the typical microflora. *Advances in Food Science*, 18: 131–144

Ouiddir, M., Bettache, G., Salas, M.L., Pawtowski, A., Donot, C., Brahimi, S., ... and Mounier, J. 2019. Selection of Algerian lactic acid bacteria for use as antifungal bioprotective cultures and application in dairy and bakery products. *Food Microbiology*, 82: 160–170.

Özogul, F., Küley, E., Küley, F., Kulawik, P. and Rocha, J.M. 2022. Impact of sumac, cumin, black pepper and red pepper extracts in the development of foodborne pathogens and formation of biogenic amines. *European Food Research and Technology* (Springer. doi: https://link.springer.com/article/10.1007/s00217-022-04006-x.

Păcularu-Burada, B., Georgescu, L.A., Vasile, M.A., Rocha, J.M. and Bahrim, G.-E. 2020. Selection of wild lactic acid bacteria strains as promoters of postbiotics in gluten-free sourdoughs. *Microorganisms (MDPI)*, 8(5): 643, Special Issue: Microbial Safety of Fermented Products; doi: https://doi.org/10.3390/microorganisms8050643; ISSN 2076-2607, Publication date: 28/04/2020.

Păcularu-Burada, B., Turturică, M., Rocha, J.M. and Bahrim, G.-E. 2021. Statistical approach to potentially enhance the postbiotication of gluten-free sourdough. *Applied Sciences (MDPI)*, 11: 11, 5306. Special Issue: Advances of Lactic Fermentation for Functional Food Production. doi: https://doi.org/10.3390/app11115306. Online ISSN 2076-3417, Publication date: 07/06/2021.

Pagani, M.A., Bottega, G. and Mariotti, M. 2013. Technology of baked goods. pp. 47–83. *In*: Gobbetti, M. and Gänzle, M. (Eds.). *Handbook on Sourdough Biotechnology*. Springer, Boston, MA. https://doi.org/10.1007/978-1-4614-5425-0_3.

Paramithiotis, S., Chouliaras, Y., Tsakalidou, E. and Kalantzopoulos, G. 2005. Application of selected starter cultures for the production of wheat sourdough bread using a traditional three-stage procedure. *Process Biochemistry*, 40(8): 2813–2819. https://doi.org/10.1016/j.procbio.2004.12.021.

Pareyt, B., Finnie, S.M., Putseys, J.A. and Delcour, J.A. . Lipids in bread making: Sources, interactions, and impact on bread quality. *J. Cereal Sci.*, 54: 266–279.

Park, J., Yun, J., Lim, J.A., Kang, D.H. and Ryu, S. 2012. Characterization of an endolysin, LysBPS13, from a *Bacillus cereus* bacteriophage. *FEMS Microbiology Letters*, 332(1): 76–83.

Patel, A. and Prajapati, J. 2013. Food and health applications of exopolysaccharides produced by Lactic acid Bacteria. *Adv. Dairy Res.*, 1: 1–7.

Pateras, I.M. 2007. Bread spoilage and staling. pp. 275–298. *In*: Cauvain, S.P. and Young, S.L. (Eds.). *Technology of Breadmaking*; Springer: Berlin/Heidelberg, Germany.

Pattison, T., Lindsay, D. and Holy, A.V. 2004. Natural antimicrobials as potential replacements for calcium propionate in bread : research article. *South African Journal of Science*, 100: 342–348.

Paul, P., Ramraj, S.K., Neelakandan, Y., et al. 2011. Production and purification of a novel exopolysaccharide from lactic acid bacterium *Streptococcus phocae* PI80 and its functional characteristics activity *in vitro*. *Bioresour Technol.*, 102: 4827–4833.

Paul Ross, R., Morgan, S. and Hill, C. 2002. Preservation and fermentation: past, present and future. *International Journal of Food Microbiology*, 79(1-2): 3–16. https://doi.org/10.1016/s0168-1605(02)00174-5.

Pawlowska, A.M., Zannini, E., Coffey, A. and Arendt, E.K. 2012. "Green preservatives": combating fungi in the food and feed industry by applying antifungal lactic acid bacteria. *Advances in Food and Nutrition Research*, 66: 217–238. doi:10.1016/b978-0-12-394597- 6.00005-7.

Pena, L. and Burton, B.K. 2012. Survey of health status and complications among propionic acidemia patients. *Am. J. Med. Genet. A.*, 158A(7): 1641–6. http://www.ncbi.nlm.nih.gov/pubmed/22078457.

Peng, Q. and Yuan, Y. 2018. Characterization of a novel phage infecting the pathogenic multidrug-resistant *Bacillus cereus* and functional analysis of its endolysin. *Applied Microbiology and Biotechnology*, 102(18): 7901–7912.

Pepe, O., Ventorino, V., Cavella, S., Fagnano, F. and Brugno, F. 2013. Prebiotic content of bread prepared with flour from immature wheat grain and selected dextran-producing lactic acid bacteria. *Appl. Environ. Microbiol.*, 79(12).

Pereira, D.M., Valentão, P., Pereira, J.A. and Andrade, P.B. 2009. Phenolics: from chemistry to biology. *Molecules*, 14: 2202–2211.

Pereira-Barros, M.A., Barroso, M.F., Martín-Pedraza, L., Vargas, L., Benedé, S., Villalba, M., Rocha, J.M., Campuzano, S. and Pingarrón, J.M. 2019. Direct PCR-free electrochemical biosensing of plant-food derived nucleic acids in genomic DNA extracts. Application to the determination of the key allergen Sola l 7 in tomato seeds. *Biosensors and Bioelectronics*, 137: 171–177. Elsevier. ISSN: 0956-5663. doi: https://doi.org/10.1016/j.bios.2019.05.011.

Pessione, E. 2012. Lactic acid bacteria contribution to gut microbiota complexity: lights and shadow. *Front. Cell. Infect. Microbiol.*, Volume 2, Article, 86: 1–15.

Petkova, M., Gotcheva, V., Dimova, M., Bartkiene, E., Rocha, J.M., Angelov, A. 2022. Screening of *Lactiplantibacillus plantarum* strains from sourdoughs for biosuppression of *Pseudomonas syringae* pv. syringae and *Botrytis cinerea* in table grapes". *Microorganisms*, 10(11): 2094. doi: https://doi.org/10.3390/microorganisms10112094. Special issue: Characterization of the Diversity of Food Microorganisms and Their Metabolites 2022, https://www.mdpi.com/journal/microorganisms/special_issues/micro_metabo2.

Pinilla, C.M.B., Thys, R.C.S. and Brandelli, A. 2019. Antifungal properties of phosphatidylcholine-oleic acid liposomes encapsulating garlic against environmental fungal in wheat bread. *International Journal of Food Microbiology*, 293: 72–78.

Plessas, S., Pherson, L., Bekatorou, A., Nigam, P. and Koutinas, A. 2005. Bread making using kefir grains as baker's yeast. *Food Chemistry*, 93(4): 585–589. https://doi.org/10.1016/j.foodchem.2004.10.034.

Prema, P., Smila, D., Palavesam, A. and Immanuel, G. 2008. Production and characterization of an antifungal compound (3-Phenyllactic Acid) produced by *Lactobacillus plantarum* Strain. *Food and Bioprocess Technology*, 3(3): 379–386. https://doi.org/10.1007/s11947-008-0127-1.

Rahman, M., Islam, R., Hasan, S., Zzaman, W., Rana, M.R., Ahmed, S., Roy, M., Sayem, A., Matin, A., Raposo, A., Zandonadi, R.P., Botelho, R.B.A. and Sunny, A.R. 2022. A comprehensive review on bio-preservation of bread: an approach to adopt wholesome strategies. *Foods*, 11(3): 319. https://doi.org/10.3390/foods11030319.

Ramos-Vivas, J., Elexpuru-Zabaleta, M., Samano, M.L., Barrera, A.P., Forbes-Hernández, T.Y., Giampieri, F. and Battino, M. 2021. Phages and enzybiotics in food biopreservation. *Molecules*, 26(17): 5138.

Rathod, N.B., Nirmal, N.P., Pagarkar, A., Özogul, F. and Rocha, J.M. 2022. Antimicrobial impacts of microbial metabolites on the preservation of fish and fishery products: a review with current knowledge. *Microorganisms (MDPI)*, 10(4): 773. doi: https://doi.org/10.3390/microorganisms10040773. ISSN 2076-2607, Publication date: 03/04/2022. Special issue (2nd edition): "Screening and characterization of the diversity of food microorganisms and their metabolites 2022": https://www.mdpi.com/journal/microorganisms/special_issues/micro_metabo2.

Raveschot, C., Cudennec, B., Coutte, F., Flahaut, C., Fremont, M., Drider, D. and Dhulster, P. 2018. Production of bioactive peptides by *Lactobacillus* species: from gene to application. *Front. Microbiol.*, 9:2354. doi: 10.3389/fmicb.2018.02354.

Regulation, H.A.T. 1993. Council regulation (EEC) No 315/93 of 8 February 1993 laying down Community procedures for contaminants in food. *Official Journal* L, 37(13/02): 0001–0003.

Reis, J.A., Paula, A.T., Casarotti, S.N. and Penna, A.L.B. 2012. Lactic acid bacteria antimicrobial compounds: characteristics and applications. *Food Engineering Reviews* 4: 124–140. https://doi.org/10.1007/s12393-012-9051-2.

Richard-Molard, D. 1994. Le gout du pain (The taste of bread. *In:* Guinet, R. (Ed.): *La Panification Francaise (The French Bakery).* Tec Collection Sciences et Techniques Agro-alimentaires, Paris, France.

198   *Novel Approaches in Biopreservation for Food and Clinical Purposes*

Rizzello, C.G., Cassone, A., Coda, R. and Gobbetti, M. 2011. Antifungal activity of sourdough fermented wheat germ used as an ingredient for bread making. *Food Chemistry*, 127(3): 952–959. https://doi.org/10.1016/j.foodchem.2011.01.063.

Rizzello, C.G., Cavoski, I., Turk, J., Ercolini, D., Nionelli, L., Pontonio, E., de Angelis, M., de Filippis, F., Gobbetti, M. and di Cagno, R. 2015. Organic cultivation of *Triticum turgidum* subsp. durum is reflected in the flour-sourdough fermentation-bread Axis. *Applied and Environmental Microbiology*, 81(9): 3192–3204. https://doi.org/10.1128/aem.04161-14.

Rocha, J.M. and Malcata, F.X. 1999. On the microbiological profile of traditional Portuguese sourdough. *Journal of Food Protection*, 62(12): 1416–1429. ISSN: 0362-028X. url: http://www.scopus.com/inward/record.url?eid=2-s2.0-0345201643&partnerID=MN8TOARS. Publication date: 16/04/1999.

Rocha, O., Ansari, K. and Doohan, F.M. 2005. Effects of trichothecene mycotoxins on eukaryotic cells: a review. *Food Additives and Contaminants*, 22(4): 369–378. https://doi.org/10.1080/02652030500058403.

Rocha, J.M., Kalo, P.J. and Malcata, F.X. 2010a. Neutral lipids in non-starch lipid and starch lipid extracts from portuguese sourdough bread. *European Journal of Lipid Science and Technology*, 112: 1138–1149. doi: 10.1002/ejlt.201000101.

Rocha, J.M., Kalo, P.J. Ollilainen, V. and Malcata, F.X. 2010b. Separation and identification of neutral cereal lipids by normal phase high-performance liquid chromatography, using evaporative light-scattering and electrospray mass spectrometry for detection. *Journal of Chromatography A*, 1217: 3013–3025. doi: 10.1016/j.chroma.2010.02.034.

Rocha, J.M. 2011. Microbiological and lipid profiles of broa: contributions for the characterization of a traditional portuguese bread. Ph.D thesis dissertation. Instituto Superior de Agronomia, Higher Institute of Agriculture, Universidade de Lisboa (University of Lisbon) (ISA-UL): Lisbon, Portugal, 705 pages. Thesis available at http://hdl.handle.net/10400.5/3876.

Rocha, J.M., Kalo, P.J. and Malcata, F.X. 2011. Neutral lipids in free, bound and starch lipid extracts of flours, sourdough and portuguese sourdough bread, determined by NP-HPLC-ELSD. *Cereal Chemistry*, 88(4): 400–408. doi: 10.1094/CCHEM-11-10-0157.

Rocha, J.M. and Malcata, F.X. 2012. Microbiological profile of maize and rye flours, and sourdough used for the manufacture of traditional Portuguese bread. *Food Microbiology*, 31: 72–88. doi: 10.1016/j.fm.2012.01.008. Publication date: 25/01/2012.

Rocha, J.M., Kalo, P.J. and Malcata, F.X. 2012a. Fatty acid composition of non-starch and starch neutral lipid extracts of portuguese sourdough bread. *Journal of the American Oil Chemists' Society*, 89(11): 2025–2045. doi: 10.1007/s11746-012-2110-2.

Rocha, J.M., Kalo, P.J. and Malcata, F.X. 2012b. Composition of neutral lipid classes and content of fatty acids throughout sourdough breadmaking. *European Journal of Lipid Science and Technology*, 114(3): 294–305. doi: 10.1002/ejlt.201100208.

Rocha, J.M. and Malcata, F.X. 2016a. Microbial ecology dynamics in portuguese broa sourdough. *Journal of Food Quality*, 39(6): 634–648. https://doi.org/10.1111/jfq.12244.

Rocha, J.M. and Malcata, F.X. 2016b. Behavior of the complex micro-ecology in maize and rye flour and mother-dough for broa throughout storage. *Journal of Food Quality*, 39: 218–233. doi: 10.1111/jfq.12183. Publication date: 06/08/2015.

Rodrigues, I. and Naehrer, K. 2012a. A three-year survey on the worldwide occurrence of mycotoxins in feedstuffs and feed. *Toxins*, 4(9): 663–675.

Rodrigues, I. and Naehrer, K. 2012b. Prevalence of mycotoxins in feedstuffs and feed surveyed worldwide in 2009 and 2010. *Phytopathologia Mediterranea*, 175–192.

Rodriguez, A., Nerin, C. and Batlle, R. 2008. New cinnamon-based active paper packaging against Rhizopusstolonifer food spoilage. *Journal of Agricultural and Food Chemistry*, 56(15): 6364–6369.

Rodriguez, M., Medina, L. and Jordano, R. 2000. Effect of modified atmosphere packaging on the shelf life of sliced wheat flour bread. *Food/Nahrung*, 44: 247–252.

Roseline, F. and Waturangi, D.E. 2021. Isolation, characterization, and application of bacteriophages against several food spoilage bacteria: *Bacillus subtilis*, *Bacillus cereus*, and *Shewanella putrefaciens*. *Bacterial Empire*, 4: 263.

Rosell, C.M. 2011. The science of doughs and bread quality. *Flour and Breads and Their Fortification in Health and Disease Prevention*, 3–14. https://doi.org/10.1016/b978-0-12-380886-8.10001-7.

Rosenquist, H. and Hansen, Å. 1998. The antimicrobial effect of organic acids, sour dough and nisin against *Bacillus subtilis* and *B. licheniformis* isolated from wheat bread. *Journal of Applied Microbiology*, 85(3): 621–631. https://doi.org/10.1046/j.1365-2672.1998.853540.x.

Russo, P., Fares, C., Longo, A., Spano, G. and Capozzi, V. 2017. *Lactobacillus plantarum* with broad antifungal activity as a protective starter culture for bread production. *Foods*, 6(12): 110. https://doi.org/10.3390/foods6120110.

Ryan, L.A., Zannini, E., dal Bello, F., Pawlowska, A., Koehler, P. and Arendt, E.K. 2011. *Lactobacillus amylovorus* DSM 19280 as a novel food-grade antifungal agent for bakery products. *International Journal of Food Microbiology*, 146(3): 276–283. https://doi.org/10.1016/j.ijfoodmicro.2011.02.036.

Ryan, L.A.M., dal Bello, F., Arendt, E.K. and Koehler, P. 2009. Detection and quantitation of 2,5-Diketopiperazines in wheat sourdough and bread. *Journal of Agricultural and Food Chemistry*, 57(20): 9563–9568. https://doi.org/10.1021/jf902033v.

Ryan, L., dal Bello, F. and Arendt, E. 2008. The use of sourdough fermented by antifungal LAB to reduce the amount of calcium propionate in bread. *International Journal of Food Microbiology*, 125(3): 274–278. https://doi.org/10.1016/j.ijfoodmicro.2008.04.013.

Sabilon, L. and Bianchini, A. 2015. From field to table: a review on the microbiological quality and safety of wheat-based products. *Cereal Chemistry*, 93(2): 105 – 115. https://doi.org/10.1094/CCHEM-06-15-0126-RW.

Sacco, C., Donato, R., Zanella, B., Pini, G., Pettini, L., Marino, M. F., Rookmin, A.D. and Marvasi, M. 2020. Mycotoxins and flours: effect of type of crop, organic production, packaging type on the recovery of fungal genus and mycotoxins. *International Journal of Food Microbiology*, 334: 108808. https://doi.org/10.1016/j.ijfoodmicro.2020.108808.

Sadeghi, A., Ebrahimi, M., Mortazavi, S.A. and Abedfar, A. 2019. Application of the selected antifungal LAB isolate as a protective starter culture in pan whole-wheat sourdough bread. *Food Control*, 95: 298–307. https://doi.org/10.1016/j.foodcont.2018.08.013.

Sadik, O.A., A.O. Aluoch, and A. Zhou. 2009. Status of biomolecular recognition using electrochemical techniques. *Biosensors and Bioelectronics*, 24(9): 2749–2765.

Salgado, P.R., Di Giorgio, L., Musso, Y.S. and Mauri, A.N. 2021. Recent developments in smart food packaging focused on biobased and biodegradable polymers. *Frontiers in Sustainable Food Systems*, 5. https://doi.org/10.3389/fsufs.2021.630393.

Salim-ur-Rehman, Paterson, A. and Piggott, J.R. 2006. Flavour in sourdough breads: a review. *Trends in Food Science &Amp; Technology*, 17(10): 557–566. https://doi.org/10.1016/j.tifs.2006.03.006.

Salovaara, H. and Valjakka, T. 1987. The effect of fermentation temperature, flour type, and starter on the properties of sour wheat bread. *Int. J. Food Sci. Technol.*, 22: 591–597.

Samapundo, S., Devlieghere, F., Vroman, A. and Eeckhout, M. 2017. Antifungal activity of fermentates and their potential to replace propionate in bread. *LWT-Food Science and Technology*, 2017;76: 101–107. DOI: 10.1016/j.lwt.2016.10.043.

Samarajeewa, U., Sen, A.C., Cohen, M.D. and Wei, C.I. 1990. Detoxification of aflatoxins in foods and feeds by physical and chemical methods1. *Journal of Food Protection*, 53(6): 489–501. https://doi.org/10.4315/0362-028x-53.6.489.

Sangmanee, P. and Hongpattarakere, T. 2014. Inhibitory of multiple antifungal components produced by *Lactobacillus plantarum* K35 on growth, aflatoxin production and ultrastructure alterations of *Aspergillus flavus* and *Aspergillus parasiticus*. *Food Control*, 40: 224–233. https://doi.org/10.1016/j.foodcont.2013.12.005.

Saranraj, P. and Sivasahthivelan, P. 2015. Microorganisms involved in spoilage of bread and its control measures. pp. 132–149 *In*: Cristina, M. Rosell, Joanna Bajerska and Aly F. El Sheikha (Eds.). *Bread and Its Fortification Nutrition and Health Benefits*. CRC Press, Taylor & Fransis Group, New York, USA https://doi.org/10.1201/b18918.

Saranraj, P. and Geetha, M. 2012. Microbial spoilage of bakery products and its control by preservatives. *International Journal of Pharmaceutical & Biological Archive*, 3(1): 38–48.

Sauer, F. 1977. Control of yeast and molds with preservatives. *Food Technol.*, 31: 66 – 67.

200 *Novel Approaches in Biopreservation for Food and Clinical Purposes*

Schaefer, L., Auchtung, T.A., Hermans, K.E., Whitehead, D., Borhan, B. and Britton, R.A. 2010. The antimicrobial compound reuterin (3-hydroxypropionaldehyde) induces oxidative stress via interaction with thiol groups. *Microbiology*, 156(6): 1589–1599. https://doi.org/10.1099/mic.0.035642-0.

Schmidt, M., Lynch, K.M., Zannini and Arendt, E.K. 2018. Fundamental study on the improvement of the antifungal activity of *Lactobacillus reuteri* R29 through increased production of phenyllactic acid and reuterin. *Food Control*, 88: 139–148.

Schnürer, J. and Magnusson, J. 2005. Antifungal lactic acid bacteria as biopreservatives. *Trends in Food Science & Technology*, 16(1–3): 70–78. https://doi.org/10.1016/j.tifs.2004.02.014.

Schuch, R., Pelzek, A.J., Nelson, D.C. and Fischetti, V.A. 2019. The PlyB endolysin of bacteriophage vB_BanS_Bcp1 exhibits broad-spectrum bactericidal activity against *Bacillus cereus* sensu lato isolates. *Applied and Environmental Microbiology*, 85(9): e00003-19.

Seiler, D.A.L. 1983. Preservation of bakery products. FMBRA Bulletin No. 4, CCFRA, Chipping Campden, UK, pp. 166–177.

Seiler, D.A.L. 1989. Modified atmosphere packaging of bakery products. pp. 119–133. *In*: Brody, A.L. (Ed.). *Controlled/modified Atmosphere/Vacuum Packaging of Foods*. Trumbell, CT: Food and Nutrition Press.

Seiler, D.A.L. 1984. Preservation of bakery products. *Institute of Food Science and Technology Proceedings*, 17: 31–39.

Sharma, H., Özogul, F., Bartkiéne, E., Rocha, J.M. 2021. Impact of lactic acid bacteria and their metabolites on the techno-functional properties and health benefits of fermented dairy products. *Critical Reviews in Food Science and Nutrition* (Taylor & Francis). doi: https://doi.org/10.1080/10408398.2021.2007844. Publication date: 2021-11-15.

Shetty, P.H. and Jespersen, L. 2006. *Saccharomyces cerevisiae* and lactic acid bacteria as potential mycotoxin decontaminating agents. *Trends in Food Science & Technology*, 17(2): 48–55. https://doi.org/10.1016/j.tifs.2005.10.004.

Singh, A., S. Poshtiban and S. Evoy. 2013. Recent advances in bacteriophage based biosensors for foodborne pathogen detection. *Sensors*, 13(2): 1763–1786.

Singh, V.P. 2018. Recent approaches in food bio-preservation-a review. *Open Veterinary Journal*, 8(1): 104–111.

Sivasankar, P., Seedevi, P., Poongodi, S., Sivakumar, M., Murugan, T., Sivakumar, L. et al. 2018. Characterization, antimicrobial and antioxidant property of exopolysaccharide mediated silver nanoparticles synthesized by *Streptomyces violaceus* MM72. *Carbohydr. Polym.*, 181: 752–759. doi: 10.1016/j.carbpol.2017. 11.082.

Sjögren, J., Magnusson, J., Broberg, A., Schnürer, J. and Kenne, L. 2003. Antifungal 3-hydroxy fatty acids from *Lactobacillus plantarum* MiLAB 14. *Appl. Environ. Microbiol.*, 69: 7554–7557.

Skrajda-Brdak, M., Konopka, I., Tańska, M. and Czaplicki, S. 2019. Changes in the content of free phenolic acids and antioxidative capacity of wholemeal bread in relation to cereal species and fermentation type. *Eur. Food Res. Technol.*, 245: 2247–2256.

Smith, J.P., Ooraikul, B., Koersen, W.J., Jackson, E.D. and Lawrence, R.A. 1986. Novel approach to oxygen control in modified atmosphere packaging of bakery products. *Food Microbiol.*, 3(4): 315–320.

Smith, J.P., Daifas, D.P., El-Khoury, W., Koukoutsis, J. and El-Khoury, A. 2004. Shelf life and safety concerns of bakery products—a review. *Critical Reviews in Food Science and Nutrition*, 44(1): 19–55. https://doi.org/10.1080/10408690490263774.

Son, B., Yun, J., Lim, J. A., Shin, H., Heu, S. and Ryu, S. 2012. Characterization of LysB4, anendolysin from the *Bacillus cereus*-infecting bacteriophage B4. *BMC Microbiology*, 12(1): 1–9.

Sperber, W. 2003. Microbiology of milled cereal grains: issues in customer specifications. *Tech. Bull IAOM*, 3:7929–7931.

Spicher, G. 1980. Zur Aufklärung der Quellen und Wege der Schimmelkontamination des Brotes im Grossbackbetreib Zentralblatt für Bakteriologie Parasitenkunde, Infektionskrankheiten und Hygiene, 1 Abt. *Original Reiheb Hygiene Betriebshygiene Preventive Medizin*. 170: 508–528.

Stevens, M., Vollenweider, S., Lacroix, C. and Zurich, E.T.H. 2011. The potential of reuterin produced by *Lactobacillus reuteri* as a broad spectrum preservative in food. *Protective Cultures, Antimicrobial Metabolites and Bacteriophages for Food and Beverage Biopreservation*, 129–160.

Stolz, P. 1999. Mikrobiologie des Sauerteiges. pp. 35–60. *In*: Spicher, G. and Stephan, H. (Eds.). *Handbuch Sauerteig: Biologie, Biochemie, Technologie* 5th ed. Hamburg: Behr's Verlag.

Stolz, P. 2003. Biological fundamentals of yeast and *Lactoba- cilli* fermentation in bread dough. pp. 37–56. *In*: Kulp, K. and Lorenz, K. (Eds.). *Handbook of Dough Fermentations*. Marcel Dekker, New York.

Stoyanova, L.G., Ustyugova, E.A. and Netrusov, A.I. 2012. Antibacterial metabolites of lactic acid bacteria: Their diversity and properties. *Applied Biochemistry and Microbiology*, 48(3): 229–243. https://doi.org/10.1134/s0003683812030143.

Stratford, M. and Anslow, P. 1998. Evidence that sorbic acid does not inhibit yeast as a classic 'weak acid preservative.' *Letters in Applied Microbiology*, 27(4): 203–206. https://doi.org/10.1046/j.1472-765x.1998.00424.x.

Stratford, M., Steels, H., Nebe-von-Caron, G., Novodvorska, M., Hayer, K. and Archer, D.B. 2013b. Extreme resistance to weak-acid preservatives in the spoilage yeast *Zygosaccharomyces bailii*. *International Journal of Food Microbiology*, 166: 126–134.

Sugihara, F.T. 2003. Commercial starters in the United States. *In*: Kulp, K. and Lorenz, K. (Eds.). *Handbook of Dough Fermentations*. Marcel Dekker, New York.

Sugihara, T.F., Kline, L. and Miller, M.W. 1971. Microorganisms of the San Francisco Sour Dough Bread Process. *Applied Microbiology*, 21(3): 456–458. https://doi.org/10.1128/am.21.3.456-458.1971.

Suhr, K.I. and Nielsen, P.V. 2004. Effect of weak acid preservatives on growth of bakery product spoilage fungi at different water activities and pH values. *International Journal of Food Microbiology*. 95: 67–78. DOI: 10.1016/j.ijfoodmicro.2004.02.004.

Suhr, K.I. and Nielsen, P.V. 2005, January. Inhibition of fungal growth on wheat and rye bread by modified atmosphere packaging and active packaging using volatile mustard essential oil. *Journal of Food Science*, 70(1): M37–M44. https://doi.org/10.1111/j.1365-2621.2005.tb09044.x.

Sun, D., Li, H., Song, D., Zhang, L., Zhao, X. and Xu, X. 2019. Genome, transcriptome and fermentation analyses of *Lactobacillus plantarum* LY-78 provide new insights into the mechanism of phenyllactate biosynthesis in lactic acid bacteria. *Biochemical and Biophysical Research Communications*, 519(2): 351–357. https://doi.org/10.1016/j.bbrc.2019.09.011.

Suo, B., Nie, W., Wang, Y., Ma, J., Xing, X., Huang, Z., Xu, C., Li, Z. and Ai, Z. 2020. Microbial diversity of fermented dough and volatile compounds in steamed bread prepared with traditional Chinese starters. *LWT*, 126: 109350. https://doi.org/10.1016/j.lwt.2020.109350.

Suomalainen, T.H. and Mäyrä-Makinen, A.M. 1999. Propionic acid bacteria as protective cultures in fermented milks and breads. *Le Lait*, 79(1): 165–174. https://doi.org/10.1051/lait:1999113.

Sweetman, J. 2016. Commercialization of foods for customers with specific dietary needs. pp. 63–77. *In*: Osborn, S. and Morley, W. (Eds.). *Developing Food Products for Consumers with Specific Dietary Needs*. Woodhead Publishing.

Taglieri, I., Macaluso, M., Bianchi, A., Sanmartin, C., Quartacci, M.F., Zinnai, A. and Venturi, F. 2020. Overcoming bread quality decay concerns: main issues for bread shelf life as a function of biological leavening agents and different extra ingredients used in formulation. A review. *Journal of the Science of Food and Agriculture*, 101(5): 1732–1743. https://doi.org/10.1002/jsfa.10816.

Talarico, T.L., Casas, I.A., Chung, T.C. and Dobrogosz, W.J. 1988. Production and isolation of reuterin, a growth inhibitor produced by *Lactobacillus reuteri. Antimicrobial Agents and Chemotherapy*, 32(12): 1854–1858. https://doi.org/10.1128/aac.32.12.1854.

Teixeira, J.S., McNeill, V. and Gänzle, M.G. 2012. Levansucrase and sucrose phoshorylase contrib- ute to raffinose, stachyose, and verbascose metabolism by lactobacilli. *Food Microbiol*, 31: 278–284.

Teixeira-Costa, B.E. and Andrade, C.T. 2021. Natural polymers used in edible food packaging—history, function and application trends as a sustainable alternative to synthetic plastic. *Polysaccharides*, 3(1): 32–58. https://doi.org/10.3390/polysaccharides3010002.

Tolpeznikaite, E., Starkute, V., Zokaityte, E., Ruzauskas, M., Pilkaityte, R., Viskelis, P., Urbonaviciene, D., Ruibys, R., Rocha, J.M. and Bartkiene, E. 2022. Effect of solid-state fermentation and ultrasonication processes on antimicrobial and antioxidant properties of algae extracts. *Frontiers in Nutrition, Section Nutrition and Food Science Technology*, 9: 990274. doi: https://doi.org/10.3389/fnut.2022.990274.

Trakselyte-Rupsiene, K., Juodeikiene, G., Alzbergaite, G., Zadeike, D., Bartkiene, E., Özogul, F., Ruller, L., Robert, J. and Rocha, J.M. 2022. Bio-refinery of plant drinks press cake permeate

202 *Novel Approaches in Biopreservation for Food and Clinical Purposes*

using ultrafiltration and lactobacillus fermentation into antimicrobials and its effect on the growth of wheatgrass *in vivo*. *Food Bioscience (Elsevier)*, 46: 101427. doi: https://doi.org/10.1016/j.fbio.2021.101427. Publication date: 01/04/2022.

U.S. Food and Drug Administration. "21 CFR 184 – Direct Food Substances Affirmed As Generally Recognized As Safe." 1 Apr. 2022, https://www.accessdata.fda.gov/scripts/cdrh/cfdocs/cfcfr/CFRSearch.cfm?CFRPart=184.

Uma Reddy, M., Gulla, S. and Nagalakshmi, A.V.D. 2010. Effect of fermentation on aflatoxin reduction in selected food products. *Food Sci Technol. Nutr.* 4(1): 93–99.

ur-Rehman, S., Nawaz, H., Hussain, S., Mushtaq Ah, M., Anjum Murt, M. and Saeed Ahma, M. 2007. Effect of sourdough bacteria on the quality and shelf life of bread. *Pakistan Journal of Nutrition*, 6(6): 562–565. https://doi.org/10.3923/pjn.2007.562.565.

Valanciene, E., Jonuskiene, I., Syrpas, M., Augustiniene, E., Matulis, P., Simonavicius, A. and Malys, N. 2020. Advances and prospects of phenolic acids production, biorefinery and analysis. *Biomolecules*, 10(6): 874. https://doi.org/10.3390/biom10060874.

Valerio, F., De Bellis, P., Di Biase, M., Lonigro, S.L., Giussani, B., Visconti, A., ... and Sisto, A. 2012. Diversity of spore-forming bacteria and identification of *Bacillus amyloliquefaciens* as a species frequently associated with the ropy spoilage of bread. *International Journal of Food Microbiology*, 156(3): 278–285

Verma, A.K., Banerjee, R., Dwivedi, H.P. and Juneja, V.K. 2014. *BACTERIOCINS* | Potential in Food Preservation. *Encyclopedia of Food Microbiology*, 180–186.

Vermeulen, N., Kretzer, J., Machalitza, H., Vogel, R.F. and Gänzle, M.G. 2006. Influence of redox-reactions catalysed by homo- and hetero-fermentative lactobacilli on gluten in wheat sourdoughs. *J. Cereal Sci.*, 43: 137–143.

Vero, S., Garmendia, G., González, M.B., Bentancur, O. and Wisniewski, M. 2012. Evaluation of yeasts obtained from Antarctic soil samples as biocontrol agents for the management of postharvest diseases of apple (*Malus* × *domestica*. *FEMS Yeast Research*, 13(2): 189–199. https://doi.org/10.1111/1567-1364.12021.

Vogel, R.F., Pavlovic, M., Ehrmann, M.A., Wiezer, A., Liesegang, H., Offschanka, S., Voget, S., Angelov, A., Böcker, G. and Liebl, W. 2011. Genomic analysis reveals *Lactobacillus sanfranciscensis* as stable element in traditional sourdoughs. *Microbiol Cell Fact*, 10(Suppl 1): S6.

Wagacha, J.M. and Muthomi, J.W. 2008. Mycotoxin problem in Africa: current status, implications to food safety and health and possible management strategies. *International Journal of Food Microbiology*, 124(1): 1–12.

Wan, X., Geng, P., Sun, J., Yuan, Z. and Hu, X. 2021. Characterization of two newly isolated bacteriophages PW2 and PW4 and derived endolysins with lysis activity against *Bacillus cereus* group strains. *Virus Research*, 302: 198489.

Wang, H., Yan, Y., Wang, J., Zhang, H. and Qi, W. 2012. Production and characterization of antifungal compounds produced by *Lactobacillus plantarum* IMAU10014. *PLoS ONE*, 7(1): e29452. https://doi.org/10.1371/journal.pone.0029452.

Wang, Y., Wu, J., Lv, M., Shao, Z., Hungwe, M., Wang, J., Bai, X., Xie, J., Wang, Y. and Geng, W. 2021. Metabolism characteristics of lactic acid bacteria and the expanding applications in food industry. *Frontiers in Bioengineering and Biotechnology*, 9. https://doi.org/10.3389/fbioe.2021.612285.

Wang, Y.T., Xue, Y.R. and Liu, C.H. 2015. A brief review of bioactive metabolites derived from deep-sea fungi. *Marine Drugs*, 13(8): 4594–4616. https://doi.org/10.3390/md13084594.

Weckx, S., van Kerrebroeck, S. and de Vuyst, L. 2019. Omics approaches to understand sourdough fermentation processes. *International Journal of Food Microbiology*, 302: 90–102. https://doi.org/10.1016/j.ijfoodmicro.2018.05.029.

Weidenbörner, M., Wieczorek, C., Appel, S. and Kunz, B. 2000. Whole wheat and white wheat flour—the mycobiota and potential mycotoxins. *Food Microbiology*, 17(1): 103–107. https://doi.org/10.1006/fmic.1999.0279.

WHO. 2006. World Health Organization. Evaluation of certain mycotoxins in food. 56th report of the joint FAO/WHO Expert Committee on Food Additives (JECFA). WHO technical Report 906. Geneva, 2002 16–26.

World Health Organization. 2020. The state of food security and nutrition in the world 2020: transforming food systems for affordable healthy diets (Vol. 2020). Food & Agriculture Org.

Wu, S., Peng, Y., Xi, J., Zhao, Q., Xu, D., Jin, Z. and Xu, X. 2022. Effect of sourdough fermented with corn oil and lactic acid bacteria on bread flavor. LWT, 155, 112935. https://doi.org/10.1016/j.lwt.2021.112935.

Xi, J., Zhao, Q., Xu, D., Jin, Y., Wu, F., Jin, Z. and Xu, X. 2021. Volatile compounds in Chinese steamed bread influenced by fermentation time, yeast level and steaming time. *LWT*, 141: 110861. https://doi.org/10.1016/j.lwt.2021.110861.

Yan, B., Zhao, J., Fan, D., Tian, F., Zhang, H. and Chen, W. 2016. Antifungal activity of *Lactobacillus plantarum* against *Penicillium roqueforti in vitro* and the preservation effect on chinese steamed bread. *Journal of Food Processing and Preservation*, 41(3): e12969. https://doi.org/10.1111/jfpp.12969.

Yan, P.S., Song, Y., Sakuno, E., Nakajima, H., Nakagawa, H. and Yabe, K. 2004. Cyclo (L-Leucyl-L-Prolyl) produced by achromobacter xylosoxidans inhibits aflatoxin production by *Aspergillus parasiticus*. *Appl. Environ. Microbiol.*, 70: 7466–7473.

Yazar, G. and Tavman, E. 2012. Functional and technological aspects of sourdough fermentation with *Lactobacillus sanfranciscensis*. *Food Engineering Reviews*, 4(3): 171–190. https://doi.org/10.1007/s12393-012-9052-1.

Yépez, A., Luz, C., Meca, G., Vignolo, G., Mañes, J. and Aznar, R. 2017. Biopreservation potential of lactic acid bacteria from Andean fermented food of vegetal origin. *Food Control*, 78: 393–400. doi: 10.1016/j.foodcont.2017.03.009.

Yilmaz, B., Bangara, S.P., Echegaray, N., Surib, S., Tomasevic, I., Lorenzo, J.M., Melekoglu, E., Rocha, J.M. and Özogul, F. 2022a. The impacts of Lactiplantibacillus plantarum on the functional properties of fermented foods: A review of current knowledge. *Microorganisms (MDPI)*, 10(4): 826. doi: https://doi.org/10.3390/microorganisms10040826. ISSN 2076-2607, Publication date: 15/04/2022. Special issue (2nd edition): "Screening and characterization of the diversity of food microorganisms and their metabolites 2022": https://www.mdpi.com/journal/microorganisms/special_issues/micro_metabo2.

Yilmaz, N., Özogul, F., Fadiloglu, E.E., Moradi, M., Simat, V. and Rocha, J.M. 2022b. Reduction of biogenic amine formation by foodborne pathogens using postbiotics in lysine-decarboxylase broth. *Journal of Biotechnology*, 358: 118–127. doi: https://doi.org/10.1016/j.jbiotec.2022.09.003.

Yu, J., Chang, P.K., Ehrlich, K.C., Cary, J.W., Bhatnagar, D., Cleveland, T.E., Payne, G.A., Linz, J.E., Woloshuk, C.P. and Bennett, J.W. 2004. Clustered pathway genes in *Aflatoxin biosynthesis*. *Applied and Environmental Microbiology*, 70(3): 1253–1262. https://doi.org/10.1128/aem.70.3.1253-1262.2004.

Zannini, E., Garofalo, C., Aquilanti, L., Santarelli, S., Silvestri, G. and Clementi, F. 2009. Microbiological and technological characterization of sourdoughs destined for bread-making with barley flour. *Food Microbiol.*, 26: 744–753.

Zannini, E., Moroni, A., Belz, M. et al. 2014. Bread. pp. 448–488. *In*: Bamforth, C.W. and Ward, R.E. (Eds.). *Oxford Handbook of Food Fermentations*. New York: Oxford University Press.

Zelada-Guillén, G.A., J. Riu, A. Dьzgьn, and F.X. Rius. 2009. Immediate detection of living bacteria at ultralow concentrations using a carbon nanotube based potentiometric aptasensor. *Angewandte Chemie International Edition*, 48(40): 7334–7337.

Zendo, T. and Sonomoto, K. 2011. Classification and diversity of bacteriocin. *In*: Sonomoto, K. and Yokota, A. (Eds.). *Lactic Acid Bacteria and Bifidobacteria: Current Progress in Advanced Research*. Caister Academic Press, Portland, USA.

Zhang, C., Brandt, M.J., Schwab, C. and Gänzle, M.G. 2010. Propionic acid production by cofermentation of *Lactobacillus buchneri* and *Lactobacillus diolivorans* in sourdough. *Food Microbiology*, 27(3): 390–395. https://doi.org/10.1016/j.fm.2009.11.019.

Zhang, L., Liu, C., Li, D., Zhao, Y., Zhang, X., Zeng, X. and Li, S. 2013. Antioxidant activity of an exopolysaccharide isolated from *Lactobacillus plantarum* C88. *International Journal of Biological Macromolecules*, 54: 270–275.

Zhou Y., Cui, Y. and Qu, X. 2019. Exopolysaccharides of lactic acid bacteria: structure, bioactivity and associations: a review. *Carbohydr. Polym.* 207: 317–332. doi: 10.1016/j.carbpol.2018.11.093.

Zinedine, A., Juan, C., Idrissi, L. and Mañes, J. 2007. Occurrence of ochratoxin A in bread consumed in Morocco. *Microchemical Journal*, 87(2): 154–158. https://doi.org/10.1016/j.microc.2007.07.004.

Zokaityte, E., Cernauskas, D., Klupsaite, D., Lele, V., Starkute, V., Zavistanaviciute, P., Ruzauskas, M., Gruzauskas, R., Juodeikiene, G., Rocha, J.M., Bliznikas, S., Viskelis, P., Ruibys, R., Bartkiene, E. 2020. Bioconversion of milk permeate with selected lactic acid bacteria strains and apple by-products into beverages with antimicrobial properties and enriched with galactooligosaccharides. *Microorganisms (MDPI)*, 8(8): 1182. doi: https://doi.org/10.3390/microorganisms8081182. ISSN 2076-2607, Publication date: 03/08/2020. Special issue "Screening and characterization of the diversity of food microorganisms and their metabolites".

CHAPTER 7

Biopreservation of Beverages

Suchi Parvin Biki[1] and *Enriqueta Garcia-Gutierrez*[2,*]

Introduction

Beverages are drinkable liquids that are not water. Humankind has consumed beverages for millennia and the beverage industry is one of the biggest in the world. There is a great variety of beverages, including milk, fruit juices, tea, sodas, coffee, beer, liquors or energy drinks, leading to different classifications according to different criteria, for example, sugar content (sugared, e.g., fruit juices; and non-sugared, e.g., tea or coffee) or presence or absence of carbon dioxide (sparkling, e.g., sodas; or not, e.g., juices) (Kalpana and Rajeswari, 2019).

Given the wide variety of beverages, it can be difficult to establish common strategies to assess microbiological risk and extend shelf life. As consumers demand safer and healthier products, new approaches need to be developed to ensure the quality of the products. Some of these approaches include the use of natural preservatives (Vara et al., 2019) or the development of active packaging (Shekarchizadeh and Nazeri, 2020; Ramos et al., 2015).

Generally, most beverages available for consumers contain flavoring, sweeteners, acidifiers, emulsifiers or coloring agents (Buglass, 2015). Ultimately, these methods can enhance flavor, and some can also slow down the antioxidant activities of bacteria and molds. The use of preservatives is the most reliable and extended method to maintain a microbiological risk within an acceptable range for a specific beverage.

In the case of alcoholic beverages, as these drinks contain certain levels of ethanol, the microbiological risk is lower and are considered microbiologically safe. However, microorganisms might develop and cause spoilage, generating economic loses. Alcoholic beverages are classified as fermented or distilled and their

---

[1] Hajee Mohammad Danesh Science and Technology University, Dinajpur-5200, Bangladesh.
[2] Department of Agricultural Engineering, Institute of Plant Biotechnology, Polytechnic University of Cartagena, Murcia, Spain.
* Corresponding author: enriqueta.garcia@upct.es

206 *Novel Approaches in Biopreservation for Food and Clinical Purposes*

classification also depends on the fermentation taking place on sugars or starches, fruits or grains (Nykänen and Nykänen, 2017; Lea et al., 2003). Thus, we can find beer, cider and wine or whiskies and liquors (Bujake, 2000; Buglass, 2011). This fermentation can be due to the natural microbiological flora or by the addition of starter cultures (Doyle and Meng, 2006).

Fermentation is a natural method that has been used worldwide for centuries to preserve foods and beverages and generate new products. Fermentation offers other advantages, such as improving the nutritional value of the products, remove undesirable components and make the product safer for the consumers.

There are four main types of fermentation processes: alcoholic, lactic acid, acetic acid and alkali (Anal, 2019). The alcoholic fermentation consists in the production of ethanol by yeasts, like in the case of wines and beers (Walker and Stewart, 2016). Lactic acid fermentation is conducted by lactic acid bacteria (LAB) on milk products, cereals, fruits and vegetables (Ziarno and Cichońska, 2021; Widyastuti and Febrisiantosa, 2014; Szutowska, 2020; Bautista-Gallego et al., 2020). *Acetobacter* species are responsible of transforming the alcohol into acetic acid for the acetic acid fermentation and alkali fermentation happens on protein-rich materials, like legumes, eggs or fish (Guillamón and Mas, 2009; Raspor and Goranovič, 2008; Singh et al., 2018; Deshpande, 2000).

In this chapter, we provide an overview of the different microorganisms associated to different beverages, either during production or of the final product, the different microbial spoilage that threat the quality of these beverages and the different biopreservation strategies used, or under development, to improve the quality and safety of beverages.

Microbial microflora of beverages

It is common to find microorganisms associated to beverages, either during their production or in the final composition. Many of them are necessary for the development of the final product, as it is the case of fermented beverages, either alcoholic or non-alcoholic, like kefir or kombucha.

Dairy fermentates typically contain a wide variety of microorganisms during the fermentation stage and the final product. Kefir, derived from milk, is fermented using kefir grains or mother cultures (Marshall and Cole, 1985). These grains, typically containing LAB, acetic acid bacteria (AAB) and yeasts, will determine the final microbial composition and properties of the kefir (Chen et al., 2009). More specifically, we can find *Lactococcus*, *Lactobacillus*, *Leuconostoc*, and *Acetobacter*, including some species with probiotic properties like *Lactobacillus acidophilus*, *Lactobacillus helveticus*, *Lactobacillus casei* or *Pediococcus pentosaceus*.

Koumiss is another fermented drink derived from milk. As kefir, its associated microflora consists in a variety of lactobacilli, mainly *L. acidophilus* and *Lactobacillus bulgaricus*, and yeasts, like *Saccharomyces* sp., *Kluyveromyces* sp. and *Candida koumiss* (de Melo Pereira et al., 2022). Similarly, production of buttermilk, made from cows' or buffalo's milk, is conducted by mesophilic LAB and acidophilus milk contains the *L. acidophilus* used for the fermentation (Batish and Sunita, 2004; Mohammadi et al., 2012). *L. casei* is the species used to

Biopreservation of Beverages 207

obtain yakult from milk and *Bifidobacterium bifidum* and *Bifidobacterium longum* the ones used to obtain bifidus milk (Hernandez et al., 2010; Kaur et al., 2019).

In alcoholic beverages, the raw material used by the microbial flora for the fermentation process is key for the final fermented beverage. Fruits are considered the ideal substrate, because they contain antioxidants, polyphenols and vitamins and offer support for the growth and survival of probiotic microorganisms, as they harbor a high concentration of nutrients and sugars (Lea et al., 2003). Some low alcoholic beverages, like kombucha or ginger beer, are examples of these fermentations. Kombucha and ginger beer fermentation typically contains LAB and yeasts, and sometimes AAB.

For products with higher alcohol content, the yeast *Saccharomyces cerevisiae* is commonly used, as it is tolerant to high concentrations of ethanol (Ding et al., 2009; Ma and Liu, 2010; Snoek et al., 2016). This characteristic makes it widely used for fermentation industry, whether alcoholic beverages or bioethanol production (Akhtar et al., 2018; Peris et al., 2018; Walker and Stewart, 2016).

Wine is produced using mainly grapes as raw materials, but can also use other fruits, such as cherries, dates, pineapple, apple, plums, peaches or strawberries. Among its common microflora, we can find *Saccharomyces* sp., the most abundant microorganism at the end of the fermentation and pivotal for the final properties of wine, a few *Candida* species and other yeasts like *Hanseniaspora uvarum*, *Torulaspora delbrueckii*, *Kloeckera apiculata*, *Kloeckera thermotolerans* or *Metschnikowia pulcherrima* (Sun et al., 2014).

According to alcohol content, ciders can be soft (1–5%), hard (5–8%) or apple wine up to 14%. They can be also still or sparkling, depending on gas content. The apple juice is inoculated with *S. cerevisiae* or *Saccharomyces bayanus* and ammonium salts are also added to reduce alcohol production by nondegradation of must amino acids (Joshi et al., 2017). Other species added to contribute to fermentation are *Metschnikowia*, *Candida*, and *Pichia* and others, such as *Dekkera* sp., *Zygosaccharomyces* sp., *Saccharomycodes* sp. or *Hanseniaspora* sp., which can develop during the maturation process (Tamang and Kailasapathy, 2010).

Alcoholic beverages based on cereal fermentation use a variety of grains, such as barley, maize, wheat, millet, oat, rye, sorghum or rice (Arendt and Zannini, 2013). These grains are heated, mashed and filtrated (Kaur et al., 2019). The purpose of making the cell wall polysaccharides available for yeast and bacterial metabolization, mainly *S. cerevisiae* and LAB (Bamforth and Cook, 2019; García et al., 2019).

Beer is an example of cereal-based fermentation beverage. Beer sugars come mainly from malted barley (but also wheat, rice, corn or rye) and can be flavored with different elements, such as hops (*Humulus lupulus*), coriander, citrus or cherries (Leake and Silverman, 1971). Raw materials undergo a series of steps, like the abovementioned mashing, which consists in crushing the malt to release the kernel and the grain is mixed with water, to activate the amylolytic enzymes that will break down the starch and produce fermentable carbohydrates and sugars (Rani and Bhardwaj, 2021). After this, the insoluble spent grains are evacuated in a tank (named lauter) in a process named lautering, to obtain the wort. The kettle boil consists in mixing the wort with the hops or its extracts, to add the characteristic aroma and flavor to beer and develop the color. After this, the wort is cooled down, filtrated

208   *Novel Approaches in Biopreservation for Food and Clinical Purposes*

and centrifuged. Finally, the yeast is added and the fermentation of the sugars lasts generally 4–7 days, when the yeast will be settled. The beer will be moved to an aging tank in cold environment and the yeast will be recovered.

In distilled alcoholic beverages, the fermented products are prepared without adding starters and the alcoholic content is separated afterwards. Different distilled beverages have different original material, e.g., rum is obtained from molasses, brandy from grapes and whisky from barley. Fermentation of barley to obtain whisky is conducted typically with *S. cerevisiae* (Walker and Hill, 2016), while the fermentation of sugarcane juice to obtain rum is conducted by *S. cerevisiae*, but can include also *Schizosaccharomyces pombe* and growth of other species like *Candida*, *Pichia* and *Kluyveromyces*.

Microbial spoilage of beverages

Given the wide variety of beverages and the native microbiota, numerous spoilage microorganisms can contaminate beverages. Thus, non-alcoholic beverages can be contaminated during the production process, either by the raw materials or the flavorings, the water or any other chemical used. In addition, the machines involved in the process and the filling lines in the factory environment can be a source of contamination, due to poor hygienic maintenance and cleaning (Lawlor et al., 2009). Packaging materials and storage conditions can also contaminate the final product (Kregiel, 2015).

Non-alcoholic beverages are characterized for having a high water activity and the presence of vitamins and minerals make them very susceptible to microbial growth, even in carbonated beverages with low pH. Yeasts are the microorganisms that cause the primary spoilage. The most common yeasts that can be found in non-alcoholic beverages include *Aureobasidium pullulans*, *Saccharomyces* sp. (*S. cerevisiae*, *Saccharomyces bayanus*, *Saccharomyces exiguous*), *Pichia* sp., *Zygosaccharomyces bailii*, *Brettanomyces*, *Hanseniaspora*, *Hansenula*, *Candida* sp. (*Candida davenportii*, *Candida parapsilosis*, *Candida tropicalis*, *Candida solani*), *Criptococcous* sp., *Debaryomyces* sp., *Rhodotorula*, *Sporidiobolus*, *Dekkera bruxellensis*, and *Sporobolomyces* (Wareing, 2016). Presence of yeasts is associated to a series of visual defects, such as surface films, presence of clouds or swollen packages, tainting and particulates, and odors like aldehyde off-flavor, vinegar or petroleum-like odor (Stratford, 2006).

Molds can also contaminate non-alcoholic beverages, growing as masses over some of them, despite that in others, like carbonated beverages, their growth can be limited. The visual defects that cause are the presence of mycelial mats and discoloration, with musty and stale odors. There is a range of species involved in non-alcoholic beverage contamination, such as *Aspergillus* sp., *Fusarium* sp., *Penicillum* sp. or *Rizhopus* sp. (Juvonen et al., 2011).

There are different types of bacteria involved in non-alcoholic beverages spoilage. LAB can produce loss of carbon dioxide, ropiness and turbidity, while inducing cheesy or sour odors (Rawat, 2015). Typical LAB involved in this type of spoilage are *Lactobacillus* sp., *Leuconostoc* sp. and *Weissella* sp. AAB, such as

Acetobacter sp., *Gluconobacter* sp. or *Gluconacetobacter* sp., which can form surface films and swollen packages, in addition to haze, roppiness and sour and vinegar off-flavors (Juvonen et al., 2011). Other bacteria are particularly difficult to remove in industrial environments, like *Alicyclobacillus* sp., which is aerobic, acidophilic, thermophilic and have the ability to form spores that are resistant to pasteurization (Chang and Kang, 2004). Their activity is very difficult to detect visually, but it provides a smoky taint. Other bacteria involved in non-alcoholic beverage spoilage are considered pathogenic, such as *Listeria monocytogenes*, *Yersinia enterocolitica*, *Escherichia coli* and *Salmonella* sp. (Shankar et al., 2021).

Spoilage microorganisms can contaminate alcoholic beverages at the same stages than non-alcoholic beverages. Some temperatures at which fermentations take place are favorable for spoilage microbial growth. Wine can be spoilaged by a wide number of yeasts, LAB and AAB bacteria (Du Toit and Pretorius, 2000). Mold contamination is rare, as they are present mostly in the raw materials and they are usually eliminated during the fermentation, due to the production of alcohol. Spoilage molds include *Aspergillus*, *Penicillium*, *Alternaria*, *Botrytis*, *Cladosporium*, *Oidium* and *Uncinula* (Campaniello and Sinigaglia, 2017; Rubio-Bretón et al., 2017). Different yeasts and bacterial species will produce different defects. Thus, yeasts like *Pichia* sp., *Hanseniaspora* sp., *Hansenula* sp., *Metschnikowia* sp., *Dekkera* sp. and *Candida* produce ester and aldehyde taints and increase volatile acidity, but *Candida* sp. and *Pichia* sp. are also involved in the production of surface slime. Other yeasts, like *Brettanomyces intermedius*, produce high levels of acetic acid, and *Saccharomyces* sp. and *Zygosaccharomyces* sp. can induce re-fermentation in the bottle, while *Schizosaccharomyces* sp. deactivates the wine (Escott et al., 2018; Martorell et al., 2007; Thomas and Davenport, 1985). Some species of *Saccharomyces* and *Pichia* can produce flocculent masses and oxidized taint from acetaldehyde (Pretorius, 2000).

Bacteria can produce a variety of defects in wine (Bartowsky, 2009). LAB can be particularly damaging for wine production (Du Toit and Pretorius, 2000). *Lactobacillus* sp. produces ethyl carbamate precursors, acidification of wine (due to the production of lactic and acetic acids), mannitol formation, mousy taints or flocculent growth. *Leuconostoc* sp. produces ropiness and bitterness (due to glycerol metabolism), *Oenococcus* sp. produces histamine and buttery flavor (due to production of diacetyl) and *Pediococcus* sp. produces bitterness from acrolein formation from glycerol and the production of polysaccharides increasing viscosity (Bartowsky, 2019). AAB like *Acetobacter* sp. and *Gluconobacter* sp. oxidates ethanol to acetaldehyde and acetic acid, the production of ethyl acetate, the production of acetoin from lactic acid, metabolism of glycerol to dihydroxyacetone and ropiness. Other sporulate bacteria, like *Bacillus* sp. and *Clostridium* sp., form sediments and increase acidity by producing butyric acid.

Beer is also affected by the activity of numerous microorganisms (Sakamoto and Konings, 2003). Yeasts of different genera, including *Saccharomyces* sp., *Brettanomyces* sp., *Candida* sp., *Debaryomyces* sp., *Hanseniaspora* sp., *Kluyveromyces* sp., *Pichia* sp., *Torulaspora* sp. or *Zygosaccharomyces*, are involved

210 *Novel Approaches in Biopreservation for Food and Clinical Purposes*

in the formation of phenolic compounds, fatty acids, and estery off-flavors, as well as hazes and turbidity (Gutiérrez et al., 2018). On the other hand, wild yeasts can produce sulphur taints, and drains like off-flavor. Molds can also produce off-flavors (*Alternaria* sp., *Cladosporium* sp., *Epicoccum* sp. and *Fusarium* sp.) and roughness (*Aspergillus* sp.) (Vaughan et al., 2005). LAB can produce sourness, creaminess and acidic taste (*Lactobacillus* sp.) and cloudiness, and buttery aroma from diacetyl (*Pediococcus* sp.). Gram-positive bacteria, like *Kocuria kristinae*, alters aroma, giving an atypical fruity one, while other Gram-negative bacteria, like *Zygomonas* sp. produce turbidity and off-odors, acidic taste and unpleasant smells (Kordialik-Bogacka, 2022).

Cider spoilage is due to organisms like *Brettanomyces* sp., *Acetobacter* sp. and *S. ludwigii* and can be summarized in mousiness, indole taint, sulphur and rotten-egg taints, discoloration, hazes and deposits (Tubia et al., 2018). "Cider sickness" or "framboisé" is due to the activity of the bacterium *Zymomonas anaerobia*, which can ferment sugars in sweet beverages and generate a microbiological stability problem that affects cider during storage (Coton and Coton, 2003; Bauduin et al., 2006). Storage stability of cider can also be affected by indoles as a result of tryptophan breakdown (Joshi et al., 2021; Lea and Drilleau, 2003).

Preservation strategies for enhanced food safety and extended shelf life in beverages

Preservatives are substances capable of inhibiting or delaying the growth of microorganisms and, therefore, extending the shelf life of a product. They can be added to food to delay the action of microorganisms or to make the product resistant to acidification, fermentation and other types of undesirable food alterations (Kalpana and Rajeswari, 2019). Food preservation has become a critical aspect in industry, due to the time lengths and characteristics of commercialization, like storage, distribution and consumption of out-of-season products.

Preservatives can be categorized into Class I (e.g., oils, glucose, salt, vinegar, etc.) and Class II preservatives (e.g., benzoic and sulfurous acids, including their salts, nitrites, nitrates, potassium and sodium and their salts, propyl or methyl), where Class I has no restrictions to be added to foodstuffs (Kalpana and Rajeswari, 2019; Ortega-Rivas and Ortega-Rivas, 2012). Preservatives can also be divided according to other criteria, such as source of origin, into natural or artificial; or according to mechanism of action, into antioxidants and antimicrobials. In any case, preservatives belonging to both divisions should extend the period of time where the product can be safely consumed without changing the form, taste, texture or nutritional value of the beverage (Bomgardner, 2014).

Natural preservatives are conventional preservatives that can be found in nature (e.g., alcohol, salt or vinegar, among others), used in raw or cooked form to extend shelf life (Chauhan, 2010). Artificial preservatives are chemical substances that are classified as food additives. Food preservatives can be referred to as safe (Generally Recognized as Safe – GRAS) and are considered the best and most efficient option for extending shelf life (e.g., benzoates, sulfites, parabens, nitrites, etc.) (Sharma, 2015). Antioxidant preservatives are substances that can prevent or retard the oxidative mechanisms that lead to food spoilage, by generating resistance to the effects of the oxygen, including discoloration, texture changes or rancidity (Silva et al., 2016). Antimicrobial preservatives are substances that stop or kill microbes present in the food products. These preservatives include parabens, benzoic acid, sorbic acid, vinegar, isobutyl and propyl (Ochiai et al., 2002). Antienzymatic preservatives will target enzymes in the food that are activated after the harvest, slowing down the deterioration, e.g., ascorbic and citric acids or ethylenediaminetetraacetic acid (EDTA) (Glevitzky et al., 2009).

Despite the safety recommendations approved for the use of preservatives in the beverage industry, these can exert some adverse effects. It has been reported that carbonated drinks containing phosphoric acids can damage the enamel of teeth, disturb the digestive system and exert effects on the kidneys, like the development of stones or the elimination of calcium along the phosphoric acid, potentially leading to osteoporosis and bone fracture (Saldana et al., 2007; Chen et al., 2020). Additionally, the high levels of sugars associated to these drinks have been linked to obesity, diabetes and hypertension (Malik et al., 2006; Malik et al., 2010). Nitrosamines have been associated to colorectal and stomach cancers (Cantwell and Elliott, 2017; Bryan et al., 2012). Other negative effects of preservatives, like sulfites, are allergic reactions or asthma (Simon, 2003; Wilson and Bahna, 2005; Anand and Sati, 2013) and headaches (Sharma, 2015; Zaeem et al., 2016).

Because of these potential negative effects, consumers are demanding more natural products and a reduction of chemical preservatives, despite that those negative effects are classified as just residual levels of carcinogenicity, teratogenicity and toxicity (Mesías et al., 2021). This growing interest in natural products goes hand in hand with the use of biopreservation techniques, which ensure food and safety from a natural approach.

Biopreservation maintains food in an acceptable status suitable for consumption in terms of food safety and nutritional value, reduces waste and ensures supply, making food prices affordable. Arguably, the most prevalent methods are traditional strategies based on the reduction of the water activity, nutrient availability, modulation of the presence or absence of oxygen and modulation of temperature and pH. The typical methods are: smoking, dehydration, heating and fermentation (Food Preservation by Reducing Water Activity, 2016; Knorr and Augustin, 2021). Biopreservation is evolving and new biotechnology-based strategies, like the use of encapsulated essential oils or bacteriophages, allow targeted removal of specific spoilage microorganisms. These methods can be applied individually or in combination. Here is a summary of these strategies.

Protective cultures

Protective cultures are bacteria that have been selected because they can exert antimicrobial activity in foods, reducing the microbial load (Hammami et al., 2019; Arbulu et al., 2022). Protective cultures are used in meat, vegetable, fish and dairy preparations. Among the organisms used in protective cultures, LAB are the most studied (Bintsis, 2018; Shi and Maktabdar, 2022). Some examples including various LAB species showed antagonistic activity against pathogens in dairy beverages (milk, yogurt), like *Lactobacillus plantarum* showed efficacy against *L. monocytogenes* (Young and O'sullivan, 2011). Inoculation of *L. lactis*, a lacticin 3347 producer, in cottage cheese and yogurt was effective against *Listeria, Bacillus* by reducing two log cycle during storage (Coelho et al., 2014; Morgan et al., 2002). Other applications showed successful use of *E. faecalis* in reducing *E. coli* of *L. monocytogenes* in grape juice (de Lima Marques et al., 2017).

Microbial products

Arguably, the most effective antimicrobial compounds produced by bacteria are bacteriocins. Bacteriocins are ribosomally-synthesized antimicrobial peptides with a wide or narrow spectrum of action depending on the compound (Cotter et al., 2005). They have different mode of actions, which makes them suitable to target Gram-positive and Gram-negative bacterial pathogens involved in deterioration and safety concerns in food (May et al., 2008). Bacteriocins are involved in different bacterial dynamics and, despite that can be found in many environments, their use for food preservation to improve microbial safety and stability is well accepted (Garcia-Gutierrez et al., 2018; Egan et al., 2017). Their main advantages for food applications involve their thermostability, their ability to exert their action at a wide range of pHs, their sensitivity to digestive proteases and their nontoxicity to humans (Cotter et al., 2005).

Among the different bacteriocins that have been investigated for food applications, nisin is the most important and the one that is most widely used. Nisin is a small peptide with several variants that has a wide range of action against Gram-positive and Gram-negative foodborne pathogens such as *Listeria* sp. (Gharsallaoui et al., 2016). In beverages, it has been used in treatments in orange juice (in combination with Pulse Electric Field (PEF)) and is effective against natural flora (Rodrigo et al., 2005; Buckow et al., 2013). In apple ciders, nisin is combined with PEF to target *E. coli* spoilage (de Souza et al., 2019). Nisin showed inhibitory activity against *Alicyclobacillus acidoterrestris, Bacillus cereus, Staphylococcus aureus* and *L. monocytogenes* in cashew, peach and mango juice (de Oliveira Junior et al., 2015). A bacteriocin considered an alternative to nisin is enterocin AS-48, which has shown to display antimicrobial activity against *A. acidoterrestris* in different natural and commercial juices (Grande et al., 2005; Abriouel et al., 2010).

Other compounds with preservative activity produced by bacteria are alcohols. Ethanol is the most common ingredient produced during fermentation and it is extensively used in various chemical, textile and food industries (Sarris and Papanikolaou, 2016). Glycerol produced by *Candida glycerinogenes* is used in beverages as sweetener, flavor enhancer and preservative (Wang et al., 2001).

Different types of enzymes are generated from microorganisms such as *Aspergillus niger, Aspergillus flavus, Bacillus subtilis* and used in preservation of beverages, like protease enzymes in wine, beer and fruit juices to prevent haze formation (Kamini et al., 1999; Gurung et al., 2013; Adrio and Demain, 2014). Several antimicrobial compounds can be extracted from microorganisms and used in food. Natamycin, produced from *Streptococcus natalensis*, is used in beverages and shows

Biopreservation of Beverages 213

a wide range of efficacy against fungus by causing leakage in the cell membrane (Singh, 2018; Davidson and Doan, 2020).

Essential oils

Essential oils are complex mixtures of volatile compounds synthesized by plants that can exert antimicrobial effects in appropriate concentrations (Hyldgaard et al., 2012; de Souza et al., 2019). These antimicrobial effects have been reported particularly strong and with a wide spectrum of action when added to different beverages, especially juices (Maldonado et al., 2013; Loeffler et al., 2014; de Sousa Guedes and de Souza, 2018). The main constituents of the essential oils are terpenes and terpenoids, and their antimicrobial activity is linked to their ability to cause leakage in the bacterial cell by disrupting the lipids in the cell membrane (Burt, 2004). Most essential oils are GRAS, which makes them suitable to preserve various beverages.

There is a wide variety of essential oils of different origins that are already part of the beverage production process. For example, clove oil is used in watermelon juice. When its concentration was of 4500- and 9000-mg/L, clove oil showed a reduction of mesophilic count up to 8 log cycle during storage (Siddiqua et al., 2015). Another experiments reported that mentha essential oils were active against *E. coli*, *L. monocytogenes*, and *Salmonella enteritidis* in pineapple and mango juice by disrupting and depolarizing the cellular membrane (de Sousa Guedes and de Souza, 2018). Lemon essential oil has been proven useful to inhibit the germination and growth of *A. acidoterrestris* spores during 11 days of refrigeration (Maldonado et al., 2013). Essential oils can be used individually or as emulsions of dual combinations to reduce problematic strains, like the acid-resistant *Z. bailii* in apple juice (Loeffler et al., 2014).

Other examples include rosemary extract, obtained from *Rosmarinus officinalis*, a Mediterranean shrub. The rosemary oil has antiviral, antimicrobial, antimycotic and antioxidant properties and its use is well extended in food industry due to its low cost and high availability. It is particularly useful to provide stability to omega 3-rich oils, which are prone to rancidity (Lopez-Garcia, 2000). It contains steroids, flavones, diterpenes and triterpenes, each of them responsible of different bioactivities. However, it has been proposed that its antifungal activity is due to monoterpenes, which impair the fungal membrane (Juglal et al., 2002). *Thymus vulgaris L.* is another Mediterranean bushy herb that contains flavonoids and phenolic antioxidants, like thymol and carvacrol (Piccaglia and Marotti, 1993). Its essential oil is considered to have very potent antimicrobial properties (Descalzo and Sancho, 2008). Another extract that is used as preservative is basil extract, obtained from leaf of Indian and Iran herb *Ocimum basilicum*. It has antimicrobial and antioxidant properties due to the presence of phenolic and aromatic compounds, including flavonoids, tannins, thymol and eugenol, among others (Hussain et al. 2008; May et al., 2008). Neem oil is also used as a preservative. It is obtained from pressing fruits and seeds of *Azadirachta indica*. The extract of neem has antibacterial, antifungal and antiprotozoal activities and it is currently commercialized into tea (Vara et al., 2019).

Natural preservatives

Natural substances can be used as a substitute for chemical preservatives in non-alcoholic beverage products. Those substances are normally extracted from plant or animal sources.

Algin

This compound is a hydrophilic polysaccharide, naturally derived from seaweed including the species *Macrocystis pyrifera*, *Ascophyllum nodosum* and some types of *Laminaria* sp. Propylene glycol alginate (PGA) is highly demanded in the beverage industry, due to its stabilization capacity, film formation and the enhancement of the viscosity, which makes it suitable to be used as a thickening agent and foam stabilizer (Fabra et al., 2008; Cheong et al., 2014). Alginates are used in milkshakes, ice cream, puddings, salad dressing, cheese and yogurts to thicken the products as well as extending the shelf life (Cheong et al., 2014; Imeson, 2012; Saha and Bhattacharya, 2010). Moreover, it can replace other compounds in the food industry, such as agar, pectin, starch or gum tragacanth (Vara et al., 2019).

Carrageenan

This compound is extracted from the variety of seaweed called Irish Moss *Chondrus crispus*. It is normally used in products whose pH is neutral, and forms a light gel structure by interacting with milk proteins. It is used in puddings, milkshakes, jellies and ice creams and it is particularly interesting for beverages, as it gives good body to them (Vara et al., 2019). It can be used to substitute fat and to thicken low- and non-fat foods, as it can recreate the feeling of the fat when consumed (Vara et al., 2019).

Guar Gum

It is a powdery product generated from Indian legume seeds of *Cyamopsis tetragonoloba* (Prem et al., 2005). It has the ability to hydrate fast in cold water. Due to its excellent thickening properties in the solution, it has been added in cocoa-based drinks, fruit juices and alcoholic beverages to enhance texture (Mudgil et al., 2014).

Honey

Honey is produced by bees from the nectar of plants and it is famous for its antimicrobial and preservation properties (Vara et al., 2019). Honey can self-preserve when it is undiluted and it is highly nutritive. It provides aroma and sweetness to beverages. In industry, honey as a prebiotic, flavor enhancer and antioxidant, but also inhibits mold and bacterial growth due to its acidic pH (Mundo et al., 2004). Honey contains a wide range of components such as fructose, glucose, enzymes, vitamins, minerals, phenolic acids, flavonoids and amino acids (Mundo et al., 2004; Molan, 1992).

Propolis

Propolis is a resin used by bees to build their hives. It has aromatic, antimicrobial and antioxidants properties (Cowan, 1999). Propolis components are mainly of phenolic nature, especially flavonoids, but also contains quinones, polyphenols, steroids, coumarins and amino acids, which makes them suitable for antibacterial and antifungal activity against, e.g., mycotoxin in apple, mandarin, orange and grape juices (Koc et al., 2007; Silici and Karaman, 2014). Propolis is also considered as GRAS (Burdock, 1998).

Citric acid

Citric acid (or citrate) is a highly soluble, weak organic acid naturally present in fruits, especially citrus ones. It is used as a flavorant, acidulant and preservative in margarine, salad dressings, canned fruit juices, soft drinks, cheese, yogurts, sausages and pickles (Brima and Abbas, 2014; Lopez-Garcia, 2000).

Erythorbic acid

It is also known as isoascorbic acid and has vegetable origin. It is mostly used as an antioxidant in beverages like soft drinks and fruit juices and works maintaining the color and the natural flavor and preventing the formation of nitrosamines (Vara et al., 2019).

Bacteriophages

Bacteriophages are found everywhere and are considered to be the most abundant organisms on Earth (Clokie et al., 2011). Bacteriophages are viruses that attack bacteria but not human, animal or plant cells, which makes them suitable for treating infections caused by both Gram-positive and Gram-negative bacteria (Hanlon, 2007; Sulakvelidze et al., 2001). Their applications have expanded to food, with the purpose of targeting specific pathogens to extend shelf life and decontaminate equipment and contact surfaces (Garcia et al., 2010). Thus, this strategy has showed successful results against *Salmonella*, *Campylobacter*, *E. coli*, *Cronobacter sakazakii*, *L. monocytogenes* and *S. aureus* biofilms (Garcia et al., 2010; Mann, 2008). A five-phage cocktail and lytic phage was effective against *Shigella* spp. in yogurt by decreasing its level about 1 log cycle (Soffer et al., 2017). A lytic phage significantly reduces *Staphylococcus aureus* in curd and decreased *L. monocytogenes* in melon and pear juice respectively (García et al., 2007; Oliveira et al., 2014).

Others

There are a number of methods not related to biopreservation, but used in combination with biopreservation strategies, which are applied to prevent microbial spoilage in beverages while maintaining nutritional and sensory quality. These methods include thermal and non-thermal treatments (Grumezescu and Holban, 2019).

216 *Novel Approaches in Biopreservation for Food and Clinical Purposes*

- Pasteurization is a thermal method that involves the application of heat, either as a combination of low-temperature long-time or high-temperature short-time. There can be modifications in the pasteurization process, like the ultra high-temperature treatment, for low-acid beverages at pH > 4.6, 130–150°C for 1–9 sec. The flash pasteurization is another variation, which applies a cycle of 75–85°C for over 1–4 min and a cycle of 90–96°C for over 30–90 sec, at a pH < 4.6, like the carbonated drinks, generally to sensitive products when the filling is aseptic (Ashurst, 2005). In-pack pasteurization can be applied to acidic beverages exposed to a high risk of spoilage.

- The high process processing (HPP) is a non-thermal pasteurization strategy where the liquid goes into high pressure, around 3300–600 MPa for around 10 min. It is generally applied to acidic juices, due to the inability of pressure-tolerant spores to survive at low pH levels. Depending on the type of juice, there are different protocols, e.g., apple juice requires 545 MPa for 1 min, whereas blueberry juices require 400–600 MPa for 15min. This technique is also applied to treat the juices before and after the fermentations for the alcoholic beverages, either to avoid microbial growth or to stop further yeast growth. It allows to inactivate microorganisms while preserving original flavor and nutritious quality (Özkan-Karabacak et al., 2019). HHP can be applied independently on the shape, size and composition, but its efficiency will depend on exposure time, temperature, pressure, packaging type and food parameters (Koutchma, 2014). The pressure can range between 200–900 MPa, although 400–600 MPa is the accepted pressure range to inactivate vegetative microorganisms (Smelt et al., 2001; Koutchma, 2014). HPP has been used for wine preservation (Van Leeuw et al., 2014; Garrido and Borges, 2013).

- Hydrodynamic cavitation produces bubbles under pressure fluctuation. It is a strategy used to inactivate *S. cerevisiae* in apple juice and eliminate LAB and *Z. bailii* (Milly, 2007; Buckow et al., 2013; Sun et al., 2020).

- Pulse electric field (PEF) is another non-thermal strategy that uses short pulses, of 1–100 µs, of high electric fields, for nano or milliseconds, and to an intensity of 10–80 kV/cm to foods located between two electrodes. PEF kills microbes by lysis, after disrupting the microbial cell membrane (Gabrić et al., 2018). Usually, PEF is used in combination with other strategies.

- Irradiation involves the use of X-rays, electron beams and gamma-rays to treat beverages without affecting their physical appearance (Pendyala et al., 2020).

- Membrane filtration is a process conducted at a low temperature by removing the undesirable particles of the beverage, generally a fruit juice, by passing it through a membrane (Girard and Fukumoto, 2000).

Regulatory framework

The regulatory framework for the use of biopreservatives in beverages is not different from the solid food. Both EFSA and FDA recognize LAB protective cultures as safe

(Mukherjee et al., 2022). In addition to LAB species, *E. faecium* strain was also got recognition as safe from Council Directive 70/524/EEC (Calo-Mata et al., 2008; Authority, 2005; Alvarez-Sieiro et al., 2016).

Regarding microbial products, nisin is considered as GRAS by the European Food Safety Authority (EFSA) and the Food and Drug Administration (FDA) (Authority, 2006). FDA approved nisin for human consumption (Gillor et al., 2008) and it can be added into dairy products as additive. Currently, this bacteriocin is widely used as preservative in more than 50 countries (Calo-Mata et al., 2008). Other bacteriocins are approved as GRAS too (Authority, 2005; Committee, 2014; Hussain et al., 2017; Lee and Chang, 2018). FDA and EFSA have approved the use of essential oils as flavoring agent in beverages, as some essential oils also are GRAS (Burt, 2004).

Both FDA and EFSA have approved the use of bacteriophages as preservatives in vegetables and fruit juices prior conditioning that the use of this phage must follow good manufacturing practice. Bacteriophages are normally considered as safe as they do not show toxicity during oral ingestion. The bacteriophage (Listex P-100) was approved in 2006 by FDA, as it follows all regulations under GRAS, to be used in raw products as preservatives (Carlton et al., 2005; Monk et al., 2010).

Conclusions

Beverage products are ubiquitously used all over the world. There are a number of preservative strategies used to prevent the development of microbial contaminations and decrease their potential pathogenic risk in beverages. However, biopreservation is increasingly gaining attention as an alternative to chemical preservation strategies. Some natural antimicrobial compounds (e.g., essential oils and plant extracts) are used in juices for flavoring and preservation purposes. Primary bacterial metabolites, such as organic acids and alcohols produced by microorganisms during fermentation are also used as natural preservatives, because they show a wide range of inhibitory effects when inoculated in beverage items. Other alternatives include the use of bacteriophages, which are considered the most innovative concept towards beverage preservation. These approaches can be used alone or in combination. Overall, biopreservation is gaining worldwide recognition and its use is expanding in industry. However, its main future projection comes from the innovative applications and developments that can be tailored for the production needs of each specific beverage, like targeting specific spoilage microorganisms without altering the properties of the product.

Acknowledgements

This work was supported by a Beatriz Galindo scholarship from the Spanish Ministry of Universities (BG22/00060) awarded to EGG.

References

Abriouel, Hikmate, Rosario Lucas, Nabil Ben Omar, Eva Valdivia and Antonio Gálvez. 2010. Potential applications of the cyclic peptide enterocin AS-48 in the preservation of vegetable foods and beverages. *Probiotics and Antimicrobial Proteins*, 2: 77–89.

Adrio, Jose L. and Arnold L. Demain. 2014. Microbial enzymes: tools for biotechnological processes. *Biomolecules*, 4: 117–39.

Akhtar, N., Karnwal, A., Upadhyay, A.K., Paul, S. and Mannan, M.A. 2018. *Saccharomyces cerevisiae* bio-ethanol production, a sustainable energy alternative. *Asian J. Microbiol. Biotechnol. Environ. Sci.*, 20: 202–06.

Alvarez-Sieiro, Patricia, Manuel Montalbán-López, Dongdong Mu and Oscar P Kuipers. 2016. Bacteriocins of lactic acid bacteria: extending the family. *Applied Microbiology and Biotechnology*, 100: 2939–51.

Anal, Anil Kumar. 2019. Quality ingredients and safety concerns for traditional fermented foods and beverages from Asia: A Review. *Fermentation*, 5: 8.

Anand, S.P. and Sati, N. 2013. Artificial preservatives and their harmful effects: looking toward nature for safer alternatives. *Int. J. Pharm. Sci. Res.*, 4: 2496–501.

Arbulu, Sara, Beatriz Gómez-Sala, Enriqueta Garcia-Gutierrez and Paul D. Cotter. 2022. Bioprotective Cultures and Bacteriocins for Food. *In: Good Microbes in Medicine, Food Production, Biotechnology, Bioremediation, and Agriculture.*

Arendt, Elke K. and Emanuele Zannini. 2013. *Cereal Grains for the Food and Beverage Industries* (Elsevier). ISBN 978-0-85709-413-1. pp. 1–61.

Ashurst, P.R. 2005. Non carbonated beverages. *Chemistry and Technology of Soft Drinks and Fruit Juices*: 129–49.

Authority, European Food Safety. 2005. Opinion of the Scientific Committee on a request from EFSA related to a generic approach to the safety assessment by EFSA of microorganisms used in food/feed and the production of food/feed additives. *EFSA Journal*, 3: 226.

Authority, European Food Safety. 2006. Opinion of the Scientific Panel on food additives, flavourings, processing aids and materials in contact with food (AFC) related to The use of nisin (E 234) as a food additive. *EFSA Journal*, 4: 314.

Bamforth, Charles W. and David J. Cook. 2019. *Food, Fermentation, and Micro-organisms* (John Wiley & Sons). ISBN: 978-1-405-19872-1. pp. 43–91.

Bartowsky, Evaline J. 2009. Bacterial spoilage of wine and approaches to minimize it. *Letters in Applied Microbiology*, 48: 149–56.

Bartowsky, Eveline. 2019. Lactic Acid Bacteria (LAB) in grape fermentations: an example of LAB as contaminants in food processing. *In: Lactic Acid Bacteria* (CRC Press). ISBN: 9780429057465. pp. 344–361.

Batish, V.K. and Grover Sunita. 2004. Fermented milk products. *Concise Encyclopedia of Bioresource Technology*: 201–09.

Bauduin, Remi, J.-M. Le Quere, E. Coton and Jo Primault. 2006. Factors leading to the expression of "framboisé" in French ciders. *Lwt-Food Science and Technology*, 39: 966–71.

Bautista-Gallego, J., Medina, E., Sánchez, B., Antonio Benítez-Cabello and Francisco Noé Arroyo-López. 2020. Role of lactic acid bacteria in fermented vegetables. *Grasas y Aceites*, 71: e358–e58.

Beech, F.W. 1972. Cider making and cider research: a review. *Journal of the Institute of Brewing*, 78: 47–91.

Bintsis, T. 2018. Lactic acid bacteria as starter cultures: an update in their metabolism and genetics. *AIMS Microbiol*, 4: 665–84.

Bomgardner, M. 2014. Extending shelf life with natural preservatives. *Chem Eng News*, 92: 13–4.

Brima, Eid I. and Anass M. Abbas. 2014. Determination of citric acid in soft drinks, juice drinks and energy drinks using titration. *Int. J. Chem. Stud.*, 1: 30–34.

Bryan, Nathan S., Dominik D. Alexander, James R. Coughlin, Andrew L. Milkowski and Paolo Boffetta. 2012. Ingested nitrate and nitrite and stomach cancer risk: an updated review. *Food and Chemical Toxicology*, 50: 3646–65.

Buckow, Roman, Sieh Ng and Stefan Toepfl. 2013. Pulsed electric field processing of orange juice: a review on microbial, enzymatic, nutritional, and sensory quality and stability. *Comprehensive Reviews in Food Science and Food Safety*, 12: 455–67.

Buglass, Alan J. 2011. *Handbook of Alcoholic Beverages, 2 Volume Set: Technical, Analytical and Nutritional Aspects* (John Wiley & Sons).

Buglass, Alan J. 2015. Chemical composition of beverages and drinks. pp. 1–62. *In*: Peter Chi Keung Cheung (Ed.). *Handbook of Food Chemistry* (Springer Berlin Heidelberg: Berlin, Heidelberg). ISBN 978-3-642-41609-5.

Bujake, John E. 2000. Beverage spirits, distilled. *Kirk-Othmer Encyclopedia of Chemical Technology*, vol. 1 pages n/a.

Burdock, G.A. 1998. Review of the biological properties and toxicity of bee propolis (propolis). *Food and Chemical Toxicology*, 36: 347–63.

Burt, Sara. 2004. Essential oils: their antibacterial properties and potential applications in foods—a review. *International Journal of Food Microbiology*, 94: 223–53.

Calo-Mata, Pilar, Samuel Arlindo, Karola Boehme, Trinidad de Miguel, Ananias Pascoal and Jorge Barros-Velazquez. 2008. Current applications and future trends of lactic acid bacteria and their bacteriocins for the biopreservation of aquatic food products. *Food and Bioprocess Technology*, 1: 43–63.

Campaniello, Daniela and Milena Sinigaglia. 2017. Wine spoiling phenomena. *In: The Microbiological Quality of Food* (Elsevier). ISBN 9780081005033. pp. 237–255.

Cantwell, Marie and Chris Elliott. 2017. Nitrates, nitrites and nitrosamines from processed meat intake and colorectal cancer risk. *J. Clin. Nutr. Diet.*, 3: 27.

Carlton, R.M., Noordman, W.H., Biswas, B., De Meester, E.D. and Martin J. Loessner. 2005. Bacteriophage P100 for control of *Listeria monocytogenes* in foods: genome sequence, bioinformatic analyses, oral toxicity study, and application. *Regulatory Toxicology and Pharmacology*, 43: 301–12.

Chang, S.S. and Kang, D.H. 2004. *Alicyclobacillus* spp. in the fruit juice industry: history, characteristics, and current isolation/detection procedures. *Crit. Rev. Microbiol.*, 30: 55–74.

Chauhan, Harmeet. 2010. Naturally Occurring Preservatives in Food and their Role in Food Preservation.

Chen, L., Liu, R., Zhao, Y. and Shi, Z. 2020. High consumption of soft drinks is associated with an increased risk of fracture: A 7-Year Follow-Up Study. *Nutrients*, 12.

Chen, T.H., Wang, S.-Y., Chen, K.-N., Liu, J.-R. and Chen, M.-J. 2009. Microbiological and chemical properties of kefir manufactured by entrapped microorganisms isolated from kefir grains. *Journal of Dairy Science*, 92: 3002–13.

Cheong, Kok Whye, Hamed Mirhosseini, Nazimah Sheikh Abdul Hamid, Azizah Osman, Mahiran Basri and Chin Ping Tan. 2014. Effects of propylene glycol alginate and sucrose esters on the physicochemical properties of modified starch-stabilized beverage emulsions. *Molecules*, 19: 8691–706.

Clokie, M.R., Millard, A.D., Letarov, A.V. and Heaphy, S. 2011. Phages in nature. *Bacteriophage*, 1: 31–45.

Coelho, M.C., Silva, C.C.G., Ribeiro, S.C., Dapkevicius, M.L.N.E. and Rosa, H.J.D. 2014. Control of *Listeria monocytogenes* in fresh cheese using protective lactic acid bacteria. *International Journal of Food Microbiology*, 191: 53–59.

Committee, EFSA Scientific. 2014. Scientific Opinion on a Qualified Presumption of Safety (QPS) approach for the safety assessment of botanicals and botanical preparations. *EFSA Journal*, 12: 3593.

Coton, E. and Coton, M. 2003. Microbiological origin of "framboisé" in French ciders. *Journal of the Institute of Brewing*, 109: 299–304.

Cotter, P.D., Hill, C. and Ross, R.P. 2005. Bacteriocins: developing innate immunity for food. *Nat. Rev. Microbiol.*, 3.

Cowan, Marjorie Murphy. 1999. Plant products as antimicrobial agents. *Clinical Microbiology Reviews*, 12: 564–82.

Davidson, Michael, P. and Craig Doan. 2020. Natamycin. *In: Antimicrobials in Food* (CRC Press). ISBN 9780429058196. pp. 338–350.

de Lima Marques, Juliana, Graciele Daiana Funck, Guilherme da Silva Dannenberg, Claudio Eduardo dos Santos Cruxen, Shanise Lisie Mello El Halal, Alvaro Renato Guerra Dias, Ângela Maria Fiorentini and Wladimir Padilha da Silva. 2017. Bacteriocin-like substances of *Lactobacillus curvatus* P99: characterization and application in biodegradable films for control of *Listeria monocytogenes* in cheese. *Food Microbiology*, 63: 159–63.

de Melo Pereira, Gilberto, V., Dão Pedro de Carvalho Neto, Bruna, L. Maske, Juliano De Dea Lindner, Alexander, S. Vale, Gabriel, R. Favero, Jéssica Viesser, Júlio, C. de Carvalho, Aristóteles Góes-Neto and Carlos, R. Soccol. 2022. An updated review on bacterial community composition of traditional fermented milk products: what next-generation sequencing has revealed so far? *Crit. Rev. Food Sci. Nutr.*, 62: 1870–89.

de Oliveira Junior, Adelson Alves, Hyrla Grazielle Silva de Araújo Couto, Ana Andréa Teixeira Barbosa, Marcelo Augusto Guitierrez Carnelossi and Tatiana Rodrigues de Moura. 2015. Stability, antimicrobial activity, and effect of nisin on the physico-chemical properties of fruit juices. *International Journal of Food Microbiology*, 211: 38–43.

de Sousa Guedes, Jossana Pereira and Evandro Leite de Souza. 2018. Investigation of damage to *Escherichia coli, Listeria monocytogenes* and *Salmonella enteritidis* exposed to *Mentha arvensis* L. and *M. piperita* L. essential oils in pineapple and mango juice by flow cytometry. *Food Microbiology*, 76: 564–71.

de Souza, Evandro Leite, Erika Tayse da Cruz Almeida and Jossana Pereira de Sousa Guedes. 2019. Emerging nonchemical potential antimicrobials for beverage preservation. *In*: *Preservatives and Preservation Approaches in Beverages* (Elsevier). Academic Press. ISBN 9780128166857. pp. 149–178.

Descalzo, Adriana María and Sancho, A.M. 2008. A review of natural antioxidants and their effects on oxidative status, odor and quality of fresh beef produced in Argentina. *Meat Science*, 79: 423–36.

Deshpande, S.S. 2000. *Fermented grain legumes, seeds and nuts: A global perspective* (Food & Agriculture Org.).

Ding, Junmei, Xiaowei Huang, Lemin Zhang, Na Zhao, Dongmei Yang and Keqin Zhang. 2009. Tolerance and stress response to ethanol in the yeast *Saccharomyces cerevisiae*. *Applied Microbiology and Biotechnology*, 85: 253–63.

Doyle, Michael P. and Jianghong Meng. 2006. Bacteria in food and beverage production. *Prokaryotes. Springer, New York*: 797–811.

Du Toit, Maret and Isak S. Pretorius. 2000. Microbial Spoilage and Preservation of Wine: Using Weapons for Nature's own Arsenal. *South African Journal of Enology & Viticulture*, 21: 1.

Egan, Kevin, R. Paul Ross and Colin Hill. 2017. Bacteriocins: antibiotics in the age of the microbiome. *Emerging Topics in Life Sciences*, Apr 21; 1(1): 55–63.

Escott, Carlos, Juan Manuel Del Fresno, Iris Loira, Antonio Morata and José Antonio Suárez-Lepe. 2018. *Zygosaccharomyces rouxii*: control strategies and applications in food and winemaking. *Fermentation*, 4: 69.

Fabra, Maria José, Pau Talens and Amparo Chiralt. 2008. Effect of alginate and λ-carrageenan on tensile properties and water vapour permeability of sodium caseinate–lipid based films. *Carbohydrate Polymers*, 74: 419–26.

Food Preservation by Reducing Water Activity. 2016. *Food Microbiology: Principles into Practice*.

Gabrić, Domagoj, Francisco Barba, Shahin Roohinejad, Seyed Mohammad Taghi Gharibzahedi, Milivoj Radojčin, Predrag Putnik and Danijela Bursać Kovačević. 2018. Pulsed electric fields as an alternative to thermal processing for preservation of nutritive and physicochemical properties of beverages: A review. *Journal of Food Process Engineering*, 41: e12638.

Garcia-Gutierrez, Enriqueta, Melinda J. Mayer, Paul D. Cotter and Arjan Narbad. 2018. Gut microbiota as a source of novel antimicrobials. *Gut microbes*: 1–21.

García, Cristina, Manuel Rendueles and Mario Díaz. 2019. Liquid-phase food fermentations with microbial consortia involving lactic acid bacteria: A review. *Food Research International*, 119: 207–20.

García, Pilar, Carmen Madera, Beatriz Martínez and Ana Rodríguez. 2007. Biocontrol of *Staphylococcus aureus* in curd manufacturing processes using bacteriophages. *International Dairy Journal*, 17: 1232–39.

Garcia, Pilar, Lorena Rodriguez, Ana Rodriguez and Beatriz Martinez. 2010. Food biopreservation: promising strategies using bacteriocins, bacteriophages and endolysins. *Trends in Food Science & Technology*, 21: 373–82.

Garrido, Jorge and Fernanda Borges. 2013. Wine and grape polyphenols—A chemical perspective. *Food Research International*, 54: 1844–58.

Gharsallaoui, A., Oulahal, N., Joly, C. and Degraeve, P. 2016. Nisin as a food preservative: Part 1: physicochemical properties, antimicrobial activity, and main uses. *Crit. Rev. Food Sci. Nutr.*, 56: 1262–74.

Gillor, O., Etzion, A. and Riley, M.A. 2008. The dual role of bacteriocins as anti-and probiotics. *Applied Microbiology and Biotechnology*, 81: 591–606.

Girard, B. and Fukumoto, L.R. 2000. Membrane processing of fruit juices and beverages: a review. *Critical Reviews in Food Science Nutrition*, 40: 91–157.

Glevitzky, M., Dumitrel, G.A., Perju, D. and Popa, M. 2009. Studies regarding the use of preservatives on soft drinks stability. *Chemical Bulletin of Politehnica University of Timisoara*, 54: 31–36.

Grande, Ma J., Lucas, R., Abriouel, H., Ben Omar, N., Maqueda, M., Martínez-Bueno, M., Martínez-Cañamero, M., Valdivia, E. and Gálvez, A. 2005. Control of *Alicyclobacillus acidoterrestris* in fruit juices by enterocin AS-48. *International Journal of Food Microbiology*, 104: 289–97.

Grumezescu, Alexandru Mihai and Alina Maria Holban. 2019. *Preservatives and Preservation Approaches in Beverages: Volume 15: The Science oof Beverages* (Academic Press).

Guillamón, José Manuel, and Albert Mas. 2009. Acetic acid bacteria. *Biology of Microorganisms on Grapes, in Must and in Wine*: 31–46.

Gurung, Neelam, Sumanta Ray, Sutapa Bose and Vivek Rai. 2013. A broader view: microbial enzymes and their relevance in industries, medicine, and beyond. *BioMed Research International*, 2013.

Gutiérrez, Alicia, Teun Boekhout, Zoran Gojkovic and Michael Katz. 2018. Evaluation of non-Saccharomyces yeasts in the fermentation of wine, beer and cider for the development of new beverages. *Journal of the Institute of Brewing*, 124: 389–402.

Hammami, R., Fliss, I. and Corsetti, A. 2019. Editorial: application of protective cultures and bacteriocins for food biopreservation. *Front Microbiol.*, 10: 1561.

Hanlon, Geoffrey William. 2007. Bacteriophages: an appraisal of their role in the treatment of bacterial infections. *International Journal of Antimicrobial Agents*, 30: 118–28.

Hernandez, L.H.H., Barrera, T.C., Mejia, J.C., Mejia, G.C., Del Carmen, M., Dosta, M., De Lara Andrade, R. and Sotres, J.A.M. 2010. Effects of the commercial probiotic *Lactobacillus casei* on the growth, protein content of skin mucus and stress resistance of juveniles of the Porthole livebearer *Poecilopsis gracilis* (Poecilidae). *Aquaculture Nutrition*, 16: 407–11.

Hussain, Abdullah Ijaz, Farooq Anwar, Syed Tufail Hussain Sherazi and Roman Przybylski. 2008. Chemical composition, antioxidant and antimicrobial activities of basil (*Ocimum basilicum*) essential oils depends on seasonal variations. *Food Chemistry*, 108: 986–95.

Hussain, Malik A., Huan Liu, Qi Wang, Fang Zhong, Qian Guo and Sampathkumar Balamurugan. 2017. Use of encapsulated bacteriophages to enhance farm to fork food safety. *Crit. Rev. Food Sci. Nutr.*, 57: 2801–10.

Hyldgaard, Morten, Tina Mygind and Rikke Louise Meyer. 2012. Essential oils in food preservation: mode of action, synergies, and interactions with food matrix components. *Frontiers in Microbiology*, 3: 12.

Imeson, Alan P. 2012. *Thickening and Gelling Agents for Food* (Springer Science & Business Media).

Joshi, V.K., Sharma, S. and Thakur, A.D. 2017. Wines: White, red, sparkling, fortified, and cider. *In*: *Current Developments in Biotechnology and Bioengineering* (Elsevier), pp. 353–406, ISBN 9780444636669.

Joshi, V.K., Somesh Sharma and Vikas Kumar. 2021. Cider: the production technology. *In*: *Winemaking* (CRC Press). ISBN 9781351034265. pp. 574–601.

Juglal, S., Govinden, R. and Odhav, B. 2002. Spice oils for the control of co-occurring mycotoxin-producing fungi. *Journal of Food Protection*, 65: 683–87.

Juvonen, Riikka, Vertti Virkajärvi, Outi Priha and Arja Laitila. 2011. Microbiological spoilage and safety risks in non-beer beverages. *VTT Tiedotteita-Research Notes*, 2599.

Kalpana, V.N. and Devi Rajeswari, V. 2019. Preservatives in beverages: perception and needs. *In*: *Preservatives and Preservation Approaches in Beverages* (Elsevier). Academic Press, pp. 1–30, ISBN 9780128166857.

222 *Novel Approaches in Biopreservation for Food and Clinical Purposes*

Kamini, N.R., Hemachander, C., Geraldine Sandana Mala, J. and Puvanakrishnan, R. 1999. Microbial enzyme technology as an alternative to conventional chemicals in leather industry. *Current Science*: 80–86.

Kaur, Prabhjot, Gargi Ghoshal and Uttam C. Banerjee. 2019. Traditional bio-preservation in beverages: fermented beverages. *In: Preservatives and Preservation Approaches in Beverages* (Elsevier). Academic Press, pp. 69–113, ISBN 9780128166857.

Knorr, D. and Augustin, M.A. 2021. Food processing needs, advantages and misconceptions. *Trends in Food Science & Technology*, 108: 103–10.

Koc, Ayse Nedret, Sibel Silici, Fatma Mutlu-Sariguzel and Osman Sagdic. 2007. Antifungal activity of propolis in four different fruit juices. *Food Technology and Biotechnology*, 45: 57–61.

Kordialik-Bogacka, Edyta. 2022. Biopreservation of beer: potential and constraints. *Biotechnology Advances*, 58: 107910.

Koutchma, Tatiana. 2014. *Adapting high hydrostatic pressure (HPP) for food processing operations* (Academic Press).

Kregiel, Dorota. 2015. Health safety of soft drinks: contents, containers, and microorganisms. *BioMed Research International*, 2015.

Lawlor, Kathleen A., James D. Schuman, Peter G. Simpson and Peter J. Taormina. 2009. Microbiological spoilage of beverages. *Compendium of the Microbiological Spoilage of Foods and Beverages*: 245–84.

Lea, Andrew G.H. and Jean-François Drilleau. 2003. Cidermaking. *Fermented Beverage Production*: 59–87.

Lea, Andrew G.H., John Raymond Piggott and John R. Piggott. 2003. *Fermented Beverage Production* (Springer Science & Business Media).

Leake, Chauncey D. and Milton Silverman. 1971. The chemistry of alcoholic beverages. *The Biology of Alcoholism: Volume 1: Biochemistry*: 575–612.

Lee, Seul Gi and Hae Choon Chang. 2018. Purification and characterization of mejucin, a new bacteriocin produced by Bacillus subtilis SN7. *LWT*, 87: 8–15.

Loeffler, Myriam, Sophia Beiser, Sarisa Suriyarak, Monika Gibis and Jochen Weiss. 2014. Antimicrobial efficacy of emulsified essential oil components against weak acid–adapted spoilage yeasts in clear and cloudy apple juice. *Journal of Food Protection*, 77: 1325–35.

Lopez-Garcia, Rebecca. 2000. Citric acid. *Kirk-Othmer Encyclopedia of Chemical Technology*: 1–25.

Ma, Menggen and Lewis Liu, Z. 2010. Mechanisms of ethanol tolerance in *Saccharomyces cerevisiae*. *Applied Microbiology and Biotechnology*, 87: 829–45.

Maldonado, Maria Cristina, Marina Paola Aban and Antonio Roberto Navarro. 2013. Chemicals and lemon essential oil effect on *Alicyclobacillus acidoterrestris* viability. *Brazilian Journal of Microbiology*, 44: 1133–37.

Malik, V.S., Schulze, M.B. and Hu, F.B. 2006. Intake of sugar-sweetened beverages and weight gain: a systematic review. *Am. J. Clin. Nutr.*, 84: 274–88.

Malik, V.S., Popkin, B.M., Bray, G.A., Després, J.P. and Hu, F.B. 2010. Sugar-sweetened beverages, obesity, type 2 diabetes mellitus, and cardiovascular disease risk. *Circulation*, 121: 1356–64.

Mann, Nicholas H. 2008. The potential of phages to prevent MRSA infections. *Research in Microbiology*, 159: 400–05.

Marshall, Valerie M. and Wendy M. Cole. 1985. Methods for making kefir and fermented milks based on kefir. *Journal of Dairy Research*, 52: 451–56.

Martorell, Patricia, Malcolm Stratford, Hazel Steels, Ma Teresa Fernández-Espinar and Amparo Querol. 2007. Physiological characterization of spoilage strains of *Zygosaccharomyces bailii* and *Zygosaccharomyces rouxii* isolated from high sugar environments. *International Journal of Food Microbiology*, 114: 234–42.

May, André, Odair Alves Bovi, Nilson Borlina Maia, Lauro Euclides Soares Barata, Rita de Cassia Zacardi de Souza, Eduardo Mattoso Ramos de Souza, Andrea Rocha Almeida de Moraes and Mariane Quaglia Pinheiro. 2008. Basil plants growth and essential oil yield in a production system with successive cuts. *Bragantia*, 67: 385–89.

Mesías, F.J., Martín, A. and Hernández, A. 2021. Consumers' growing appetite for natural foods: Perceptions towards the use of natural preservatives in fresh fruit. *Food Research International*, 150: 110749.

Milly, Paul Jesse. 2007. Utilizing Hydrodynamic Cavitation and Ultraviolet Irradiation to Improve the Safety of Minimally Processed Fluid Food. University of Georgia.

Mohammadi, Reza, Sara Sohrabvandi and Amir Mohammad Mortazavian. 2012. The starter culture characteristics of probiotic microorganisms in fermented milks. *Engineering in Life Sciences*, 12: 399–409.

Molan, Peter C. 1992. The antibacterial activity of honey: 1. The nature of the antibacterial activity. *Bee World*, 73: 5–28.

Monk, A.B., Rees, C.D., Barrow, P., Hagens, S. and Harper, D.R. 2010. Bacteriophage applications: where are we now? *Letters in Applied Microbiology*, 51: 363–69.

Morgan, S.M., O'sullivan, L., Ross, R.P. and Hill, C. 2002. The design of a three strain starter system for Cheddar cheese manufacture exploiting bacteriocin-induced starter lysis. *International Dairy Journal*, 12: 985–93.

Mudgil, Deepak, Sheweta Barak and Bhupendar Singh Khatkar. 2014. Guar gum: processing, properties and food applications—a review. *Journal of Food Science and Technology*, 51: 409–18.

Mukherjee, A., B. Gómez-Sala, E.M. O'Connor, J.G. Kenny and P.D. Cotter. 2022. Global regulatory frameworks for fermented foods: a review. *Front. Nutr.*, 9: 902642.

Mundo, Melissa A., Olga I. Padilla-Zakour and Randy W. Worobo. 2004. Growth inhibition of foodborne pathogens and food spoilage organisms by select raw honeys. *International Journal of Food Microbiology*, 97: 1–8.

Nykänen, Lalli and Irma Nykänen. 2017. Distilled beverages. *In*: *Volatile compounds in foods and beverages* (Routledge). ISBN 9780203734285. pp. 537–571. Taylor and Francis.

Ochiai, Nobuo, Kikuo Sasamoto, Masahiko Takino, Satoru Yamashita, Shigeki Daishima, Arnd C. Heiden and Andreas Hoffmann. 2002. Simultaneous determination of preservatives in beverages, vinegar, aqueous sauces, and quasi-drug drinks by stir-bar sorptive extraction (SBSE) and thermal desorption GC–MS. *Analytical and Bioanalytical Chemistry*, 373: 56–63.

Oliveira, M., Viñas, I., Colàs, P., Anguera, M., Usall, J. and Abadias, M. 2014. Effectiveness of a bacteriophage in reducing *Listeria monocytogenes* on fresh-cut fruits and fruit juices. *Food Microbiology*, 38: 137–42.

Ortega-Rivas, Enrique, and Enrique Ortega-Rivas. 2012. Classification of food processing and preservation operations. *Non-thermal Food Engineering Operations*: 3–10.

Özkan-Karabacak, Azime, Bige İnceda yı and Ömer Utku Çopur. 2019. Preservation of beverage nutrients by high hydrostatic pressure. *In*: *Preservatives and Preservation Approaches in Beverages* (Elsevier). Academic Press, pp. 309–337, ISBN 9780128166857.

Pendyala, Brahmaiah, Ankit Patras, Ramaswamy Ravi, Vybhav Vipul Sudhir Gopisetty and Michael Sasges. 2020. Evaluation of UV-C irradiation treatments on microbial safety, ascorbic acid, and volatile aromatics content of watermelon beverage. *Food and Bioprocess Technology*, 13: 101–11.

Peris, David, Roberto Pérez-Torrado, Chris Todd Hittinger, Eladio Barrio and Amparo Querol. 2018. On the origins and industrial applications of *Saccharomyces cerevisiae* × *Saccharomyces kudriavzevii* hybrids. *Yeast*, 35: 51–69.

Piccaglia, R. and Marotti, M. 1993. Characterization of several aromatic plants grown in northern Italy. *Flavour and Fragrance Journal*, 8: 115–22.

Prem, Deepak, Subhadra Singh, Padam Prakash Gupta, Jaivir Singh and Sish Pal Singh Kadyan. 2005. Callus induction and *de novo* regeneration from callus in Guar (*Cyamopsis tetragonoloba*). *Plant cell, Tissue and Organ Culture*, 80: 209–14.

Pretorius, Isak S. 2000. Tailoring wine yeast for the new millennium: novel approaches to the ancient art of winemaking. *Yeast*, 16: 675–729.

Ramos, Marina, Arantzazu Valdés, Ana Cristina Mellinas and María Carmen Garrigós. 2015. New trends in beverage packaging systems: a review. *Beverages*, 1: 248–72.

Rani, Heena and Rachana D. Bhardwaj. 2021. Quality attributes for barley malt: The backbone of beer. *Journal of Food Science*, 86: 3322–40.

Raspor, Peter and Dušan Goranovič. 2008. Biotechnological applications of acetic acid bacteria. *Critical Reviews in Biotechnology*, 28: 101–24.

Rawat, Seema. 2015. Food Spoilage: microorganisms and their prevention. *Asian Journal Of Plant Science and Research*, 5: 47–56.

224   *Novel Approaches in Biopreservation for Food and Clinical Purposes*

Rodrigo, D., Sampedro, F., Martínez, A., Rodrigo, M. and Barbosa-Cánovas, G.V. 2005. Application of PEF on orange juice products. pp. 131–144. *In*: *Novel Food Processing Technologies* (CRC Press).

Rubio-Bretón, Pilar, Teresa Garde-Cerdán, Juana Martínez, Ana Gonzalo-Diago, Eva P. Pérez-Álvarez and Matteo Bordiga. 2017. Wine aging and spoilage. pp. 112–158. *In*: *Post-Fermentation and-Distillation Technology* (CRC Press).

Saha, Dipjyoti and Suvendu Bhattacharya. 2010. Hydrocolloids as thickening and gelling agents in food: a critical review. *Journal of Food Science and Technology*, 47: 587–97.

Sakamoto, Kanta and Wil N. Konings. 2003. Beer spoilage bacteria and hop resistance. *International Journal of Food Microbiology*, 89: 105–24.

Saldana, T.M., O. Basso, R. Darden and D.P. Sandler. 2007. Carbonated beverages and chronic kidney disease. *Epidemiology*, 18: 501–6.

Sarris, Dimitris and Seraphim Papanikolaou. 2016. Biotechnological production of ethanol: biochemistry, processes and technologies. *Engineering in Life Sciences*, 16: 307–29.

Shankar, Vijayalakshmi, Shahid Mahboob, Khalid A Al-Ghanim, Zubair Ahmed, Norah Al-Mulhm and Marimuthu Govindarajan. 2021. A review on microbial degradation of drinks and infectious diseases: A perspective of human well-being and capabilities. *Journal of King Saud University-Science*, 33: 101293.

Sharma, Sanjay. 2015. Food preservatives and their harmful effects. *International Journal of Scientific and Research Publications*, 5: 1–2.

Shekarchizadeh, Hajar and Fatemeh Sadat Nazeri. 2020. Active nanoenabled packaging for the beverage industry. *In*: *Nanotechnology in the Beverage Industry* (Elsevier). pp. 587–607, ISBN 9780128199411.

Shi, Ce and Maryam Maktabdar. 2022. Lactic acid bacteria as biopreservation against spoilage molds in dairy products – a review. *Frontiers in Microbiology*, 12.

Siddiqua, S., Anusha, B.A., Ashwini, L.S. and Negi, P.S. 2015. Antibacterial activity of cinnamaldehyde and clove oil: effect on selected foodborne pathogens in model food systems and watermelon juice. *Journal of Food Science and Technology*, 52: 5834–41.

Silici, Sibel and Kevser Karaman. 2014. Inhibitory effect of propolis on patulin production of *Penicillium expansum* in apple juice. *Journal of Food Processing and Preservation*, 38: 1129–34.

Silva, Natalia Kellen Vieira da, Luiz Bruno de Sousa Sabino, Luciana Siqueira de Oliveira, Lucicléia Barros de Vasconcelos Torres and Paulo Henrique Machado de Sousa. 2016. Effect of food additives on the antioxidant properties and microbiological quality of red guava juice. *Revista Ciência Agronômica*, 47: 77–85.

Simon, Ronald A. 2003. Adverse reactions to food additives. *Current Allergy and Asthma Reports*, 3: 62–66.

Singh, Jagriti, Akanksha Rastogi, Debajyoti Kundu, Mohan Das and Rintu Banerjee. 2018. A new perspective on fermented protein rich food and its health benefits. *Principles and Applications of Fermentation Technology*: 417–36.

Singh, Veer Pal. 2018. Recent approaches in food bio-preservation-a review. *Open Veterinary Journal*, 8: 104–11.

Smelt, Jan P., Johan C. Hellemons and Margaret Patterson. 2001. Effects of high pressure on vegetative microorganisms. *Ultra High Pressure Treatments of Foods*: 55–76.

Snoek, Tim, Kevin J. Verstrepen and Karin Voordeckers. 2016. How do yeast cells become tolerant to high ethanol concentrations? *Current Genetics*, 62: 475–80.

Soffer, Nitzan, Joelle Woolston, Manrong Li, Chythanya Das and Alexander Sulakvelidze. 2017. Bacteriophage preparation lytic for Shigella significantly reduces Shigella sonnei contamination in various foods. *PLoS One*, 12: e0175256.

Stratford, Malcolm. 2006. Food and beverage spoilage yeasts. *Yeasts in Food and Beverages*: 335–79.

Sulakvelidze, Alexander, Zemphira Alavidze and J. Glenn Morris Jr. 2001. Bacteriophage therapy. *Antimicrob Agents Chemother*, 45: 649–59.

Sun, Shu Yang, Han Sheng Gong, Xiao Man Jiang and Yu Ping Zhao. 2014. Selected non-Saccharomyces wine yeasts in controlled multistarter fermentations with *Saccharomyces cerevisiae* on alcoholic fermentation behaviour and wine aroma of cherry wines. *Food Microbiology*, 44: 15–23.

Sun, Xun, Jingting Liu, Li Ji, Guichao Wang, Shan Zhao, Joon Yong Yoon and Songying Chen. 2020. A review on hydrodynamic cavitation disinfection: the current state of knowledge. *Science of the Total Environment*, 737: 139606.

Szutowska, Julia. 2020. Functional properties of lactic acid bacteria in fermented fruit and vegetable juices: a systematic literature review. *European Food Research and Technology*, 246: 357–72.

Tamang, Jyoti Prakash and Kasipathy Kailasapathy. 2010. *Fermented Foods and Beverages of the World* (CRC press).

Thomas, D.S. and Davenport, R.R. 1985. *Zygosaccharomyces bailii*—a profile of characteristics and spoilage activities. *Food Microbiology*, 2: 157–69.

Tubia, Imanol, Karthik Prasad, Eva Pérez-Lorenzo, Cristina Abadín, Miren Zumárraga, Iñigo Oyanguren, Francisca Barbero, Jacobo Paredes and Sergio Arana. 2018. Beverage spoilage yeast detection methods and control technologies: a review of Brettanomyces. *International Journal of Food Microbiology*, 283: 65–76.

Van Leeuw, Robin, Claire Kevers, Joël Pincemail, Jean-Olivier Defraigne and Jacques Dommes. 2014. Antioxidant capacity and phenolic composition of red wines from various grape varieties: Specificity of Pinot Noir. *Journal of Food Composition and Analysis*, 36: 40–50.

Vara, Saritha, Manoj Kumar Karnena and Bhavya Kavitha Dwarapureddi. 2019. Natural preservatives for nonalcoholic beverages. *In*: *Preservatives and Preservation Approaches in Beverages* (Elsevier). Academic Press, pp. 179–201, ISBN 9780128166857.

Vaughan, Anne, Tadhg O'Sullivan and Douwe Van Sinderen. 2005. Enhancing the microbiological stability of malt and beer—a review. *Journal of the Institute of Brewing*, 111: 355–71.

Walker, Graeme M. and Annie E. Hill. 2016. *Saccharomyces cerevisiae* in the production of whisk (e) y. *Beverages*, 2: 38.

Walker, Graeme M. and Graham G. Stewart. 2016. *Saccharomyces cerevisiae* in the production of fermented beverages. *Beverages*, 2: 30.

Wang, Zhengxiang, Jian Zhuge, Huiying Fang and Bernard A. Prior. 2001. Glycerol production by microbial fermentation: a review. *Biotechnology Advances*, 19: 201–23.

Wareing, Peter. 2016. Microbiology of soft drinks and fruit juices. *Chemistry and Technology of Soft Drinks and Fruit Juices*: 290–309.

Widyastuti, Yantyati and Andi Febrisiantosa. 2014. The role of lactic acid bacteria in milk fermentation. *Food and Nutrition Sciences*, 2014.

Wilson, Brian G. and Sami L. Bahna. 2005. Adverse reactions to food additives. *Annals of Allergy, Asthma & Immunology*, 95: 499–507.

Young, N.W.G. and O'sullivan, G.R. 2011. The influence of ingredients on product stability and shelf life. *In*: *Food and beverage stability and shelf life* (Elsevier). Woodhead Publishing, pp. 132–183, ISBN 9781845697013.

Zaeem, Zoya, Lily Zhou and Esma Dilli. 2016. Headaches: a review of the role of dietary factors. *Current Neurology and Neuroscience Reports*, 16: 1–11.

Ziarno, Małgorzata and Patrycja Cichońska. 2021. Lactic acid bacteria-fermentable cereal- and pseudocereal-based beverages. *Microorganisms*, 9: 2532.

CHAPTER **8**

Biopreservation in Medicines

Sanjogta Thapa Magar

Introduction

Medicines are formulations used for the administration and delivery of pharmaceutical drugs, via a variety of routes. Formulations range from simple aqueous solutions or dry powders to complex combinations of ingredients and physicochemical characteristics. This chemical and physicochemical complexity can favor the survival and multiplication of microorganisms that might contaminate these medicines during the manufacturing process, causing damage in the product and posing a threat for the consumer's safety. The Good Manufacturing Practices (GMPs) should prevent these contaminations of the raw materials during the processing and packaging processes. Many pharmaceutical products contain preservatives to extend their shelf life. It is essential to add preservatives to such products, particularly those with high water content, to prevent alteration and degradation by microorganisms during storage (Dodge et al., 2015; Moser and Meyer, 2011). However, there is little official list for recommendations for specific medicines, as there is non-mandatory public disclosure of preservative content in medicines.

Ideal preservatives would work at low concentrations against all types of microorganisms, be nontoxic and compatible with other constituents of the preparation, and remain stable for the preparation's shelf-life. Sometimes, one type of preservation alone is not enough to ensure maintenance of the product and two or more preservatives are combined for better efficiency. Preservatives are widely used in pharmaceutical products such as emulsions, suspensions, semisolid and parenteral preparations for oral, dental, dermal, nasal, vaccines and rectal and ophthalmic products (Shaikh et al., 2016; Kumari et al., 2019; Charnock and Otterholt, 2012).

Sun International Incorporated Pvt. Ltd (Baskin Robbins), Jyoti Bhawan, Jamal, Kathmandu, Nepal.

Microbial and pathogenic microorganisms of concern. Consequences of microbial contamination

Microorganisms can exert a range of metabolic activities that ultimately are hazards for the stability of the product and the safety of the consumer. This will depend on the type of microorganism contaminating the medicine, the infective dose, the route of administration and the immune status of the consumer (Hiom, 2013).

The most common microbial hazards in liquid medicines are *Pseudomonas* sp., while dry powders, tablets and capsules are frequently contaminated by Gram-negative rods, bacterial and fungi spores. In the past, there have been a number of outbreaks due to contaminated medicines. Thus, salmonellosis was associated to contaminated tablets, while eye infections have been associated several times with contaminated ophthalmic solutions and even tetanus was linked to talc powders (Tremewan, 1946; Kallings, 1966; Templeton III et al., 1982). These events led to an increased regulatory control in many countries.

Medicines for eye health constitute a sensitive area for microbiological control. Multidose eye drop containers are sensitive to generate pathogenic contamination (Taşli and Coşar, 2001; Teuchner et al., 2015; da Costa et al., 2020) and there are a limited number of preservatives that do not harm the eye. Eye-treatment contamination is not only generated by bacteria, but also by protozoa, like *Acanthamoeba* sp. (Lindsay et al., 2007; Shimmura-Tomita et al., 2018; Fanselow et al., 2021).

Another medicine area susceptible of microbial contamination are injections (Meers et al., 1973; Landman et al., 2000; Tabatabaei et al., 2017). Most of the infections derived from parenteral infusions come from the administration devices, like the catheters and cannula (Tebbs et al., 1996). Parenteral infusions are usually formed by sterile components but the use of preservatives is more limited (Floyd, 1999; Lowe, 2018; Chapman, 2019).

If patients are immunocompromised, weakened by chemotherapy or other conditions, opportunistic microbial infections might inflict more damage than if they were in normal conditions (Millership et al., 1986).

Other untreatable diseases have been transmitted via contaminated medicines, like the human immunodeficiency virus (HIV), hepatitis C or the Creutzfeldt-Jakob disease (Brown et al., 1995; Gomperts, 1996; Swerdlow et al., 2003). Even products as ultrasound gel and antiseptic solutions are susceptible of bacterial contaminations, like the ones caused by *Burkholderia cepacia* and *Pseudomonas* spp. (Arjunwadkar et al., 2001; Hutchinson et al., 2004).

Preservation strategies

There are different strategies based on the formulation for the maintenance of medicine stability and safety against microbial contamination, either by modification of the physicochemical properties or by modulating the efficacy of any present preservative.

Formulation effects on microbial growth

Many ingredients in the medicine formulations are biodegradable and cover the nutritional requirements for most microbial contaminants. However, for them to survive it would require the appropriate physicochemical conditions, like the available water (Aw) or pH, which can be manipulated for the purpose of making more difficult this growth and acting as a preservative strategy.

Live microorganisms require a certain level of available water to grow and maintain their metabolism. As a general rule, microorganisms can live in the range of 0.91–0.6 Aw, where no microorganism is expected to grow below that level (Allen Jr., 2018). Therefore, reducing the Aw level would minimize the ability of certain organisms to grow on tablets, capsules or powders. However, this Aw reduction needs to be maintained throughout the life of the product, even during the distribution in the case of humid climates (Fassihi et al., 1978; Whiteman, 1995). One of the Aw reduction strategies involves the addition of water-binding solutes of low molecular weight, such as sucrose, sorbitol, glycerol, urea, polyacrylamide hydrogels or alcoholic formulations, which already have antimicrobial properties (LINTNER, 1997; O'Malley and Maibach, 1997; Morris and Leech, 2017). Nevertheless, despite reducing the Aw in the formulation, microbial contamination can occur by accumulation of moisture on the surfaces of the containers, leading to formation of bacterial biofilms and fungal growth (Beveridge and Bendell, 1988).

Microorganisms require a certain pH to grow, ranging wider or narrower depending on the species (Jin and Kirk, 2018). pH control is also a strategy to limit the colonization and development of microbial spoilage, but these modifications need to have into consideration the drug requirements, to avoid inactivation. Therefore, the application of this strategy is limited, allowing other microorganisms to take over.

Modification of the temperature can also reduce growth. Typically, a temperature reduction will inhibit the growth or even kill some of the microorganisms. However, most of the activity of the microbiological contaminants in medicines can be recovered by increasing temperature again, to 25 degrees in the case of fungi and 30 degrees in the case of bacteria (Symonds et al., 2016).

Formulation effects on preservative efficacy

Choosing one preservative or another depends on several factors, such as the intrinsic properties of the agent, the extrinsic environment (product formulation) and the nature of the product usage. Other factors will influence the efficacy of the antimicrobial preservatives, such as the concentration, temperature and environmental pH.

The concentration of the preservative has an exponential effect on the microbial death, but will depend on the type of preservative that is used (Hugo and Denyer, 1987; Russell and McDonnell, 2000). Moreover, the activity of a preservative will depend on the free concentration of the active form of the molecule in the aqueous phase and will be affected by partitioning in complex multiphase formulations and pH effects.

The efficacy of preservatives can be affected when they are incorporated into multiphase formulations (Van Doorne, 1990; Attwood, 2012; Dempsey, 2017).

The partitioning of preservatives between oil and water phases will depend on the oil:water ratio, influencing the migration of the preservative and potentially concentrating them in specific areas (Attwood, 2012). Additionally, the affinity interactions with other components of the formulation can affect the availability and effectivity of the preservatives, and the molecules bound to other ingredients are not considered to act contributively (Dempsey, 2017).

Environmental pH can affect the efficacy and potency of the preservatives, mainly by inducing ionic changes and it will depend on the preservative's mode of action. Similarly, temperature can affect preservative activity and, therefore, recommended storage temperatures are important to follow.

The effectivity of preservatives is believed to be lower at low Aw values (e.g., powders or tablets), despite the lack of evidence (Hiom, 2004). However, low levels of electrolytes can enhance the effect of phenolic disinfectants, similarly to the effect of los Aw, due to the addition of sucrose, glycerol and other glycols (Waites et al., 2009; Hiom, 2013). Low Aw can also difficult the sterilizing effects of ethylene oxide and high temperatures (Hiom, 2004).

Other elements that can interact with preservatives are the suspending and thickening agents, either by enhancing or diminish their activity. Some of these reactive reagents include cyclodextrins, alginates and polyethylene glycols, among others (Kumar et al., 2018). Moreover, the low solubility of some preservatives might cause precipitation in the form of salts or complexes, impairing their function.

Preservative efficacy can also be reduced by adsorption to surfaces and absorption into the solids, especially in plastic containers (Sahnoune et al., 2021; Lang et al., 2022). Efficacy can also be reduced by the inactivation of the molecules that interact with the microorganisms and, therefore, are not available to exert their function anymore.

Preservatives

The ideal properties of preservatives

Ideally, a good preservative would have the following characteristics (Shaikh et al., 2016; Oni et al., 2013):

(1) It should not be toxic and should not cause any irritation;

(2) Effective at low concentration against a wide variety of microorganisms;

(3) Soluble in the formulation at the required concentration;

(4) Non-toxic and non-sensitizing externally and internally at the concentration required;

(5) Compatible with a wide variety of drug solubilizing and dispensing agents;

(6) Free from objectionable odor, taste or color;

(7) Active with long-term stability over a wide range of pH and temperature.

Classification of preservatives

A. Based on mechanisms of action

A.1. Antimicrobial preservatives

These preservatives are added to the products to kill or inhibit the growth of microorganisms that are inadvertently introduced during production or use (Fahelelbom and El-Shabrawy, 2007). Antimicrobial preservatives are used in a product that contains water and is highly susceptible to contamination, such as solutions, suspensions, and emulsions administered orally, solutions applied externally creams and sterile preparations repeatedly used (e.g., injectable multi-dose preparations and eye-drops).

There are two main types of antimicrobial preservatives: antifungal preservatives and antibacterial preservatives (Shaikh et al., 2016). Antifungal preservatives include compounds such as benzoic and ascorbic acids and their salts, and phenolic compounds such as methyl, ethyl, propyl and butyl p-hydroxybenzoate (parabens) (Dodge et al., 2015). Preservatives that inhibit bacteria include quaternary ammonium salts, alcohols, phenols and mercurials (Kumari et al., 2019).

A.2. Antioxidants

These preservatives are added to the products to prevent the oxidation of active pharmaceutical ingredients, which are sensitive and degrade in presence of oxygen.

The antioxidants can be divided into three categories:

The first category includes true antioxidants, or anti-oxygen, which probably inhibit oxidation by reacting with free radicals blocking the chain reaction. Examples are alkygallates butylated hydroxy anisole, butylated hydroxytoluene, nordihydroguaiaretic acid and tocopherols (Himoudy, 2016).

The second category includes reducing agents that have lower redox potentials than the drug or adjuvant which they are intended to protect, and are, therefore, more readily oxidized. Reducing agents may act also by reacting with free radicals. Examples are ascorbic acid and the potassium and sodium salts of sulfurous acid (Himoudy, 2016).

Those in the third group are antioxidant synergists, which usually exhibit little antioxidant activity themselves but may enhance the action of first-group antioxidants by interacting with heavy metal ions that catalyze oxidation. Examples of antioxidant synergists are citric acid, lecithin, and tartaric acid (Himoudy, 2016).

A.3. Chelating agents

These preservatives are added in an attempt to prevent the degradation of pharmaceutical formulations by forming a complex with the pharmaceutical ingredient. Such preservatives include disodium ethylenediamine tetraacetic acid (EDTA), polyphosphates, and citric acid (George and Brady, 2020).

B. Based on source

B.1. Natural preservatives

These preservatives are natural substances that provide intrinsic protection against microbial growth in products. They include neem oil, salt (sodium chloride), lemon, honey, sugar, vinegar, diatomaceous, etc. (Pawar et al., 2011).

B.2. Artificial preservatives

These preservatives are derived from chemical synthesis and are active against various microorganisms when used in small concentrations. They include benzoates, sodium benzoate, sorbates, propionates, and nitrites (Fahelelbom and El-Shabrawy, 2007).

Characteristics of the most common preservatives in medicines

There is a number of natural and artificial preservatives that are commonly used in medicines. We have listed some of their characteristics. Table 1 summarizes examples of how these preservatives are typically present in medicines and Tables 2, 3, 4 and 5 compile their recommended concentrations for liquid oral, parenteral, ophthalmic/nasal and creams.

1. Benzoic acid: It is a colorless, crystalline compound that includes sodium and potassium benzoates. Bacteriostatic and fungistatic properties are commonly associated with benzoic acid and its salt form, sodium benzoate. The pharmaceutical industry uses it as an antimicrobial preservative, antifungal, and tablet and capsule lubricant. A combination of benzoic acid and salicylic acid is used to treat athletes' feet and ringworm with salicylic acid ointment (Chipley, 2020).

2. Methylparaben: Methylparaben belongs to a group of chemicals known as parabens. It is a white crystalline powder with a characteristic odor, freely soluble in water and alcohol. In the pharmaceutical industry, methylparaben has been used as a preservative for over 50 years to prevent pathogen growth and unwanted chemical changes (Lincho et al., 2021).

3. Benzyl alcohol: Benzyl alcohol (α-hydroxytoluene) is aromatic alcohol commonly used as a preservative in multidose medication vials and parenteral solutions to inhibit the growth of bacteria. It is also used as an anti-parasite medication in some products (Lincho et al., 2021).

4. Sorbic acid: Sorbic acid is a white crystalline compound slightly soluble in water, and soluble in organic solvents that are used as a preservative in pharmaceuticals. It is non-toxic to humans and inhibits yeasts, molds, and some bacteria from growing. It is safe to use sorbic acid in a wide range of drugs. Pharmaceutical products are often preserved with sorbic acid and its salts, sodium sorbate, potassium sorbate, and calcium sorbate (Anurova et al., 2019).

5. Ethylenediaminetetraacetic Acid (EDTA): In many pharmaceutical formulations, EDTA is used as an antioxidant, chelating agent for heavy metals,

232 *Novel Approaches in Biopreservation for Food and Clinical Purposes*

Table 1. Some common examples of preservatives used in different pharmaceutical products. Adapted from (Horowitz, 2010; Anand and Sati, 2013).

Category	Product	Preservatives
Oral	Tablets, capsules, suspension, syrup	Methyl, ethyl, propyl parabens and their combinations, sodium benzoate, benzoic acid, calcium lactate, potassium sorbate, calcium sorbate, sodium sorbate, sorbic acid, propionic acid.
Dermal	Creams, lotion, ointment, soap, gel	Benzalkonium chloride, cetrimide, EDTA, benzoic acid, thiomersal, imidurea, chlorhexidine, chlorocresol, phenyl salicylate.
Dental	Toothpaste, mouthwash, gargle	Sodium benzoate, benzoic acid, potassium sorbate, sodium phosphate, triclosan, cetyl, pyridinium chloride, methyl and ethyl parabens.
Ophthalmic	Eye drops, ointments, contact lenses and solutions amino	Benzalkonium chloride, EDTA, benzoic acid, thiomersal, imidurea, chlorhexidine, polyamine propyl biguanide, sodium perborate, boric acid, thimerosal, phenylmercuric salt.
Nasal	Nasal drops, sprays, aerosols	Benzalkonium chloride, phenylcarbinol, potassium sorbate, chlorobutanol, chlorocresol, EDTA.
Rectal	Suppositories, enema	Benzyl alcohol, benzoic acid, sodium benzoate, methyl hydroxybenzoate, chlorhexidine, gluconate.
Parenteral	Vaccines, injections	Methyl, ethyl, propyl, butyl parabens and their combinations, benzyl alcohol, chlorobutanol, chlorhexidine, thiomersal, formaldehyde, benzoic acid, sorbic acid, phenol, 3-cresol, thimerosal, phenylmercuric salt.

and preservative. EDTA is used in eye wash and ophthalmic solutions that have bactericidal properties to enhance the activity of preservatives and antibacterial ingredients, as well as stabilize their antioxidant effect of them (Kabara, 1997).

6. Sodium chloride: Sodium chloride mineral exists in the form of white crystalline powder or colorless crystals with a saline taste. Many pharmaceutical products contain sodium chloride for producing isotonic solutions. There are several uses for saline, including nasal sprays, intravenous lock flushes, and eye washes or solutions (Horowitz, 2010).

7. Ascorbic acid: Vitamin C is a water-soluble vitamin found in citrus fruits and vegetables, as well as dietary supplements. It is colorless to light-yellow, non-hygroscopic, odorless, crystalline powder or crystals. Aqueous pharmaceutical formulations contain ascorbic acid as an antioxidant. For injectable solutions and oral liquids, ascorbic acid has been used to adjust pH. Besides injections, ascorbic acid is available as tablets, capsules, chewable tablets, and capsules with ascorbic acid in them (Shaikh et al., 2016).

Biopreservation in Medicines   233

Table 2. Some of the preservatives used in pharmaceutical formulations and their concentrations for liquid oral preparation. Adapted from (Oni et al.; SHAQRA and AL-SHAWAGFEH, 2012; Boukarim et al., 2009; Niazi, 2019b; Tiwari et al., 2018).

Preservative	Recommended concentration (%)
Benzoic acid	0.1–0.2
Benzethonium chloride	0.01–0.02
Benzyl alcohol	3.0
Benzylkonium chloride	0.002–0.02
Bronidol	0.001–0.05
Bronopol	0.01–0.1
Butyl paraben	0.1–0.4
Chlorobutanol	0.5
Chloro cresol	0.2
Ethyl paraben	0.1–0.25
Methyl paraben	0.25
Meta cresol	0.15–0.3
Phenol	0.1–0.5
Phenylmercuric nitrate	0.002–0.1
Propylene glycol	15–30
Propyl paraben	0.5–0.25
Sodium benzoate	0.1–0.2
Sorbic acid	0.1–0.2
Thiomersal	0.1

Table 3. Some of the preservatives used in pharmaceutical formulations and their concentrations for parenteral preparation. Adapted from (Shaikh et al., 2016; Kumari et al., 2019; Meyer et al., 2007).

Preservative	Recommended concentration (%)
Benzethonium chloride	0.01
Benzyl alcohol	0.5–10
Benzylkonium chloride	0.01
Butyl paraben	0.015
Chlorobutanol	0.25–0.5
Chloro cresol	0.1–0.18
Ethyl paraben	0.005–0.02
Methyl paraben	0.01–0.5
Meta cresol	0.1–0.25
Phenol	0.065–0.02
Phenylmercuric nitrate	0.002
Propylparaben	0.005–0.002
Thimerosal	0.01
Chlorobutanol	0.5
Methyl p-hydroxybenzoate & Propyl p-hydroxybenzoate	0.18 0.02 (in combination)

234 *Novel Approaches in Biopreservation for Food and Clinical Purposes*

Table 4. Some of the preservatives used in pharmaceutical formulations and their concentrations for ophthalmic/nasal preparation. Adapted from (Kumari et al., 2019; Fahelelbom and El-Shabrawy, 2007; Anand and Sati, 2013; Niazi, 2019a).

Preservative	Recommended concentration (%)
Methyl paraben	0.1
Ethyl paraben	0.1
Propylparaben	0.1
Butyl paraben	0.1
Chlorobutanol	0.5
Thiomersal	0.01
Phenylmercuric nitrate	0.004
Benzylkonium chloride	0.004-0.02
Benzethonium chloride	0.004-0.01

Table 5. Some of the preservatives used in pharmaceutical formulations and their concentrations for ointments and creams preparation. Adapted from (Kumari et al., 2019; Fahelelbom and El-Shabrawy, 2007; Anand and Sati, 2013; Niazi, 2019a).

Preservatives	Recommended concentration (%)
Methyl paraben	0.001–0.2
Ethyl paraben	0.001–0.2
Propylparaben	0.001–0.2
Butyl paraben	0.001–0.2
Chlorobutanol	0.5
Phenol	0.25–0.5
Meta cresol	0.1–0.3
Chloro cresol	0.1–0.3
Thiomersal	0.01
Phenylmercuric nitrate	0.002
Benzylkonium chloride	0.01
Benzethonium chloride	0.01

Factors affecting the efficiency of preservatives in medicines products

The various factors play important roles in the working capacity of preservatives which are as follows:

Interaction with formulation components

It is important to consider potential interactions between a preservative and its preparation even if it is known to be effective. The preservatives used in pharmaceutical drugs interact with different active ingredients, excipients and other components such as emulgents, solubility in oil, suspended solids, etc. ((FDA) 2020).

Several hydrocolloids, such as methylcellulose, alginates and tragacanth, can interact with preservatives and render them inactive (Charnock and Otterholt, 2012). When added to a formulation, it interacts and deteriorates the preservative properties resulting in the efficiency of preservatives.

Properties of preservatives

It is important to know about the properties of all preservatives to be used in the formulation. Preservatives should have properties like a broad spectrum of activity, rapid action, non-sensitizing, non-toxic, non-ionic, synergy with other preservatives, low irritation and chemical stability. However other factors like pH, compatibility with other preservatives, homogeneous mixture, and reactive action of preservatives should be considered, e.g., chorobutol may get hydrolyzed on storage if pH is not maintained in the formulation. On the other hand, the preservative must be compatible with the container used during formulation as it may react and decrease the efficiency of the preservative. As a result, all the above factors must be considered to halt the decrease in the efficiency of preservatives.

Effect of containers or packaging

The containers used to make and store pharmaceutical products including preservatives play an important role in retaining the efficiency of preservatives. The containers frequently used are plastics, glass, rubber, aluminum for tablets, etc. Preservatives used in pharmaceutical drugs can penetrate plastics, rubber, and aluminum and form a reaction. This reaction may cause decreased efficiency of preservatives. As a result, it becomes prone to microbial contamination.

Solubility

Several preservatives are insoluble in their natural forms, but become soluble in their sodium salt forms. The substitution of sodium benzoate for benzoic acid is one example.

Influence of product pH

pH is an important factor that affects the chemical stability and activity of preservatives. For example, benzoic acid at pH 3.5 is 10 times more effective than at pH 7. At low pH of 2, the effectiveness is 100 times because at low pH, the un-dissociated molecules of the acid are more and it is the unassociated molecules that create the effect.

Regulatory framework

The European Pharmacopoeia (Ph. Eur.) Commission is a decision-making body responsible for the technical decisions and the elaboration and maintenance of the European Pharmacopoeia. Thus, the Ph. Eur. Commission sets the limits for the presence of microorganisms in medicines, and this will depend on the product

236 *Novel Approaches in Biopreservation for Food and Clinical Purposes*

and its purposeful use. The legistative instruments are the European Community (EC) Directives, which are meant to be implemented by the national legislations (Permanand, 2006).

In Europe, pharmaceutical companies must obtain a marketing authorization before commercializing a product, which must contain a detail study of quality, safety and efficacy data of all the materials, including the preservatives. At European-wide level, approval can be obtained via the Committee for Medicinal Products for Human Use (CHMP) Scientific Committee, under the supervision of the European Medicine Agency (EMA). In the USA, medicines approval operates under the Federal Food, Drug and Cosmetic Act (Title 21 USC, 350 et seq.), through the Food and Drug Administration (FDA). Each country has its own Regulatory body, like the regulatory authority in the UK, the Medicines and Healthcare Products Regulatory Agency (MHRA), the Health Products Regulatory Authority (HPRA) in Ireland, the Spanish Agency of Medicines and Medical Devices (AEMPS) in Spain, Health Canada in Canada or the Pharmaceutical and Food Safety Bureau in Japan. These regulatory bodies are responsible of granting licenses to manufacturers, ensuring that clinical trials are conducted appropriately and there is an adequate pharmacovigilance after adverse reactions.

Pharmaceutical companies are asked to justify the safety, stability and effectiveness of a medicine, including the strategies taken to minimize microbial contamination and spoilage risks, like preservatives, which must demonstrate suitable efficacy over the product life and demonstrate safety.

It is rare that lists of approved preservatives are issued, although some official compilation might contain indications of the potential preservatives depending on the potential purposes. The use of a known preservative in similar medicines facilitates the toxicological data required by the licensing authorities, which ultimately leads to slow the development and application of newer or less established preservatives, due to the high cost of developing new toxicological testing.

Additionally, it is also required the development of detailed environmental impact assessments for preservatives, to assess the effect of the different components on the biosphere.

Preservative efficacy testing

Pharmaceutical products undergo a preservative efficacy test to determine the type of preservative and the minimum effective concentration. Pharmaceuticals are frequently tested for microbial contamination using antimicrobial effectiveness tests of the preservatives.

The preservation efficiency test is done in the following steps:

1. Product category
2. Test organism
3. Media culture preparation
4. Preparation of inoculum
5. Procedure
6. Interpretation

Biopreservation in Medicines 237

Various bacteria, yeasts, and molds are individually introduced into the product over 28 days, then the calculated recovery of viable microorganisms is compared with the initial inoculum density (Vu et al., 2014; 7.0 2005). Microorganisms are then compared against established acceptance criteria based on their recovery rate as shown in Table 6.

1. Antimicrobial Effectiveness Test (USP): Pharmaceutical products are primarily covered by this compendia. Product testing is categorized into four types: injections, topical products, oral products based on aqueous bases, and antacids.

Table 6. Overview of the different tests and specifications required for Antimicrobial Effectiveness Testing (AET), including Japan, Europe (including Great Britain) and the United States (Moser and Meyer, 2011; 7.0, 2005; Sutton and Porter, 2002).

	USP	JP	BP
Bacteria	*E. coli, S. aureus, P. aeruginosa*	*E. coli, S. aureus, P. aeruginosa*	*E. coli, S. aureus, P. aeruginosa*
Yeast	*C. albicans*	*C. albicans*	*C. albicans*
Mold	*A. niger*	*A. niger*	*A. niger*
Test duration	28 days	28 days	28 days
Types of Product	1 Injection 2 Topical 3 Oral (non-antacids) 4 Antacids	1A Injections, 1B Topical, 1C Oral 1D Antacids II Non-Aqueous	C-1 Parenteral and Ophthalmic preparations C-2 Oral preparations C-3 Topical preparations C-4 Ear preparations
Growth Media	-Bacteria: soybean-casein digest -Yeast and mold: Sabouraud dextrose agar	-Bacteria: soybean-casein Digest -Yeast and mold: Sabouraud dextrose, glucose-peptone agar and potato dextrose agar	-Bacteria: soybean-casein digest -Yeast and mold: Sabouraud dextrose agar
Harvest fluid	-Bacteria: saline -Yeast: saline -Mold: saline with 0.05% PS80	-Bacteria: saline or 0.1% peptone water -Yeast: saline or 0.1% peptone water -Mold: saline with 0.05% PS80 or 0.1% peptone water	-Bacteria: saline -Yeast: saline -Mold: saline with 0.05% PS80
Incubation period	-Bacteria: 18–24 h -Yeast: 44–52 h -Mold: 6–10 days or until good sporulation	-Bacteria: 18–24 h -Yeast: 40–48 h -Mold: 1 week or until good sporulation	-Bacteria: 18–24 h -Yeast: 48 h -Mold: 1 week or until good sporulation
Standardization level of inoculum	1×10^8 CFU/ml	10^8 CFU/ml	10^8 CFU/ml
Enumeration Time Intervals (Bacteria)	0, 7 d, 14 d & 28 d	0, 7 d, 14 d & 28 d	0, 6 h, 24 h, 7 d, 14 d & 28 d
Enumeration Time Intervals (Mold and Yeast)	0, 7 d, 14 d & 28 d	0, 7 d, 14 d & 28 d	0, 6 h, 24 h, 7 d, 14 d & 28 d

In USP <51> test, *Staphylococcus aureus*, *E. coli*, *Pseudomonas aeruginosa*, *Candida albicans*, and *Aspergillus brasiliensis* (*Aspergillus niger*) are tested using Standard microbiological techniques. Microorganism count and log reduction are calculated at each interval (7, 14, and 28 days). A test product's antimicrobial preservative effectiveness is determined by USP <51> acceptance criteria (Moser and Meyer, 2011).

2. Efficacy of Antimicrobial Preservation (EP or BP): British Pharmacopeia has three categories: Parenteral and Ophthalmic, Topical and Oral Preparations. This method uses only four microorganisms. The intervals evaluated depend on the categories of the product. In this method, *S. aureus*, *P. aeruginosa*, *C. albicans,* and *A. brasiliensis* (*A. niger*). Microorganism count and log reduction are calculated at each interval (6 h, 24 h, 7 days, 14 days, and 28 days).

3. Preservation Effectiveness Test (JP): This method is very similar to USP <51> however its microbial preparation is different than the USP. JP lists the use of five challenge organisms for sterile multi-dose parenteral. These organisms are *S. aureus* (Gram-positive coccus), *P. aeruginosa* (Gram-negative bacillus), and *E. coli* (Gram-negative bacillus), *C. albicans* (yeast), and *A. brasiliensis* (mold). Microorganism count and log reduction are calculated at each interval of 14 days: reduction of 0.1% of inoculum count or less and 28 days: same or less than.

Conclusions and future prospects

The development of microbial contaminations in medicines can pose a threat to the consumers and, therefore, it is necessary to establish protocols to minimize and control them. There are different strategies to reduce and control the presence of microorganisms in medicines. Some of these preservation strategies include the manipulation of the physicochemical characteristics of the medicine formulations, while others include the use of preservatives, which will difficult the establishment and development of contaminant microorganisms in the medicines. However, there are different approaches attending to different regulatory institutions. New strategies would involve the production of smart containers, micro- and nanoencapsulation and use of tailored antimicrobials.

References

7.0, European Pharmacopoeia v. 2005. 5.1.3. Efficacy of antimicrobial preservation: 447–49.

Allen Jr. and Loyd, V. 2018. Quality control: water activity considerations for beyond-use dates. *International Journal of Pharmaceutical Compounding*, 22: 288–93.

Anand, S.P. and N. Sati. 2013. Artificial preservatives and their harmful effects: looking toward nature for safer alternatives. *Int. J. Pharm. Sci. Res*, 4: 2496–501.

Anurova, M.N., Bakhrushina, E.O., Demina, N.B. and Panteleeva, E.S. 2019. Modern preservatives of microbiological stability. *Pharmaceutical Chemistry Journal*, 53: 564–71.

Arjunwadkar, V.P., Bal, A.M., Joshi, S.A., Kagal, A.S. and Bharadwaj, R.S. 2001. Contaminated antiseptics—an unnecessary hospital hazard. *Indian Journal of Medical Sciences*, 55: 393–98.

Attwood, David. 2012. *Surfactant Systems: Their Chemistry, Pharmacy and Biology* (Springer Science & Business Media).

Beveridge, E.G. and Bendell, D. 1988. Water relationships and microbial biodeterioration of some pharmaceutical tablets. *International Biodeterioration*, 24: 197–203.

Boukarim, Chawki, Jaoude, S.A., Rita Bahnam, Roula Barada and Soula Kyriacos. 2009. Preservatives in liquid pharmaceutical preparations. *J. Appl. Res.*, 9: 14–17.

Brown, Larry K., Janet R. Schultz, Rod A. Gragg and Hemophilia Behavioral Intervention Evaluation Project. 1995. HIV-infected adolescents with hemophilia: Adaptation and coping. *Pediatrics*, 96: 459--63.

Chapman, Derek G. 2019. Parenteral products. *Pharmacy Practice E-Book*, 195.

Charnock, C. and Otterholt, E. 2012. Evaluation of preservative efficacy in pharmaceutical products: the use of psychrotolerant, low-nutrient preferring microbes in challenge tests. *Journal of Clinical Pharmacy and Therapeutics*, 37: 558–64.

Chipley, John R. 2020. Sodium benzoate and benzoic acid. pp. 41–88. *In*: Branen, A.L., Sofos, J.N. and Davidson, P.M. (Eds.). *Antimicrobials in Food* (CRC Press).

da Costa, Alexandre Xavier, Maria Cecilia Zorat Yu, Denise de Freitas, Priscila Cardoso Cristovam, Lauren C LaMonica, Vagner Rogerio Dos Santos and José Alvaro Pereira Gomes. 2020. Microbial cross-contamination in multidose eyedrops: the impact of instillation angle and bottle geometry. *Translational Vision Science & Technology*, 9: 7–7.

Dempsey, G. 2017. The effect of container materials and multiple-phase formulation components on the activity of antimicrobial agents. pp. 87–98. *In*: Baird, R. and Sally, F. (Eds.). Bloomfield *Microbial Quality Assurance in Cosmetics, Toiletries and Non-sterile Pharmaceuticals* (CRC Press).

Dodge, Laura, E., Katherine E. Kelley, Paige L. Williams, Michelle A. Williams, Sonia Hernández-Díaz, Stacey A. Missmer and Russ Hauser. 2015. Medications as a source of paraben exposure. *Reproductive Toxicology*, 52: 93–100.

Fahelelbom, Khairi, M.S. and Yasser El-Shabrawy. 2007. Analysis of preservatives in pharmaceutical products. *Pharm. Rev*, 5: 1–55.

Fanselow, Nicholas, Nadia Sirajuddin, Xiao-Tang Yin, Andrew J.W. Huang, and Patrick M. Stuart. 2021. *Acanthamoeba keratitis*, pathology, diagnosis and treatment. *Pathogens*, 10: 323.

Fassihi, A.R., Parker, M.S. and Dingwall, D. 1978. The preservation of tablets against microbial spoilage. *Drug Development and Industrial Pharmacy*, 4: 515–27.

(FDA), Food and Drug Administration. 2020. Pharmaceutical microbiology manual. *Office of Regulatory Science*: 1–92.

Floyd, Alison G. 1999. Top ten considerations in the development of parenteral emulsions. *Pharmaceutical Science & Technology Today*, 2: 134–43.

George, Tom and Mark F. Brady. 2020. Ethylenediaminetetraacetic acid (EDTA).

Gomperts, Edward D. 1996. Gammagard® and reported hepatitis C virus episodes. *Clinical therapeutics*, 18: 3–8.

Himoudy, Iman. 2016. Preservatives and their role in Pharma and Clinical Research Int. J. Pharm. Sci. & Scient. Res. 2: 4, 134–151. Copyright:© 2016 Iman Himoudy. This is an open-access article distributed under the terms of the Creative Commons Attribution License, which permits unrestricted use, distribution, and reproduction in any medium, provided the original author and source are credited. *Received June*, 26.

Hiom, Sarah J. 2004. Preservation of medicines and cosmetics. *Russell, Hugo & Ayliffe's: Principles and Practice of Disinfection, Preservation and Sterilization*: 388–407.

Horowitz, Sala. 2010. Salt cave therapy: rediscovering the benefits of an old preservative. *Alternative and Complementary Therapies*, 16: 158–62.

Hugo, W.B. and Denyer, S.P. 1987. The concentration exponent of disinfectants and preservatives (biocides). *Preservatives in the Food, Pharmaceutical and Environmental Industries*: 281–91.

Hutchinson, Jim, Wendy Runge, Mike Mulvey, Gail Norris, Marion Yetman, Nelly Valkova, Richard Villemur and Francois Lepine. 2004. *Burkholderia cepacia* infections associated with intrinsically contaminated ultrasound gel: the role of microbial degradation of parabens. *Infection Control & Hospital Epidemiology*, 25: 291–96.

Jin, Qusheng and Matthew F. Kirk. 2018. pH as a primary control in environmental microbiology: 1. thermodynamic perspective. *Frontiers in Environmental Science*, 6: 21.

Kabara, Jon J. 1997. Chelating agents as preservative. *Preservative-Free and Self-Preserving Cosmetics and Drugs: Principles and Practices*: 209.

240 *Novel Approaches in Biopreservation for Food and Clinical Purposes*

Kallings, L.O., Ringertz, O. and Silverstolpe, L. 1966. Microbial contamination of medical preparations. 1965. Report to the Swedish National Board of Health. *Acta Pharmaca Suecica*, 3(3): 219–228.

Kumar, Manish, Rohit Bhatia and Ravindra K. Rawal. 2018. Applications of various analytical techniques in quality control of pharmaceutical excipients. *Journal of Pharmaceutical and Biomedical Analysis*, 157: 122–36.

Kumari, P.K., Akhila, S., Srinivasa Rao, Y. and Rama Devi, B. 2019. Alternative to artificial preservatives. *Syst. Rev. Pharm.*, 10: 99–102.

Landman, W.J.M., Veldman, K.T., Mevius, D.J. and Doornenbal, P. 2000. Contamination of Marek's disease vaccine suspensions with *Enterococcus faecalis* and its possible role in amyloid arthropathy. *Avian Pathology*, 29: 21–25.

Lang, Yvonne, Sandra O'Neill and Jenny Lawler. 2022. Considerations for the pharmaceutical industry regarding environmental and human health impacts of microplastics. pp. 61–78. *In*: Yvonne Lang (Ed.). *Influence of Microplastics on Environmental and Human Health* (CRC Press).

Lincho, João, Rui C. Martins, and João Gomes. 2021. Paraben compounds—Part I: an overview of their characteristics, detection, and impacts. *Applied Sciences*, 11: 2307.

Lindsay, Richard G., Grant Watters, Richard Johnson, Susan E. Ormonde and Grant R. Snibson. 2007. *Acanthamoeba keratitis* and contact lens wear. *Clinical and Experimental Optometry*, 90: 351–60.

LINTNER, K. 1997. Physical methods of preservation. *Inside Cosmetics*, 2: 23–29.

Lowe, Robert. 2018. Parenteral drug delivery. *Aulton's Pharmaceutics: The Design and Manufacture of Medicines. 5th ed. Amsterdam: Elsevier*: 638–52.

Meers, P.D., Calder, M.W., Mazhar, M.M. and Gerald M. Lawrie. 1973. Intravenous infusion of contaminated dextrose solution: the Devonport incident. *The Lancet*, 302: 1189–92.

Meyer, Brian K., Alex Ni, Binghua Hu and Li Shi. 2007. Antimicrobial preservative use in parenteral products: past and present. *Journal of Pharmaceutical Sciences*, 96: 3155–67.

Millership, S.E., Patel, N. and Chattopadhyay, B. 1986. The colonization of patients in an intensive treatment unit with gram-negative flora: the significance of the oral route. *Journal of Hospital Infection*, 7: 226—35.

Morris, C. and Leech, R. 2017. Natural and physical preservative systems. *In*: *Microbial Quality Assurance in Cosmetics, Toiletries and Non-sterile Pharmaceuticals* (CRC Press).

Moser, Cheryl L. and Brian K. Meyer. 2011. Comparison of compendial antimicrobial effectiveness tests: a review. *Aaps Pharmscitech*, 12: 222–26.

Niazi, Sarfaraz K. 2019a. *Handbook of Pharmaceutical Manufacturing Formulations: Volume One, Compressed Solid Products* (CRC press).

Niazi, Sarfaraz K. 2019b. *Handbook of Pharmaceutical Manufacturing Formulations: Volume Two, Uncompressed Solid Products* (CRC press).

O'Malley, Homero Penagos G. Michael, and Howard I. Maibach. 1997. American contact dermatitis society annual meeting abstract Allergic contact der. *American Journal of Contact Dermatitis*, 1: 62.

Oni, M.O., Adeyemo, I.A. and Agbolade, J.O. 2013. Potency of Preservatives in Selected Drug Mixtures in Ibadan, Oyo State, Nigeria.

Pawar, H.A., Shenoy, A.V., Narawade, P.D., Soni, P.Y., Shanbhag, P.P. and Rajal, V.A. 2011. Preservatives from nature: a review. *Int. J. Pharm. Phytopharmacol. Res.*, 1: 78–88.

Permanand, Govin. 2006. *EU pharmaceutical regulation: the politics of policy-making* (Manchester University Press).

Russell, A.D. and McDonnell, G. 2000. Concentration: a major factor in studying biocidal action. *The Journal of Hospital Infection*, 44: 1–3.

Sahnoune, Meriem, Nicolas Tokhadzé, Julien Devémy, Alain Dequidt, Florent Goujon, Philip Chennell, Valérie Sautou and Patrice Malfreyt. 2021. Understanding and characterizing the drug sorption to PVC and PE materials. *ACS Applied Materials & Interfaces*, 13: 18594–603.

Shaikh, Sabir M., Rajendra C. Doijad, Amol S. Shete and Poournima S. Sankpal. 2016. A review on: preservatives used in pharmaceuticals and impacts on health. *PharmaTutor*, 4: 25–34.

SHAQRA, QASEM M. ABU, and MAISA AL-SHAWAGFEH. 2012. Assessment of the harmonized preservatives efficacy test in oral liquid pharmaceutical preparations using reference and none reference test microorganisms. *Asian Journal of Pharmaceutical and Clinical Research*, 5: 141–44.

Shimmura-Tomita, Machiko, Hiroko Takano, Nozomi Kinoshita, Fumihiko Toyoda, Yoshiaki Tanaka, Rina Takagi, Mina Kobayashi and Akihiro Kakehashi. 2018. Risk factors and clinical signs of severe Acanthamoeba keratitis. *Clinical Ophthalmology*: 2567–73.

Sutton, Scott, V.W. and David Porter. 2002. Development of the antimicrobial effectiveness test as USP chapter <51>. *PDA Journal of Pharmaceutical Science and Technology*, 56: 300–11.

Swerdlow, A.J., Higgins, C.D., Adlard, P, Jones, M.E. and Preece, M.A. 2003. Creutzfeldt-Jakob disease in United Kingdom patients treated with human pituitary growth hormone. *Neurology*, 61: 783–91.

Symonds, I.D., Martin, D.L. and Davies, M.C. 2016. Facility-based case study: a comparison of the recovery of naturally occurring species of bacteria and fungi on semi-solid media when incubated under standard and dual temperature conditions and its impact on microbial environmental monitoring approach. *European Journal of Parenteral & Pharmaceutical Sciences*, 21.

Tabatabaei, Seyed Mehdi, Alireza Salimi Khorashad, Sahar Shahraki and Feiz Mohammad Elhami. 2017. A study on bacterial contamination of multidose vaccine vials in Southeast of Iran. *International Journal of Infection*, 4.

Taşli, H. and Coşar, G. 2001. Microbial contamination of eye drops. *Central European Journal of Public Health*, 9: 162–64.

Tebbs, S.E., Ghose, A. and Elliott, T.S.J. 1996. Microbial contamination of intravenous and arterial catheters. *Intensive Care Medicine*, 22: 272–73.

Templeton III, William C., Richard A. Eiferman, James W. Snyder, Julio C. Melo and Martin J. Raff. 1982. *Serratia keratitis* transmitted by contaminated eyedroppers. *American Journal of Ophthalmology*, 93: 723–26.

Teuchner, Barbara, Julia Wagner, Nikolaos E. Bechrakis, Dorothea Orth-Höller and Markus Nagl. 2015. Microbial contamination of glaucoma eyedrops used by patients compared with ocular medications used in the hospital. *Medicine*, 94.

Tiwari, Neha, U.V.S. Teotia and Yogendra Singh. 2018. Optimization of preservative system for liquid dosage form using box behnken design. *World J. Pharm. Res*, 7: 556–68.

Tremewan, H.C. 1946. Tetanus neonatorum in New Zealand. *New Zealand Medical Journal*, 45: 312–13.

Van Doorne, H. 1990. Interactions between preservatives and pharmaceutical components. *Guide to Microbiological Control in Pharmaceuticals. New York: Ellis Horwood.*

Vu, N., Nguyen, K. and Kupiec, T.C. 2014. The essentials of United States Pharmacopeia Chapter <51> antimicrobial effectiveness testing and its application in pharmaceutical compounding. *Int. J. Pharm. Compd.*, 18: 123–30.

Waites, Michael J., Neil L. Morgan, John S. Rockey and Gary Higton. 2009. *Industrial Microbiology: An Introduction* (John Wiley & Sons).

Whiteman, M. 1995. Evaluating the performance of tablet coatings. *Manufacturing Chemist*, 66: 24–5.

CHAPTER 9

Biopreservation of Cells and Tissues

Enriqueta Garcia-Gutierrez

Introduction

In vitro and *in vivo* testing are of paramount importance for many research studies, both for clinical and commercial applications. Therefore, cell and tissue culture has become a key requirement for the assessment of studies in different disciplines and industries, such as virology, immunology, hematology, molecular genetics and pharmacology, to cite a few (Pham, 2018; Uysal et al., 2018). Cell and tissue culture has also been pivotal for the production and commercialization of a wide variety of products, such as vaccines, monoclonal antibodies, cytokines and other recombinant proteins (Acker, 2005; Morgan et al., 2019, Kyriakidis et al., 2021). Therefore, maintenance and preservation of the cell lines becomes of high importance to obtain reliable, standardized and reproducible studies and commercial products.

Cell culture contamination is a very common event and can potentially constitute a major problem, as contamination can easily become chronic and difficult to eliminate, reducing the reliability of experimental results and conclusions. Contamination can be of different origins: physical, chemical or biological. Arguably, the biological one is considered the greatest threat, as it can spread via cross-contamination and produce a series of by-products that can further alter the physiology of the cell line, reducing its homogeneity and, therefore, altering future experimental results (Geraghty et al., 2014, Fusenig et al., 2017, Hirsch and Schildknecht, 2019).

Besides contamination, another issue that has been highlighted in the cell and tissue preservation is related to the inhibition or elimination of the natural repair process, as the cells are separated from their external natural environment, the

---

Department of Agricultural Engineering, Institute of Plant Biotechnology, Polytechnic University of Cartagena, Murcia, Spain.
Email: enriqueta.garcia@upct.es

surrounding cells to whom they stablish communication with. Therefore, these cells can become biologically altered and different from the same cell lines within the human body. Thus, biopreservation of cells and tissue aims to maintain the integrity and functionality of cells and tissues that are held outside their natural environment. Additionally, biopreservation focuses on developing strategies to safely and efficiently increase their storage time (Acker, 2005).

Currently, four main strategies are accepted for biopreservation to ensure the viability, biological stability and functionality of the cells and tissues. In this chapter, we will explore the basic concepts behind these four strategies.

Contamination affecting cell and tissue culture

We define contamination as a component that induces a negative impact on a cell or tissue culture system, ultimately affecting stability and reproducibility. We can differentiate three categories of contaminants, according to their nature: physical, chemical or biological (Coriell, 1984; Ryan, 1994; Lincoln and Gabridge, 1998).

Physical contamination

Contaminants of physical nature are external phenomena involved in environmental conditions, such as temperature, irradiation (i.e., fluorescent or UV light) or vibration. They can induce degradation and metabolism perturbations in the physiology of the cells and tissues, i.e., cryopreservative agent glycerol, when exposed to light, can produce a by-product, acrolein, which is toxic for the cells (Segeritz and Vallier, 2017). To minimize this type of contamination, it is important to appropriately store media and other components, reducing light exposition and extreme temperatures, and far from equipment that can create mechanical vibrations, like centrifuges. Additionally, equipment should be maintained and functioning according to quality control standards established in the workplace to avoid undesirable effects on the cells and tissues and ensure maximum reproducibility (Geraghty et al., 2014).

Chemical contamination

Chemical contaminants can also have a negative impact on the cell and tissue culture. Contaminants of chemical origin are introduced in the system through different channels, i.e., via reagents, water, serum, media and their supplements, raw materials or even the containers. Thus, as the water expands when it freezes, it can create fractures in an inappropriate storage vessel, compromising its purity. On the other hand, purified water can incorporate metal ions and organic and inorganic compounds from tubes and glassware (Ryan, 1994). Animal sera used to prepare the cell culture media can also introduce chemical contaminants in the system. In fact, the majority of the chemical contaminants are introduced via cell culture media, including the reagents, supplements and additives (Yao and Asayama, 2017). Therefore, it is of high importance that media and media-related reagents are properly stored and handled, ensuring a high quality control. Similarly, glassware, plastic ware and tubing might

244 *Novel Approaches in Biopreservation for Food and Clinical Purposes*

have residual amounts of toxic compounds as a result of the manufacturing process. Thus, the implementation of cleaning procedures would result in minimizing the chemical residues that might be deposited in the items (Adams, 1990; Lincoln and Gabridge, 1998).

Biological contamination

The biological contamination in cell and tissue cultures can have their origin in any organism that can get into the culture system, i.e., insects that can get into the cell culture medium or plates (Ryan, 1994). Microorganisms, such as protozoa, molds, yeasts, viruses and bacteria can also contaminate cell and tissue culture systems through the reagents (Segeritz and Vallier, 2017).

Biological contamination of single-cell parasitic protozoa, like amoebas, is not very common, but it can be damaging. Their harmful effects upon the cells mimic a viral contamination, which can lead to wrong safety assessments (Hay and Ikonomi, 2006; Yaeger, 2011). Contamination by viruses is extremely difficult to detect because of their small size. However, they do not usually represent a serious problem for tissue culture if they do not interfere in cell processes. They can be responsible of cell death when cultures self-destruct without apparent reason, but this can be due to mycoplasmas and chemical contamination too, so it is difficult to assess (Barone et al., 2020). Bacterial contaminations, on the other hand, are usually easily recognizable, due to the turbidity in cultures, and protocols for their control are standardized and well-studied (Langdon, 2004).

The main biological threat in cell and tissue cultures are the wall-less bacteria known as mycoplasmas, a genus, belonging to class Mollicutes, known for being the smallest self-replicating bacteria (Razin, 1996). The risk of contamination will depend on the environment, the methodology, the equipment, the supplies and the aseptic technique of the person manipulating the culture. However, despite many precautions, the small size of mycoplasmas (0.3–0.8 pm) allows them to pass through filtration membranes (Lung et al., 2021). A mycoplasma contamination is likely to have its origin in another contaminated cell culture (Lincoln and Gabridge, 1998). The majority of mycoplasmas isolated from contaminated cell and tissue culture have a human origin, typically *Mycoplasma orale*, *Mycoplasma fermentans*, *Mycoplasma salivarum*, *Mycoplasma buccale*, *Mycoplasma faucium*, *Mycoplasma genitalium*, *Mycoplasma hominis* and *Mycoplasma pirum* (Lincoln and Gabridge, 1998).

Mycoplasmas can alter the metabolism of the cells by consuming the nutrients available in the medium and generating by-products that can alter their metabolism, by modifying or suppressing morphology, antigenicity or biochemistry (Phelan, 2007). The degree of damage will depend on the role of the cell culture in the research or production system, the identity of the contaminant and the duration of the contamination (Lincoln and Gabridge, 1998). A comprehensive summary of the effects of mycoplasma on cell culture was made by Lincoln and Gabridge, and is adapted in Table 1.

Biopreservation of Cells and Tissues   245

Table 1. Effects of mycoplasmas on cell culture (from Lincoln and Gabridge, 1998).

Category	Possible consequence
Cytogenetics	• Natural contamination in some amniotic fluid • Increased sister chromatid exchange • Increased random chromosomal breakage and aberrations • Production of extra chromosomes • Generates confusion in interpretation of amniocentesis culture data • Reduced chromosomal number and increased aberrations • Infected sperm show reduced fertilization capability
Virology	• Decreased transcriptase activity • Interference with isolation of some retroviruses • Decreased infectivity of fowl pox and altered plague morphology • Induction of interferon production in human mononuclear cells and mouse spleen cells • Inhibition of adenovirus production • False plaque formation in cytomegalovirus and herpes-zoster stocks • Contamination of certain veterinary vaccines • Decreased thymidine incorporation in adenovirus and SV 40 • Inhibition of interferon induction by Newcastle disease virus (NDV) • Decreased production of adenovirus and herpes simplex virus • Increased yields of poliovirus, vaccinia viruses • Cosedimentation with viruses in sucrose gradients
Metabolism	• Decreased ornithine decarboxylase production • Decreased protein and RNA synthesis • Cytotoxicity from arginine depletion • Altered uridinehracil ratios • Decreased DNA and RNA synthesis • Induction of collagenase in 3T3 cells • Increased production of aryl hydrocarbon hydroxylase by carcinogens • Increased thymidine degradation • Increased proliferation of lymphoid cells • Altered nucleotide levels and energy charge ratio • Decreased purine synthesis • Diminished ATP levels • Reduced levels of dehydrogenases • Decreased oxygen consumption • Cytotoxicity, with reduced plating efficiency multiplication rates, and nuclear labeling • Inhibition of thymidine incorporation
Bioresponse	• Transformation of murine lymphocyte cell lines • Release of tumor necrosis factors • Suppression of cytotoxic responses • Induction of cytotoxicity factors for natural killer cells • Exchange of membrane antigens between host and parasite cells • Mimicry of lymphokine activity • Altered production of immunoglobulin by lymphoblastoid cells • Polyclonal antibody inadvertently raised against mycoplasma instead of HeLa cells • Inhibition of mitosis in certain lymphocyte cultures • Oxygen generation and chemiluminescence polymorphonuclear leukocytes • Cytotoxicity for thymocytes • Mitogenicity for lymphocytes • Activation of cultured murine macrophages
In hybridoma cultures	• Failure to produce viable fusion products • Monoclonals directed to the mycoplasma instead of target antigen • Interference in screening due to microbial thymidine consumption • Loss of hybrids during Hypoxanthine, Aminopterin, Thymidine (HAT) selection process

Preservation strategies

The preservation strategies involve the actions taken to both prevent and remove potential contamination and to maintain the quality and integrity of the cell and tissue culture.

Aseptic cell culture practice

Adhesion to aseptic techniques in the cell culture lab can help to reduce the extent and frequency of contaminations and, therefore, reduce the loss of cells and resources. It has been reported that the main sources of contamination are laboratory staff, the environment, and the culture medium (Segeritz and Vallier, 2017). There are extensive bibliography and handbooks on the topic, but all actions can be classified in a few common areas, like avoiding the entrance of microorganisms in the cell culture via contaminated items, such as equipment, media, vessels, etc or non-sterile aerosols suspended in the air (Coté, 2001). In this context, the use of the biosafety cabinet is crucial to avoid an important proportion of the contaminations. Furthermore, the use of 70% ethanol to spray and wipe both safety cabinet and all items entering its clean space, reduces drastically potential contaminations (Coté, 2001). The laboratory installations also require regular cleaning of floors and surfaces, and water baths can be treated with commercial solutions to avoid the growth of microorganisms.

Laboratory staff must adhere to the use of a clean lab coat, washing hands and the use of disposable gloves sprayed with 70% ethanol. Finally, the use of sterile commercial media and supplementary cell culture products can be reinforced by filter-sterilizing them, when possible, using a 0.22 µM polyethersulfone low-binding filter system and the addition of antibiotics to limit the potential growth of bacteria in media bottles once opened. In summary, an active cleaning strategy will prevent many undesired and costly contaminations.

Antimicrobials

There are different strategies to eliminate mycoplasma from cell cultures, including the use of antisera, exposition to mice macrophages, trypsin and 5-bromouracil (Lincoln and Gabridge, 1998). However, the most common strategy is the use of antibiotics. Since mycoplasmas do not have a cell wall, antibiotics targeting it, like cephalosporins, carbapenams and penicillins, will not be effective against them. The most effective method involves the isolation and determination of the antibiotic susceptibility of the organism causing the contamination (Del Giudice and Gardella, 1996; Lincoln and Gabridge, 1998). Traditionally, cell and tissue cultures are exposed to cocktails of antibiotics, with a minimum of two antibiotics that have shown effectivity against the contaminant. It is accepted that some level of mycoplasma contamination will happen, but the goal is to reduce its incidence (Lincoln and Gabridge, 1998).

Advances in in vitro *culture*

We understand as *in vitro* culture the process of preserving the normal phenotypic properties of a cell population or tissue for extended times at physiological temperatures by replicating *ex vivo* the native environment (Birch, 1989). This recreation of the native environment, of great importance in the physiological regulation of cells and tissues, includes the presence of soluble factors, extracellular matrix proteins and cell interactions (Andrews et al., 1994; Koller and Papoutsakis, 1995; Gumbiner, 1996; Chen et al., 1997; Iyer et al., 1999). Typically, cells preserved for *in vitro* culture can be either primary cultures (homogeneous cells lines perpetuated over generations) or transformed or continuous cells lines (when preserved indefinitely).

There are numerous areas of expansion and improvement of *in vitro* cell culture. One of the main areas of interest is related to a productivity increase in the large-scale cell culture. The strategies for this focus on the addition of nutrients, the elimination of waste, the mixing of the components and the oxygenation of the cultured cells. Thus, we can find continuous culture of cells in fed-batch bioreactors (Pörtner et al., 1996; Chang et al., 2014), fluidized bed reactors (Warnock and Al-Rubeai, 2006; Mendonça da Silva et al., 2020), hollow-fiber perfusion systems (Dowd et al., 2003; Schmelzer and Gerlach, 2016; Storm et al., 2016), microgravity culture systems (Freed et al., 1999; Imura et al., 2019) and development of microcarriers (Chen et al., 2020; Koh et al., 2020).

Another area of improvement is related to the development of the cellular microenviroment by understanding the role of soluble factors and cell and matrix interactions, which will improve cell proliferation and differentiation (Hesse and Wagner, 2000; Sinacore et al., 2000). Cell-cell and cell-matrix interactions are critical elements in the modulation of the cell physiology. New strategies, such as microfabrication technologies to construct extracellular matrices have been used for *in vitro* preservation of neurons and hepatocytes (Bhatia et al., 1999; Sorribas et al., 2002; Sohn et al., 2020).

Additionally, there are a number of natural and engineered tissues that have been recently developed and whose correct preservation depends on a combination of techniques. Thus, we find artificial skin, using human keratinocytes, fibroblasts, melanocytes and Langerhans cells which, combined, can produce different artificial skin constructs and multi-layered epidermis (Noordenbos et al., 1999; Werner and Smola, 2001; Chester and Papini, 2004; Acker, 2005). Hematopoietic cells are an example of cells requiring a complex biopreservation. Hematopoietic cells can potentially proliferate and differentiate into all blood cell types. However, their availability is quite low, so there is an interest in developing *in vitro* culture methods. The understanding of the different factors that regulate this differentiation and how to preserve each cell lineage is key to develop a good methodology that maintains the phenotypic properties of each cell type (Nielsen, 1999; McNiece and Briddell, 2001; Hawley et al., 2006).

Despite the wide use of *in vitro* culture for long-term preservation, this strategy is not ideal, due to the high cost of the components and the requirements to maintain

248 *Novel Approaches in Biopreservation for Food and Clinical Purposes*

cell proliferation and differentiation. Additionally, cells are prone to phenotypic and genetic drift, compromising reproducibility. They are difficult to storage in large volumes and they are very sensitive to contamination problems (Lincoln and Gabridge, 1998).

Hypothermic storage

Hypothermic preservation consists on suppressing biochemical and molecular reactions by reducing the temperature below a normal physiological range, but higher than the freezing point of the storage solution. This approach has allowed to extend the storage of red blood cells, platelets, hepatocytes, pancreatic islets, corneas, native and engineered skin, even organs, increasing transfusion and transplants success rate (Trent and Kirsner, 1998; Högman, 1999; Gulliksson, 2000; St Peter et al., 2002; Wigg et al., 2003; Acker, 2005). This versatility, combined with its inexpensive requirements, makes hypothermic storage an attractive technique for industry and research. However, hypothermic preservation can also present limitations. Cell damage can occur if the appropriate steps and protocols are not conducted correctly, producing a phenomenon known as hypothermia-induced injury. Additionally, although slowed down, cellular metabolism is not fully suppressed and eventually biological products can accumulate, ultimately inducing cell damage and death (Freitas-Ribeiro et al., 2019; Tam et al., 2020).

Cryopreservation

Cryopreservation, as hypothermic storage, aims to suppress biochemical and molecular reactions and preserve the biological structure and/or function of living systems, using freezing and storage at ultralow temperatures (as low as $-150°C$–$-196°C$) in combination with cryoprotectants (typically glycerol or dimethyl sulfoxide) in contrast to the non-freezing temperatures of hypothermic storage (Alm et al., 2014). This requires minimizing the cell injury and damage over freezing and thawing processes, known as "cryoinjury" (Fowler and Toner, 2006). Cryoinjuries are associated to a number of physical and chemical events related to the velocity of the freezing, the movement of water across the membrane due to osmotic forces and the intracytoplasmic and extracellular solute concentrations that can produce toxicity (Lovelock, 1957; Levitt, 1962; Mazur et al., 1972; Fowler and Toner, 2006). Cell responses to low temperatures and cryoinjuries can be mathematically modelled and, therefore, predicted for each type of cell (Muldrew and McGann, 1994; Gao and Critser, 2000; Karlsson, 2001; Ross-Rodriguez et al., 2010). This is particularly important, as the loss of cell viability and the damage to the extracellular matrix can impair tissue function (Taylor, 2006).

Despite that cryopreservation is used extensively for long-term storage, there are areas that require further improvement. Logistically, cryopreservation can be difficult and costly and requires training and special equipment. Additionally, post-thaw removal processes are costly, as some chemical cryoprotectants used to minimize the freezing-thawing effects, can affect patients who receive a transplant, i.e., dimethyl sulfoxide can produce gastrointestinal and cardiovascular side effects (De La Torre, 1983; Shu et al., 2014). Additionally, not all cell types are fully operational after being cryopreserved, like platelets, hepatocytes, granulocytes, spermatozoa and oocytes (Acker, 2005; Meneghel et al., 2020). Protocols must consider not only the preservation of the cellular function, but also the importance of the structure and composition for the tissue function. Thus, different methods would be considered for cell and tissue biopreservation (Fuller et al., 2004; Costa et al., 2012).

Vitrification is introduced to overcome some of these limitations. Vitrification is the process by which the aqueous solution bypasses ice formation and becomes and amorphous and glassy solid (Fahy and Wowk, 2021). Thus, if the ice formation can be avoided, it would be possible to eliminate its detrimental effects. The vitrification process requires high concentration of cryoprotectants and/or ultra-rapid cooling rates. There are some alternatives to reduce the use of high concentrations of potentially problematic cryoprotectants, such as the use of high hydrostatic pressure, synthetic ice blocking agents and the use of natural antifreeze proteins (Dou et al., 2021; Wu et al., 2021; Gore et al., 2022).

Desiccation and dry storage

Desiccation is an extended strategy in natural systems, where different organisms, like plants, bacteria or fungi are able to survive in a dry state by a series of physiological adaptations (Leopold et al., 1992). Organisms use a series of processes that can be adapted for cell and tissue preservation, like mechanisms for protection from reactive oxygen species (ROS), downregulation of metabolism and accumulation of proteins, disaccharides and amphiphilic solutes (Acker, 2005).

ROS can damage phospholipids, DNA and proteins, ultimately impairing metabolism and physiology. It has been observed that the synthesis of antioxidant molecules can balance the damaging effects of reactive oxygen (Hendry et al., 1996; Kranner et al., 2002). Since the formation of ROS is dependent on the oxygen concentration, another option is to reduce it by synthesizing oxygen-binding proteins (Potts et al., 2005).

One of the best-characterized adaptations to protect biological structures during dehydration processes is the synthesis of disaccharides, due to their role in stabilization of membranes and other cellular structures (Leopold, 1986; Alpert, 2006; Tapia et al., 2015). This is the basis behind the addition of sugars into preservation freeze-drying medium for dehydration and storage of a variety of biological entities, such as pharmaceutical agents, bacteria, yeast, viruses or liposomes (Pehkonen et al., 2008; Rockinger et al., 2021; Wolkers and Oldenhof, 2021). Additionally, the intracellular presence of sugars, like trehalose, has enhanced recovery of different cell lines, such as fibroblasts, embryonic kidney cells and mesenchymal stem cells (Guo et al., 2000; Oliver et al., 2004; Sánchez-González et al., 2012).

Downregulation of the metabolism is also used as a protective mechanism in the form of quiescence and diapause, methods used by organisms to enter a dormant state that increases tolerance to desiccation, freezing and anoxia (MacRae, 2010; MacRae, 2016; Lennon et al., 2021). Identifying the molecular mechanisms that allow to entry and exit the metabolic downregulation and the development of dormant states would

250 *Novel Approaches in Biopreservation for Food and Clinical Purposes*

be interesting to induce mammalian cells into this dehydrated state and increase preservation possibilities.

Regulatory framework

There is a lack of uniform regulation across the regulatory agencies worldwide regarding the use human cells, tissues, and cellular and tissue-based products (HCT/Ps) (Anon, 2020). In the case of USA, the topic is under the evaluation of the U.S. Food & Drug Administration (FDA). The FDA published in 2017 a comprehensive regenerative medicine policy framework to incentivate the innovation and access to safe and functional regenerative medicine products. In September of 2020, this framework was revisited to address certain aspects of 21 CFR 1271 and help manufacturers to understand how their product can be classified (Fang and Vangsness Jr., 2020).

Conclusions

Preservation of cell viability and function is an essential component of cell and tissue culture that can be affected by physical, chemical and biological contaminants, but also by the cell and tissue maintenance strategy. Preserving the cell and tissue culture standards requires different strategies that have been developed to ensure the viability and function after storage. Depending on the intended application, either clinical or industrial, each technique offers advantages and new challenges for improvement. There is extensive research in the area to understand the biological mechanisms that can bring cells into metabolic dormancy in order to improve their recovery and maintain their functionality.

Acknowledgements

This work was supported by a Beatriz Galindo scholarship from the Spanish Ministry of Universities (BG22/00060).

References

Acker, J.P. 2005. Biopreservation of cells and engineered tissues. *Tissue Engineering II*, 157–187.
Adams, R.L.P. 1990. *Cell Culture for Biochemists*: Elsevier.
Alm, J.J., Qian, H. and Le Blanc, K. 2014. Clinical grade production of mesenchymal stromal cells. *Tissue Engineering.* Elsevier, 427–469.
Alpert, P. 2006. Constraints of tolerance: why are desiccation-tolerant organisms so small or rare? *Journal of Experimental Biology*, 209: 1575–1584.
Andrews, R.G., Briddell, R.A., Appelbaum, F.R. and Mcniece, I.K. 1994. Stimulation of hematopoiesis *in vivo* by stem cell factor. *Curr. Opin. Hematol.*, 1: 187–96.
Anon, 2020. A worldwide overview of regulatory frameworks for tissue-based products. *Tissue Engineering Part B: Reviews*, 26: 181–196.
Barone, P.W., Wiebe, M.E., Leung, J.C., Hussein, I., Keumurian, F.J., Bouressa, J., Brussel, A., Chen, D., Chong, M. and Dehghani, H. 2020. Viral contamination in biologic manufacture and implications for emerging therapies. *Nature Biotechnology*, 38: 563–572.

Bhatia, S.N., Balis, U.J., Yarmush, M.L. and Toner, M. 1999. Effect of cell-cell interactions in preservation of cellular phenotype: cocultivation of hepatocytes and nonparenchymal cells. *Faseb. j.*, 13: 1883–900.

Birch, J.R. 1989. Culture of animal cells: a manual of basic technique: By R. I. Freshney, Alan R. Liss, Inc., New York, 1987. pp. xviii + 397, price $59.50. ISBN 0-8451-4241-0. *Journal of Chemical Technology & Biotechnology*, 45: 330–330.

Chang, H.N., Jung, K., Lee, J.C. and Woo, H.-C. 2014. Multi-stage continuous high cell density culture systems: a review. *Biotechnology Advances*, 32: 514–525.

Chen, C.S., Mrksich, M., Huang, S., Whitesides, G.M. and Ingber, D.E. 1997. Geometric control of cell life and death. *Science*, 276: 1425–8.

Chen, X.-Y., Chen, J.-Y., Tong, X.-M., Mei, J.-G., Chen, Y.-F. and Mou, X.-Z. 2020. Recent advances in the use of microcarriers for cell cultures and their *ex vivo* and *in vivo* applications. *Biotechnology Letters*, 42: 1–10.

Chester, D. and Papini, R. 2004. Skin and skin substitutes in burn management. *Trauma*, 6: 87–99.

Coriell, L.L. 1984. Establishing and characterizing cells in culture. *Eukaryotic Cell Cultures*, 1–11.

Costa, P.F., Dias, A.F., Reis, R.L. and Gomes, M.E. 2012. Cryopreservation of cell/scaffold tissue-engineered constructs. *Tissue Engineering Part C: Methods*, 18: 852–858.

Coté, R.J. 2001. Aseptic technique for cell culture. *Curr. Protoc. Cell Biol.*, Chapter 1, Unit 1.3.

De La Torre, J.C. 1983. *Biological Actions and Medical Applications of Dimethyl Sulfoxide*: New York Academy of Sciences.

Del Giudice, R.A. and Gardella, R.S. 1996. Antibiotic treatment of mycoplasmainfected cell cultures. *Molecular and Dagnostic Procedures in Mycoplasmology. Academic Press, San Diego*, 439-443.

Dou, M., Lu, C. an Rao, W. 2021. Bioinspired materials and technology for advanced cryopreservation. *Trends in Biotechnology*.

Dowd, J.E., Jubb, A., Kwok, K.E. and Piret, J.M. 2003. Optimization and control of perfusion cultures using a viable cell probe and cell specific perfusion rates. *Cytotechnology*, 42: 35–45.

Fahy, G.M. and Wowk, B. 2021. Principles of ice-free cryopreservation by vitrification. *Cryopreservation and Freeze-Drying Protocols.* Springer, 27–97.

Fang, W.H. and Vangsness Jr, C.T. 2020. Governmental regulations and increasing food and drug administration oversight of regenerative medicine products: what's new in 2020? *Arthroscopy: The Journal of Arthroscopic & Related Surgery*, 36: 2765–2770.

Fowler, A. and Toner, M. 2006. Cryo-injury and biopreservation. *Annals of the New York Academy of Sciences*, 1066: 119–135.

Freed, L.E., Pellis, N., Searby, N., De Luis, J., Preda, C., Bordonaro, J. and Vunjak-Novakovic, G., 1999. Microgravity cultivation of cells and tissues. *Gravit. Space Biol. Bull.*, 12: 57–66.

Freitas-Ribeiro, S., Carvalho, A.F., Costa, M., Cerqueira, M.T., Marques, A.P., Reis, R.L. and Pirraco, R.P. 2019. Strategies for the hypothermic preservation of cell sheets of human adipose stem cells. *PLoS One*, 14: e0222597.

Fuller, B.J., Lane, N. and Benson, E.E. 2004. *Life in the Frozen State*: CRC press.

Fusenig, N.E., Capes-Davis, A., Bianchini, F., Sundell, S. and Lichter, P. 2017. The need for a worldwide consensus for cell line authentication: experience implementing a mandatory requirement at the International Journal of Cancer. *PLoS Biol.*, 15: e2001438.

Gao, D. an Critser, J.K. 2000. Mechanisms of cryoinjury in living cells. *ILAR Journal*, 41: 187–196.

Geraghty, R.J., Capes-Davis, A., Davis, J.M., Downward, J., Freshney, R.I., Knezevic, I., Lovell-Badge, R., Masters, J.R., Meredith, J., Stacey, G.N., Thraves, P. and Vias, M. 2014. Guidelines for the use of cell lines in biomedical research. *Br. J. Cancer*, 111: 1021–46.

Gore, M., Narvekar, A., Bhagwat, A., Jain, R. and Dandekar, P. 2022. Macromolecular cryoprotectants for the preservation of mammalian cell culture: lessons from crowding, overview and perspectives. *Journal of Materials Chemistry B*.

Gulliksson, H. 2000. Additive solutions for the storage of platelets for transfusion. *Transfus Med.*, 10: 257–64.

Gumbiner, B.M. 1996. Cell adhesion: the molecular basis of tissue architecture and morphogenesis. *Cell*, 84: 345–57.

Guo, N., Puhlev, I., Brown, D.R., Mansbridge, J. and Levine, F. 2000. Trehalose expression confers desiccation tolerance on human cells. *Nature Biotechnology*, 18: 168–171.

252 *Novel Approaches in Biopreservation for Food and Clinical Purposes*

Hawley, R.G., Ramezani, A. and Hawley, T.S. 2006. Hematopoietic stem cells. *Methods Enzymol.*, 419: 149–79.

Hay, R.J. and Ikonomi, P. 2006. Detection of microbial and viral contaminants in cell lines. *Cell Biology.* Elsevier, 49–65.

Hendry, G., Khan, M., Greggains, V. and Leprince, O. 1996. Free radical formation in non-photosynthetic plant tissues—an overview. *Biochemical Society Transactions*, 24: 484–488.

Hesse, F. and Wagner, R. 2000. Developments and improvements in the manufacturing of human therapeutics with mammalian cell cultures. *Trends Biotechnol.*, 18: 173–80.

Hirsch, C. and Schildknecht, S. 2019. *In vitro* research reproducibility: keeping up high standards. *Frontiers in Pharmacology*, 10.

Högman, C.F. 1999. Liquid-stored red blood cells for transfusion. A status report. *Vox Sang*, 76: 67–77.

Imura, T., Otsuka, T., Kawahara, Y. and Yuge, L. 2019. "Microgravity" as a unique and useful stem cell culture environment for cell-based therapy. *Regenerative Therapy*, 12: 2–5.

Iyer, V.R., Eisen, M.B., Ross, D.T., Schuler, G., Moore, T., Lee, J.C., Trent, J.M., Staudt, L.M., Hudson, J., Jr., Boguski, M.S., Lashkari, D., Shalon, D., Botstein, D. and Brown, P.O. 1999. The transcriptional program in the response of human fibroblasts to serum. *Science*, 283: 83–7.

Karlsson, J.O. 2001. A theoretical model of intracellular devitrification. *Cryobiology*, 42: 154–69.

Koh, B., Sulaiman, N., Fauzi, M.B., Law, J.X., Ng, M.H., Idrus, R.B.H. and Yazid, M.D. 2020. Three dimensional microcarrier system in mesenchymal stem cell culture: a systematic review. *Cell & Bioscience*, 10: 1–16.

Koller, M.R. and Papoutsakis, E.T. 1995. Cell adhesion in animal cell culture: physiological and fluid-mechanical implications. *Bioprocess Technol.*, 20: 61–110.

Kranner, I., Beckett, R.P., Wornik, S., Zorn, M. and Pfeifhofer, H.W. 2002. Revival of a resurrection plant correlates with its antioxidant status. *The Plant Journal*, 31: 13–24.

Kyriakidis, N.C., López-Cortés, A., González, E.V., Grimaldos, A.B. and Prado, E.O. 2021. SARS-CoV-2 vaccines strategies: a comprehensive review of phase 3 candidates. *NPJ Vaccines*, 6: 28.

Langdon, S.P. 2004. Cell culture contamination. *Cancer Cell Culture*, 309–317.

Lennon, J.T., Den Hollander, F., Wilke-Berenguer, M. and Blath, J. 2021. Principles of seed banks and the emergence of complexity from dormancy. *Nature Communications*, 12: 1–16.

Leopold, A., Bruni, F. and Williams, R. 1992. Water in dry organisms. *Water and life.* Springer, 161–169.

Leopold, A.C. 1986. *Membranes, Metabolism, and Dry Organisms*.

Levitt, J. 1962. A sulfhydryl-disulfide hypothesis of frost injury and resistance in plants. *Journal of Theoretical Biology*, 3: 355–391.

Lincoln, C.K. and Gabridge, M.G. 1998. Cell culture contamination: sources, consequences, prevention, and elimination. *Methods Cell Biol.*, 57: 49–65.

Lovelock, J.E. 1957. The denaturation of lipid-protein complexes as a cause of damage by freezing. *Proceedings of the Royal Society of London. Series B-Biological Sciences*, 147: 427–433.

Lung, O., Candlish, R., Nebroski, M., Kruckiewicz, P., Buchanan, C. and Moniwa, M. 2021. High-throughput sequencing for species authentication and contamination detection of 63 cell lines. *Scientific Reports*, 11: 1–9.

Macrae, T.H. 2010. Gene expression, metabolic regulation and stress tolerance during diapause. *Cellular and Molecular Life Sciences*, 67: 2405–2424.

Macrae, T.H. 2016. Stress tolerance during diapause and quiescence of the brine shrimp, Artemia. *Cell Stress and Chaperones*, 21: 9–18.

Mazur, P., Leibo, S. and Chu, E. 1972. A two-factor hypothesis of freezing injury: evidence from Chinese hamster tissue-culture cells. *Experimental Cell Research*, 71: 345–355.

Mcniece, I. and Briddell, R. 2001. *Ex vivo* expansion of hematopoietic progenitor cells and mature cells. *Experimental Hematology*, 29: 3–11.

Mendonça Da Silva, J., Erro, E., Awan, M., Chalmers, S.-A., Fuller, B. and Selden, C. 2020. Small-scale fluidized bed bioreactor for long-term dynamic culture of 3D cell constructs and *in vitro* testing. *Frontiers in Bioengineering and Biotechnology*, 8.

Meneghel, J., Kilbride, P. and Morris, G.J. 2020. Cryopreservation as a key element in the successful delivery of cell-based therapies—a review. *Frontiers in Medicine*, 7.

Morgan, H., Tseng, S.-Y., Gallais, Y., Leineweber, M., Buchmann, P., Riccardi, S., Nabhan, M., Lo, J., Gani, Z., Szely, N., Zhu, C.S., Yang, M., Kiessling, A., Vohr, H.-W., Pallardy, M., Aswad, F. and

Turbica, I. 2019. Evaluation of *in vitro* assays to assess the modulation of dendritic cells functions by therapeutic antibodies and aggregates. *Frontiers in Immunology*, 10.

Muldrew, K. and Mcgann, L.E. 1994. The osmotic rupture hypothesis of intracellular freezing injury. *Biophys. J.*, 66: 532–41.

Nielsen, L.K. 1999. Bioreactors for hematopoietic cell culture. *Annu. Rev. Biomed. Eng.*, 1: 129–52.

Noordenbos, J., Doré, C. and Hansbrough, J.F. 1999. Safety and efficacy of TransCyte for the treatment of partial-thickness burns. *J. Burn Care Rehabil.*, 20: 275–81.

Oliver, A.E., Jamil, K., Crowe, J.H. and Tablin, F. 2004. Loading human mesenchymal stem cells with trehalose by fluid-phase endocytosis. *Cell Preservation Technology*, 2: 35–49.

Pehkonen, K., Roos, Y., Miao, S., Ross, R. and Stanton, C. 2008. State transitions and physicochemical aspects of cryoprotection and stabilization in freeze-drying of *Lactobacillus rhamnosus* GG (LGG). *Journal of Applied Microbiology*, 104: 1732–1743.

Pham, P.V. 2018. Medical biotechnology: techniques and applications. *Omics Technologies and Bio-Engineering*. Elsevier, 449–469.

Phelan, M.C. 2007. Techniques for mammalian cell tissue culture. *Current Protocols in Neuroscience*, 38, A. 3B. 1-A. 3B. 19.

Pörtner, R., Schilling, A., Lüdemann, I. and Märkl, H. 1996. High density fed-batch cultures for hybridoma cells performed with the aid of a kinetic model. *Bioprocess Engineering*, 15: 117–124.

Potts, M., Slaughter, S.M., Hunneke, F.-U., Garst, J.F. and Helm, R.F. 2005. Desiccation tolerance of prokaryotes: application of principles to human cells. *Integrative and Comparative Biology*, 45: 800–809.

Razin, S. 1996. Mycoplasmas. *Medical Microbiology. 4th edition*.

Rockinger, U., Funk, M. and Winter, G. 2021. Current approaches of preservation of cells during (freeze-) drying. *Journal of Pharmaceutical Sciences*.

Ross-Rodriguez, L.U., Elliott, J.A. and Mcgann, L.E. 2010. Investigating cryoinjury using simulations and experiments. 1: TF-1 cells during two-step freezing (rapid cooling interrupted with a hold time). *Cryobiology*, 61: 38–45.

Ryan, J.A. 1994. *Understanding and managing cell culture contamination*: Corning Incorporated.

Sánchez-González, D.J., Méndez-Bolaina, E. and Trejo-Bahena, N.I. 2012. Platelet-rich plasma peptides: key for regeneration. *International Journal of Peptides*, 2012.

Schmelzer, E. and Gerlach, J.C. 2016. Multicompartmental hollow-fiber-based bioreactors for dynamic three-dimensional perfusion culture. *Methods Mol. Biol.*, 1502: 1–19.

Segeritz, C.-P. and Vallier, L. 2017. Cell culture: growing cells as model systems *in vitro*. *Basic Science Methods for Clinical Researchers*. Elsevier, 151–172.

Shu, Z., Heimfeld, S. and Gao, D. 2014. Hematopoietic SCT with cryopreserved grafts: adverse reactions after transplantation and cryoprotectant removal before infusion. *Bone Marrow Transplant*, 49: 469–76.

Sinacore, M.S., Drapeau, D. and Adamson, S.R. 2000. Adaptation of mammalian cells to growth in serum-free media. *Mol. Biotechnol.*, 15: 249–57.

Sohn, L.L., Schwille, P., Hierlemann, A., Tay, S., Samitier, J., Fu, J. and Loskill, P. 2020. How can microfluidic and microfabrication approaches make experiments more physiologically relevant? *Cell Syst*, 11: 209–211.

Sorribas, H., Padeste, C. and Tiefenauer, L. 2002. Photolithographic generation of protein micropatterns for neuron culture applications. *Biomaterials*, 23: 893–900.

St Peter, S.D., Imber, C.J. and Friend, P.J. 2002. Liver and kidney preservation by perfusion. *Lancet*, 359: 604–13.

Storm, M.P., Sorrell, I., Shipley, R., Regan, S., Luetchford, K.A., Sathish, J., Webb, S. and Ellis, M.J., 2016. Hollow Fiber Bioreactors for *In vivo*-like Mammalian Tissue Culture. *J. Vis. Exp.*, (111): 53431.

Tam, E., Mcgrath, M., Sladkova, M., Almanaie, A., Alostaad, A. and De Peppo, G.M. 2020. Hypothermic and cryogenic preservation of tissue-engineered human bone. *Ann. N Y Acad. Sci.*, 1460: 77–87.

Tapia, H., Young, L., Fox, D., Bertozzi, C.R. and Koshland, D. 2015. Increasing intracellular trehalose is sufficient to confer desiccation tolerance to *Saccharomyces cerevisiae*. *Proceedings of the National Academy of Sciences*, 112: 6122–6127.

Taylor, M.J. 2006. Biology of cell survival in the cold: the basis for biopreservation of tissues and organs. *CRC-Taylor and Francis*, 15–62.

Trent, J.F. and Kirsner, R.S. 1998. Tissue engineered skin: apligraf, a bi-layered living skin equivalent. *Int. J. Clin. Pract.*, 52: 408–13.

Uysal, O., Sevimli, T., Sevimli, M., Gunes, S. and Sariboyaci, A.E. 2018. Cell and tissue culture: the base of biotechnology. *Omics Technologies and Bio-Engineering.* Elsevier, 391–429.

Warnock, J.N. and Al-Rubeai, M. 2006. Bioreactor systems for the production of biopharmaceuticals from animal cells. *Biotechnology and Applied Biochemistry*, 45: 1–12.

Werner, S. and Smola, H. 2001. Paracrine regulation of keratinocyte proliferation and differentiation. *Trends Cell Biol.*, 11: 143–6.

Wigg, A.J., Phillips, J.W. and Berry, M.N. 2003. Maintenance of integrity and function of isolated hepatocytes during extended suspension culture at 25°C. *Liver International*, 23: 201–211.

Wolkers, W.F. and Oldenhof, H. 2021. Principles underlying cryopreservation and freeze-drying of cells and tissues. *Cryopreservation and Freeze-Drying Protocols.* Springer, 3–25.

Wu, X., Yao, F., Zhang, H. and Li, J. 2021. Antifreeze proteins and their biomimetics for cell cryopreservation: Mechanism, function and application-A review. *International Journal of Biological Macromolecules*, 192: 1276–1291.

Yaeger, R.G. 2011. Protozoa: Structure, Classification, Growth, and Development.

Yao, T. and Asayama, Y. 2017. Animal-cell culture media: history, characteristics, and current issues. *Reprod. Med. Biol.*, 16: 99–117.

CHAPTER **10**

Biopreservation of Gametes and Embryos against Microbiological Risk

José Luis Girela López[1],* *and Enriqueta Garcia-Gutierrez*[2]

Introduction

Cryopreservation, or the process of freezing cells or tissues at very low temperatures, has been a valuable tool in the field of reproductive medicine for several decades. This technique allows for the preservation of human gametes, such as oocytes and sperm, as well as embryos, maintaining their biological properties for long periods. Cryopreservation of Gametes and Embryos is currently a central part of the Assisted Reproduction Techniques (ART) laboratory (Bosch et al., 2020). Significant advances like gamete donation, deferred embryo transfer and preimplantation genetic testing (PGT) are available thanks to the ability to cryopreserve sperm, oocytes, and embryos.

We must look back to the XVIIIth century to find the origins of these techniques. In 1776, Lazaro Spallanzani, while performing experiments to study the role of spermatozoa on the fertilisation process, observed that while cooling them using snow, they seemed to lose motility regained after warming them again (Capanna, 1999). Out of this anecdotic observation, the origin of cryopreservation of sperm cells

---

[1] Department of Biotechnology, University of Alicante, 03690 San Vicente del Raspeig, Spain.
[2] Department of Agricultural Engineering, Institute of Plant Biotechnology, Polytechnic University of Cartagena, Murcia, Spain.
* Corresponding author: girela@ua.es

Novel Approaches in Biopreservation for Food and Clinical Purposes

can be established with the identification of cryopreservation properties of glycerol by Christopher Polge, Audrey Smith, and Alan Parkes in 1949 (Polge et al., 1949). A few years later, Bunge and Sherman reported the first pregnancy and live birth using cryopreserved spermatozoa (Bunge and Sherman, 1953).

Cryopreservation of embryos precedes oocyte freezing. While the firsts animal embryos that survived the cryopreservation process were reported in the early '70s of the past century, the first human embryo cryopreserved was achieved in 1983 (Trounson and Mohr, 1983), with the first live birth reported one year later (Zeilmaker et al., 1984). All these achievements were obtained using the slow-freezing methodology, but in 1995 vitrification of human embryos was possible changing the paradigm and improving the embryo cryopreservation process (Kasai, 1997).

Cryopreservation of human oocytes was reported for the first time in 1988 (Chen, 1988), and the first human birth using a cryopreserved oocyte was achieved nearly ten years later (Porcu et al., 1997), due to the special nature of these cells, which are detained in the middle of the meiotic division. In 1999 was informed the first pregnancy with vitrified human oocytes (Cha, 1999), becoming the preferred technique used in this type of cells (Iussig et al., 2019).

Nowadays, cryopreservation techniques in ART are widely used, representing a part of nearly 80% of all treatments in the US (Services) and a similar trend in Europe (De Geyter et al., 2020). Vitrification of oocytes and embryos has been determinant of this increased use of cryopreservation in ART, due to the improvement in the results they provide (Rienzi et al., 2017). Nevertheless, there are concerns about the long-term health issues that can affect those born after cryopreservation, so further studies need to be performed on that topic (Sargisian et al., 2022).

Technical evolution in the cryopreservation of gametes and embryos

Two methods have been mainly used in the cryopreservation of gametes and embryos. These are the slow cooling method and the vitrification process. As was established in the introduction, the first procedure developed was the slow cooling one due to the technical limitations and the advancement of the knowledge on the effects of cryopreservation in biological structures.

a. Slow cooling

Slow cooling consists of the gradual dehydration of cells in a media with a relatively low concentration of cryoprotectants (CPA) to avoid ice formation. In most protocols, penetrant cryoprotectants (P-CPA) (i.e., glycerol, ethylene glycol, dimethyl sulfoxide (DMSO)) are used around 1.5 M concentration. Media can also consist of non-permeable cryoprotectants (NP-CPA) (i.e., sucrose, trehalose, PVP) at a concentration around 0.25–0.5 M. The addition of the freezing medium can be done in one or several steps, depending on the sensitivity of the cells to osmotic variations (Gook and Edgar, 2007).

In general, slow cooling protocols comprise the following steps:

(1) Cooling phase (20°C to –7°C): the freezing medium is still above its specific freezing point. Gradual dehydration of the cell occurs due to the presence of CPAs at low concentrations. The cooling rate at this stage is around 2°C/minute (Jang et al., 2017).

(2) Supercooling phase (–7°C): the freezing medium begins the change of state below its freezing point, which usually occurs in a controlled manner through the procedure known as seeding. It involves inducing crystallization at –6°C or –7°C, which can be performed easily by contacting the outside of the freezing device with metal tweezers previously submerged in liquid nitrogen. The objective of this step is to avoid the risks of excessive supercooling: the change of state of the freezing medium from liquid to solid is associated with the release of the latent heat of fusion, increasing the temperature and thus modifying the osmotic balance between the intra- and extracellular compartments. If this increase in temperature arrives close to that of melting, the phenomenon of recrystallization occurs, modifying the size of the ice crystals, which could affect the membranes of the gametes/embryos (Gao and Critser, 2000).

(3) Equilibrium phase after inducing the change of state (–7°C): After seeding, the formation of ice crystals in the solution produces the concentration of the liquid phase. Thus, as the temperature is reduced and the formation of new ice increases, the residual non-frozen phase becomes more and more concentrated, creating a hypertonic environment, which causes the cells to dehydrate by osmosis (Leibo and Pool, 2011). In cryomicroscopic studies on the freezing of oocytes and embryos at slow speeds (–1°C/min), It is observed that once the freezing of the extracellular medium begins, a flow of water from the cell is produced to re-establish the osmotic balance. The consequent progressive volume loss seems to reach its maximum at –40°C, where the absence of intracellular ice has been demonstrated.

(4) Freezing phase up to the temperature before immersion in liquid nitrogen (–7°C to –32°C). Once the point of maximum cell volume reduction is reached, it is possible to increase the freezing rate hundreds of times without intracellular ice formation. The quantity of water that leaves the cell depends on the cooling rate: at slow cooling rates, cells stay in equilibrium with external solutions longer. As the cooling rate increases, the time for the water to exit the cell is reduced. The optimal cooling rate can be obtained from the balance between these two properties. At slower cooling rates, long periods of exposure to hypertonic conditions cause cell death, while at faster rates cell death is associated with intracellular ice formation. Also, several factors can determine the optimal cooling rate, like cell volume and surface area, water permeability of cell membranes, and type and concentration of CPA (Leibo and Pool, 2011).

b. Vitrification

Vitrification is a freezing process that prevents the formation of ice crystals in the intracellular and extracellular space (Kuwayama et al., 2005). The glassy state maintains the molecular and ionic spatial distribution of the liquid state and can be considered an extremely dense supercooled liquid. In the slow freezing method, when the temperature decreases and the cell has already been sufficiently dehydrated, the cytoplasm, together with the freezing medium concentrated in the extracellular space, becomes a glassy solid without ice formation. In contrast, in the vitrification procedure, cells are dehydrated before the cooling begins by exposure to CPA concentrations high enough to reach the glassy state in the intra- and extracellular space. Thus, this transformation is achieved with high concentrations of CPA (4–8 M) and extremely high cooling rates.

The main risk faced during vitrification is the possible cytotoxicity that the high concentration of CPA required could entail. However, strategies to limit this potential toxicity are (Saragusty and Arav, 2011):

- The combined use of different types of CPA (permeable and non-permeable), thus decreasing the relative concentration of each one.
- Reducing time that the cells are exposed to vitrification solutions. In general, the addition of the CPAs is carried out in two stages: the first one is at room temperature (25°C) during an equilibration period of 3–10 min with concentrations of 5–7.5 M of CPAs. The second is extremely brief (30°C 60 s) since the concentration of CPAs amounts to 15 M.
- Reducing the temperature at which the vitrification solutions are added to the cell suspension to 20°C–25°C.

To achieve higher cooling speeds, the smallest possible volume of solution is used in successful specific vitrification supports, which can be differentiated as open (if contact with liquid nitrogen is direct) or closed (if contact with liquid nitrogen is indirect). At the moment, two vitrification systems are widely used for gametes and embryos: the Open Pulled Straw (Cryotop®—CryoTech Lab, Cryoloop) and the Closed Pulled Straw (Cryotip®—Irvine Scientific) (Morató et al., 2008; Liu et al., 2008; Gonzalez-Plaza et al., 2022; Hochi, 2022; Bonetti et al., 2011), along with a variety of cryoprotectant media and straws and other devices (Joaquim et al., 2017). Several studies have been conducted to identify which method provides better protection for gametes and embryos. However, there is a lack of conclusive and comparative studies, and the percentage of survival rates remains similar between both systems.

Microbial risk factors in the cryopreservation of gametes and embryos

There are several risks of contamination of gametes and embryos that are cryopreserved. One risk is that bacteria or other microorganisms could contaminate the gametes or embryos, potentially leading to infection. Additionally, cryopreserved gametes and embryos could be accidentally mixed up or mislabelled, leading to the wrong gametes or embryos being used during fertilization or implantation. This could result in the wrong baby being born or the failure of the fertilization or implantation process altogether. Cryopreservation facilities must follow strict protocols to minimize contamination risks and ensure the safety of the gametes and embryos they are storing (Rinehart, 2021).

Many different microorganisms can potentially contaminate cryopreserved embryos, including bacteria, viruses, and fungi. Bacteria are the most common type of microorganism that can contaminate cryopreserved embryos, and they can come from a variety of sources, such as the air, water, and surfaces in the cryopreservation facility. Some common types of bacteria that can contaminate cryopreserved embryos include *Escherichia coli*, *Staphylococcus* sp., and *Pseudomonas* sp. However, there are reports of contaminations in semen by other species like *Enterobacter* sp., *Acinetobacter calcoaceticus* and *Flavobacterium* sp. (Bielanski, 2012). Despite this, it is the bacterium *Stenotrophomonas maltophilia* that has been identified as one of the most typical contaminants in cryostored samples (Bielanski et al., 2003). Viruses, such as HIV, Zika (ZIKV) and hepatitis B (HBV), can also potentially contaminate cryopreserved embryos if proper precautions are not taken (Almeida et al., 2020; Mocanu et al., 2021; Washington et al., 2016). Fungi, such as yeasts and moulds, can also contaminate cryopreserved embryos, although this is less common (Borges et al., 2020). In this regard, It is important for cryopreservation facilities to follow strict protocols to prevent contamination and ensure the safety of the embryos they are storing.

Measures to reduce microbiological risk: devices and prophylactic measures in gamete and embryo cryobanking

There are a few methods considered effective against contamination risks in gamete and embryo cryobanking. Here, we summarize the most widespread.

Quarantine

Gametes should be quarantined until the tests performed to the donor guarantee that the samples are free of potentially harmful biological hazards, which also will improve the risk of contaminations later on. This is particularly relevant in the case of semen, as it is more prone to incorporate a high presence of microorganisms, which constitutes a further reason to store semen separately from oocytes and embryos (Larman et al., 2014).

Washing of gametes and embryos

In order to reduce microbiological load, semen samples are advised to be washed and performed a swim-up, while multiple washing to potential microbial pathogens are also recommended for oocytes and embryos (Maertens et al., 2004; Larman et al., 2014).

Decontamination of cryotanks

Development of routine management of the equipment to ensure it is well maintained and to avoid potential contaminations (Rinehart, 2021). Cryotanks and shippers require periodic cleaning following validated protocols (Bielanski and Vajta, 2009).

260   *Novel Approaches in Biopreservation for Food and Clinical Purposes*

Liquid nitrogen sterilization

As most contaminations are due to liquid nitrogen inappropriately stored and distributed, ensuring liquid nitrogen suitability and tackling potential contaminations at this stage will be translated into a reduction of decontaminant costs and risks for the gametes and embryos. There are a few strategies to generate and maintain sterile liquid nitrogen:

- Ultraviolet (UV) light sterilization. It has been demonstrated that this type of sterilization can be applied to small volumes of liquid nitrogen as an effective way against bacteria and fungi, such as *Pseudomonas aeruginosa, E. coli, S. maltophilia* and *Aspergillus niger* (Joaquim et al., 2017; Parmegiani et al., 2009; Parmegiani et al., 2008; Parmegiani et al., 2012). Additionally, it has been shown that, when dose of radiation and volume of liquid nitrogen are adjusted, UV light can also prevent the growth of other microorganisms, like viruses, such as the hepatitis B or dengue viruses, inactivating most of them generally at a dose of 200,000 μW/cm^2 (Parmegiani et al., 2009). Nonetheless, other viruses, like the Zika virus, present an increased resistance to those radiation levels and need to be further investigated (Lahon et al., 2016).

  The use of UV light for sterilization can generate some downside effects, like the generation of ozone, creating a potential threat for gametes and embryos. Despite this, it has been reported that this ozone formation could be irrelevant, as the liquid nitrogen unlikely contains oxygen, and the generation of ozone by UV radiation happens in presence of oxygen (Parmegiani et al., 2011).

- Sterile filtration. Consists in passing the liquid nitrogen through a 0.22 μm filter that will retain any biological microorganisms that could be present in the liquid nitrogen (McBurnie and Bardo, 2002).

Both strategies can be applied individually or combined, although they cannot completely remove viral contamination and cannot prevent cross-contamination during the storage, as the laboratories are not sterile and liquid nitrogen security cannot be guaranteed indefinitely. However, there is no evidence that gametes and embryos can get contaminated during cryobanking and transmit a disease (Pomeroy et al., 2010). Despite this, it has been proposed that the use of alternatives not requiring direct contact with liquid nitrogen would minimize the problems arisen from these situations (Larman et al., 2014).

Avoiding direct contact with liquid nitrogen

There have been published protocol modifications to generate embryo vitrification without liquid nitrogen contact, by generating an atmosphere of cooled air around the vial (around −190°C inside it), where survival and embryo development were not affected (Larman et al., 2006; Joaquim et al., 2017). This strategy would eliminate the risk of contamination during the vitrification process. However, long-term storage in liquid nitrogen can alter the integrity of the cryovials and generate some leakage resulting in contamination (Larman et al., 2014).

Alternatively, sealed straws have been used extensively in slow freezing and storage of human gametes and embryos in liquid nitrogen and have been modified for aseptic vitrification (Isachenko et al., 2005; Vanderzwalmen et al., 2003) and, combined with cooled air, offer another strategy to avoid direct contact with liquid nitrogen (Larman et al., 2014).

Closed systems are also used. They focus on avoiding direct contact with liquid nitrogen and have demonstrated to prevent contamination despite the presence of pathogens (Criado et al. 2011; Kuleshova and Shaw 2000).

Conclusion and future perspectives

Cryobanking for gametes and embryos is a wide-extended practice in assisted human reproduction. As with many other human activities, cryobanking can be affected by the activity of microorganisms, with potential risks of cross-contamination and disease. Consequently, several strategies have been developed in order to preserve the quality and safety of the gametes and embryos. However, these strategies are not perfect and they keep evolving alongside with technology. Thus, moving from direct contact with liquid nitrogen seems to be the most widespread accepted at the moment, and the improvement of vitrification systems will determine the minimization of microbiological risk.

Acknowledgements

This work was supported by a Beatriz Galindo scholarship from the Spanish Ministry of Universities (BG22/00060) awarded to EGG.

References

Almeida, Raquel das Neves, Heloisa Antoniella Braz-de-Melo, Igor de Oliveira Santos, Rafael Corrêa, Gary P. Kobinger and Kelly Grace Magalhaes. 2020. The cellular impact of the ZIKA virus on male reproductive tract immunology and physiology. *Cells*, 9: 1006.

Bielanski, A., Bergeron, H., Lau, P.C. and Devenish, J. 2003. Microbial contamination of embryos and semen during long term banking in liquid nitrogen. *Cryobiology*, 46: 146–52.

Bielanski, A. and Vajta, G. 2009. Risk of contamination of germplasm during cryopreservation and cryobanking in IVF units. *Hum Reprod*, 24: 2457–67.

Bielanski, A. 2012. A review of the risk of contamination of semen and embryos during cryopreservation and measures to limit cross-contamination during banking to prevent disease transmission in ET practices. *Theriogenology*, 77: 467–82.

Bonetti, A., Cervi, M., Tomei, F., Marchini, M., Ortolani, F. and Manno, M. 2011. Ultrastructural evaluation of human metaphase II oocytes after vitrification: closed versus open devices. *Fertil Steril*, 95: 928–35.

Borges, E.D., Berteli, T.S., Reis, T.F., Silva, A.S. and Vireque, A.A. 2020. Microbial contamination in assisted reproductive technology: source, prevalence, and cost. *J. Assist. Reprod. Genet.*, 37: 53–61.

Bosch, E., De Vos, M. and Humaidan, P. 2020. The Future of cryopreservation in assisted reproductive technologies. *Frontiers in Endocrinology*, 11: 67.

Bunge, R.G. and Sherman, J.K. 1953. Fertilizing capacity of frozen human spermatozoa. *Nature*, 172: 767–8.

Capanna, E. 1999. Lazzaro Spallanzani: at the roots of modern biology. *J. Exp. Zool.*, 285: 178–96.

Cha, K.Y. 1999. Pregnancy and implantation from vitrified oocytes following *in vitro* fertilization (IVF) and *in vitro* culture (IVC). *Fertil Steril*, 72: S2.

262 *Novel Approaches in Biopreservation for Food and Clinical Purposes*

Chen, C. 1988. Pregnancies after human oocyte cryopreservation. *Ann. N Y Acad. Sci.*, 541: 541–9.

Criado, E., Moalli, F., Polentarutti, N., Albani, E., Morreale, G., Menduni, F. and Levi-Setti, P.E. 2011. Experimental contamination assessment of a novel closed ultravitrification device. *Fertil Steril*, 95: 1777–9.

De Geyter, Ch., Wyns, C., Calhaz-Jorge, C., de Mouzon, J., Ferraretti, A.P., Kupka, M., Nyboe, A. Andersen, Nygren, K.G. and Goossens, V. 2020. 20 years of the european IVF-monitoring consortium registry: what have we learned? A comparison with registries from two other regions. *Human Reproduction*, 35: 2832–49.

Gao, D. and Critser, J.K. 2000. Mechanisms of cryoinjury in living cells. *Ilar j.*, 41: 187–96.

Gonzalez-Plaza, Alejandro, Josep M. Cambra, Inmaculada Parrilla, Maria A. Gil, Emilio A. Martinez, Cristina A. Martinez and Cristina Cuello. 2022. The open cryotop system is effective for the simultaneous vitrification of a large number of porcine embryos at different developmental stages. *Frontiers in Veterinary Science*, 9.

Gook, D.A. and Edgar, D.H. 2007. Human oocyte cryopreservation *Hum. Reprod Update*, 13: 591–605.

Hochi, Shinichi. 2022. Cryodevices developed for minimum volume cooling vitrification of bovine oocytes. *Animal Science Journal*, 93: e13683.

Isachenko, V., Montag, M., Isachenko, E., Zaeva, V., Krivokharchenko, I., Shafei, R. and van der Ven, H. 2005. Aseptic technology of vitrification of human pronuclear oocytes using open-pulled straws. *Hum. Reprod*, 20: 492–6.

Iussig, B., Maggiulli, R., Fabozzi, G., Bertelle, S., Vaiarelli, A., Cimadomo, D., Ubaldi, F.M. and Rienzi, L. 2019. A brief history of oocyte cryopreservation: arguments and facts. *Acta Obstet. Gynecol. Scand*, 98: 550–58.

Jang, T.H., Park, S.C., Yang, J.H., Kim, J.Y., Seok, J.H., Park, U.S., Choi, C.W., Lee, S.R. and Han, J. 2017. Cryopreservation and its clinical applications. *Integr. Med. Res.*, 6: 12–18.

Joaquim, D.C., Borges, E.D., Viana, I.G.R., Navarro, P.A. and Vireque, A.A. 2017. Risk of contamination of gametes and embryos during cryopreservation and measures to prevent cross-contamination. *Biomed. Res. Int.*, 2017: 1840417.

Kasai, Magosaburo. 1997. Vitrification: refined strategy for the cryopreservation of mammalian embryos. *Journal of Mammalian Ova Research*, 14: 17–28.

Kuleshova, L.L. and Shaw, J.M. 2000. A strategy for rapid cooling of mouse embryos within a double straw to eliminate the risk of contamination during storage in liquid nitrogen. *Hum. Reprod*, 15: 2604–9.

Kuwayama, M., Vajta, G., Kato, O. and Leibo, S.P. 2005. Highly efficient vitrification method for cryopreservation of human oocytes. *Reprod Biomed Online*, 11: 300–8.

Lahon, A., Arya, R.P., Kneubehl, A.R., Vogt, M.B., Dailey Garnes, N.J. and Rico-Hesse, R. 2016. Characterization of a Zika virus isolate from Colombia. *PLoS Negl. Trop. Dis.*, 10: e0005019.

Larman, M.G., Sheehan, C.B. and Gardner, D.K. 2006. Vitrification of mouse pronuclear oocytes with no direct liquid nitrogen contact. *Reprod. Biomed. Online*, 12: 66–9.

Larman, M.G., Hashimoto, S., Morimoto, Y. and Gardner, D.K. 2014. Cryopreservation in ART and concerns with contamination during cryobanking. *Reprod. Med. Biol.*, 13: 107–17.

Leibo, S.P. and Pool, T.B. 2011. The principal variables of cryopreservation: solutions, temperatures, and rate changes. *Fertil Steril*, 96: 269–76.

Liu, Y., Du, Y., Lin, L., Li, J., Kragh, P.M., Kuwayama, M., Bolund, L., Yang, H. and Vajta, G. 2008. Comparison of efficiency of open pulled straw (OPS) and Cryotop vitrification for cryopreservation of *in vitro* matured pig oocytes. *Cryo Letters*, 29: 315–20.

Maertens, A., Bourlet, T., Plotton, N., Pozzetto, B. and Levy, R. 2004. Validation of safety procedures for the cryopreservation of semen contaminated with hepatitis C virus in assisted reproductive technology. *Hum. Reprod.*, 19: 1554–7.

McBurnie, L. and Barry Bardo. 2002. Validation of sterile filtration of liquid nitrogen. *Pharmaceutical Technology*, 26: 74–83.

Mocanu, E., Drakeley, A., Kupka, M.S., Lara-Molina, E.E., Le Clef, N., Ombelet, W., Patrat, C., Pennings, G., Semprini, A.E., Tilleman, K., Tognon, M., Tonch, N. and Woodward, B. 2021. ESHRE guideline: medically assisted reproduction in patients with a viral infection/disease. *Hum. Reprod. Open*, 2021: hoab037.

Biopreservation of Gametes and Embryos against Microbiological Risk 263

Morató, R., Izquierdo, D., Paramio, M.T. and Mogas, T. 2008. Cryotops versus open-pulled straws (OPS) as carriers for the cryopreservation of bovine oocytes: effects on spindle and chromosome configuration and embryo development. *Cryobiology*, 57: 137–41.

Parmegiani, L., Cognigni, G.E., Bernardi, S., Ciampaglia, W., Infante, F., Pocognoli, P., de Fatis, C.T., Troilo, E. and Filicori, M. 2008. Freezing within 2 h from oocyte retrieval increases the efficiency of human oocyte cryopreservation when using a slow freezing/rapid thawing protocol with high sucrose concentration. *Hum. Reprod.*, 23: 1771–7.

Parmegiani, L., Cognigni, G.E. and Filicori, M. 2009. Ultra-violet sterilization of liquid nitrogen prior to vitrification. *Hum. Reprod.*, 24: 2969.

Parmegiani, L., Cognigni, G.E., Bernardi, S., Cuomo, Ciampaglia, W., Infante, F.E., Tabarelli de Fatis, C., Arnone, A., Maccarini, A.M. and Filicori, M. 2011. Efficiency of aseptic open vitrification and hermetical cryostorage of human oocytes. *Reprod. Biomed. Online*, 23: 505–12.

Parmegiani, Lodovico, Antonio Accorsi, Silvia Bernardi, Alessandra Arnone, Graciela Estela Cognigni, and Marco Filicori. 2012. A reliable procedure for decontamination before thawing of human specimens cryostored in liquid nitrogen: three washes with sterile liquid nitrogen (SLN2). *Fertility and Sterility*, 98: 870–75.

Polge, C., Smith, A.U. and Parkes, A.S. 1949. Revival of spermatozoa after vitrification and dehydration at low temperatures. *Nature*, 164: 666.

Pomeroy, K.O., Harris, S., Conaghan, J., Papadakis, M., Centola, G., Basuray, R. and Battaglia, D. 2010. Storage of cryopreserved reproductive tissues: evidence that cross-contamination of infectious agents is a negligible risk. *Fertil Steril*, 94: 1181–88.

Porcu, E., Fabbri, R., Seracchioli, R., Ciotti, P.M., Magrini, O. and Flamigni, C. 1997. Birth of a healthy female after intracytoplasmic sperm injection of cryopreserved human oocytes. *Fertil Steril*, 68: 724–6.

Rienzi, L., Gracia, C., Maggiulli, R., LaBarbera, A.R., Kaser, D.J., Ubaldi, F.M., Vanderpoel, S. and Racowsky, C. 2017. Oocyte, embryo and blastocyst cryopreservation in ART: systematic review and meta-analysis comparing slow-freezing versus vitrification to produce evidence for the development of global guidance. *Hum. Reprod. Update*, 23: 139–55.

Rinehart, L.A. 2021. Storage, transport, and disposition of gametes and embryos: legal issues and practical considerations. *Fertil Steril.*, 115: 274–81.

Saragusty, J. and Arav, A. 2011. Current progress in oocyte and embryo cryopreservation by slow freezing and vitrification. *Reproduction*, 141: 1–19.

Sargisian, N., Lannering, B., Petzold, M., Opdahl, S., Gissler, M., Pinborg, A., Henningsen, A.A., Tiitinen, A., Romundstad, L.B., Spangmose, A.L., Bergh, C. and Wennerholm, U.B. 2022. Cancer in children born after frozen-thawed embryo transfer: a cohort study. *PLoS Med.*, 19: e1004078.

Services. Department of Health & Human. Centers for Disease Control and U.S.P. (2022) 2019 National ART Summary. *Retrieved from http://www.cdc.gov/art/reports/2019/national-ART-summary.html.*

Trounson, A. and L. Mohr. 1983. Human pregnancy following cryopreservation, thawing and transfer of an eight-cell embryo. *Nature*, 305: 707–9.

Vanderzwalmen, P., Bertin, G., Ch Debauche, Standaert, V., Bollen, N., van Roosendaal, E., Vandervorst, M., Schoysman, R. and Zech, N. 2003. Vitrification of human blastocysts with the Hemi-Straw carrier: application of assisted hatching after thawing. *Hum. Reprod.*, 18: 1504–11.

Washington, Chantel I., Sara Haque, James H. Segars, Nabal Bracero, Fernando Rodriguez, David G. Ball, and Owen K. Davis. 2016. Keeping the Zika virus out of the assisted reproductive technology laboratory. pp. 293–98. *In*: Semin Reprod Med 2016; 34(05): 293–298. DOI: 10.1055/s-0036-1592067. Thieme Medical Publishers 333 Seventh Avenue, New York, NY 10001, USA.

Zeilmaker, G.H., Alberda, A.T., van Gent, I., Rijkmans, C.M. and Drogendijk, A.C. 1984. Two pregnancies following transfer of intact frozen-thawed embryos. *Fertil Steril,*, 42: 293–6.

Index

A

Alcoholic beverages 205, 207–209, 214, 216
Assisted reproduction techniques (ART) 255, 256

B

Bacteriocins 2, 4, 5, 11, 76, 81, 88, 212, 217
Bacteriophages 2, 12, 14, 211, 215, 217
Biopreservation 1–3, 12, 13, 66–68, 73, 74,
    81–87, 130–132, 134, 135, 141, 142, 145,
    154, 157, 159, 161, 171, 172, 177–182, 242,
    243, 247, 249
Biopreservatives 33–35, 42
Bioprotective cultures 99, 104, 118, 121
Breadmaking 131, 132, 134, 141, 144, 157, 158,
    163, 164, 171, 177, 180

C

Cell culture 242–247
Chemical preservation 132, 141, 142, 159, 166,
    172, 176–178, 180, 181
Clean labels 141, 172, 173
Cryopreservation 13, 248, 255, 256, 258, 259

D

Desiccation 249
Diapause 249

E

Embryos 255–261
Essential oils 99, 106, 108
European Medicine Agency (EMA) 236

F

Fermentation 3, 4, 8, 11, 44, 24, 26–28, 30, 31,
    35, 36
Fish biopreservation 104
Fish spoilage 99, 100, 107, 119
Food 1–3, 11–13

Food and Drug Administration (FDA) 234, 236
Food safety 131, 142, 163, 177, 180, 183
Food security 130, 131, 180
Foodborne pathogens 131

G

Gametes 255–261
Good Manufacturing Practices (GMPs) 226

H

Hunger eradication 130, 131
Hypothermic storage 248

I

in vitro culture 247

L

Lactic acid bacteria 24, 42–45, 67, 70, 71, 80,
    81, 88, 131, 134, 140–142, 144, 146, 147,
    152, 183

M

Meat 66–76, 78, 80–84, 87, 88
Medicines 1, 13, 226–228, 231, 234–236, 238
Mycoplasma 244–246

N

Nisin 212, 217

O

Oocytes 255–257, 259

P

Packaging 137, 163, 166–170, 173, 177, 178,
    181–183
Plant extracts 106–108, 118
Protective cultures 2, 3, 11, 13

Q

Quality 68, 73, 81, 83, 88
Quiescence 249

R

Regulatory framework 33

S

Safety 66, 68, 69, 73, 80–83, 85, 87, 88
Semen 259
Shelf life 205, 210, 214, 215

Sourdough fermentation 131, 134, 138–140,
146–148, 153–155, 159, 168, 181
Spoilage microorganisms 131, 137, 153,
166–168, 170, 181, 182
Starters 11

T

Tissue culture 242–244, 246, 250

Y

Yeasts 131, 132, 136–140, 144–149, 152,
155–159, 166, 179–181